Frontiers in Mathematics

Advisory Editors

William Y. C. Chen, Nankai University, Tianjin, China

Laurent Saloff-Coste, Cornell University, Ithaca, NY, USA

Igor Shparlinski, The University of New South Wales, Sydney, NSW, Australia

Wolfgang Sprößig, TU Bergakademie Freiberg, Freiberg, Germany

This series is designed to be a repository for up-to-date research results which have been prepared for a wider audience. Graduates and postgraduates as well as scientists will benefit from the latest developments at the research frontiers in mathematics and at the "frontiers" between mathematics and other fields like computer science, physics, biology, economics, finance, etc. All volumes are online available at SpringerLink.

M. L. Brown

Shafarevich-Tate Groups

M. L. Brown
Institut Fourier
Université de Grenoble
Saint Martin d'Hères Cedex, France

ISSN 1660-8046   ISSN 1660-8054 (electronic)
Frontiers in Mathematics
ISBN 978-3-031-71892-2   ISBN 978-3-031-71893-9 (eBook)
https://doi.org/10.1007/978-3-031-71893-9

© Springer Nature Switzerland AG 2025

This work is subject to copyright. All rights are solely and exclusively licensed by the Publisher, whether the whole or part of the material is concerned, specifically the rights of translation, reprinting, reuse of illustrations, recitation, broadcasting, reproduction on microfilms or in any other physical way, and transmission or information storage and retrieval, electronic adaptation, computer software, or by similar or dissimilar methodology now known or hereafter developed.

The use of general descriptive names, registered names, trademarks, service marks, etc. in this publication does not imply, even in the absence of a specific statement, that such names are exempt from the relevant protective laws and regulations and therefore free for general use.

The publisher, the authors and the editors are safe to assume that the advice and information in this book are believed to be true and accurate at the date of publication. Neither the publisher nor the authors or the editors give a warranty, expressed or implied, with respect to the material contained herein or for any errors or omissions that may have been made. The publisher remains neutral with regard to jurisdictional claims in published maps and institutional affiliations.

This book is published under the imprint Birkhäuser, www.birkhauser-science.com by the registered company Springer Nature Switzerland AG

The registered company address is: Gewerbestrasse 11, 6330 Cham, Switzerland

If disposing of this product, please recycle the paper.

Preface

This book determines the structure and demonstrates the finiteness of Shafarevich-Tate groups over global fields under suitable hypotheses. The methods of Euler systems and Kolyvagin systems, which are here essential, are developed for a Heegner system of which special cases are Heegner points on classical modular curves and Drinfeld modular curves as well as CM points on Shimura curves.

Some selection has had to be made for this text. A prolongation of this book would cover these further subjects, which are at most only briefly mentioned here:

(1) Iwasawa theory of abelian varieties over number fields.
(2) The Birch Swinnerton-Dyer conjecture for global fields of any characteristic.
(3) The final Chaps. 7 and 8, notably because of the hypotheses of Sect. 8.1, essentially only apply to elliptic curves; nevertheless, these can be extended to abelian varieties over global fields.

Paris, France
October 2023

M. L. Brown

Contents

Introduction		ix
1	**Preliminaries**	1
	1.1 Rings and Fields	1
	1.2 Groups	4
	1.3 Continuous Cochain Cohomology	5
	1.4 Orders in Imaginary Quadratic Fields	17
	1.5 Ring Class Fields	17
	1.6 Group Representations over a Local Principal Ideal Ring	20
2	**Elliptic Curves; Drinfeld Modules; Shimura Curves; Elliptic Spaces**	23
	2.1 Elliptic Curves	23
	2.2 Drinfeld Modules	26
	2.3 Shimura Curves	28
	2.4 Elliptic Spaces	36
	2.5 Moduli Schemes of Elliptic Spaces	37
	2.6 Modularity of Elliptic Curves over Global Fields	41
	2.7 Torsion Points on Elliptic Curves	42
3	**Complex Multiplication, Heegner Points, CM Points**	47
	3.1 Complex Multiplication of Elliptic Spaces	47
	3.2 Heegner Points	48
	3.3 CM Points on Shimura Curves	51
	3.4 Heegner Points and CM Points on Elliptic Curves	55
4	**Local Fields and Local Conditions**	69
	4.1 R-Representations, l-adic Representations	69
	4.2 Local Duality	72
	4.3 Local Conditions	79
	4.4 A Normal Form for Endomorphisms	83
	4.5 The Finite-Singular Comparison Map	97

5 Global Fields and Selmer Structures ... 105
- 5.1 Selmer Structures, Selmer Modules ... 105
- 5.2 Modification of Selmer Structures ... 107
- 5.3 Global Duality ... 107
- 5.4 Local to Global Principles for Rational Points ... 110
- 5.5 The Classical Selmer Structure of an Abelian Variety ... 117
- 5.6 Finiteness of Classical Selmer Modules of Abelian Varieties ... 123

6 Euler Systems ... 135
- 6.1 General Euler Systems ... 135
- 6.2 Euler Systems ... 139
- 6.3 Euler Systems Are General Euler Systems ... 141
- 6.4 Morphisms ... 144
- 6.5 Heegner Systems ... 152
- 6.6 The Heegner Point General Euler System: Classical and Drinfeld Modular Curves ... 157
- 6.7 The CM Point General Euler System: Shimura Curves ... 160
- 6.8 The Heegner Point and CM Point Euler Systems ... 165
- 6.9 Appendix: Further Examples of Euler Systems ... 169

7 Kolyvagin Systems ... 177
- 7.1 Kolyvagin Systems ... 177
- 7.2 Euler Systems Induce Kolyvagin Systems ... 188
- 7.3 Preliminaries on Heegner Point and CM Point Kolyvagin Systems ... 190
- 7.4 Derived Cohomology Classes ... 192
- 7.5 Properties of the Derived Classes ... 198
- 7.6 Heegner Point and CM Point Kolyvagin Systems ... 206

8 Selmer Groups and Kolyvagin Systems ... 217
- 8.1 Hypotheses H.0–H.5 for Selmer Triples ... 218
- 8.2 The Cassels-Tate Pairing ... 230
- 8.3 Kolyvagin Systems over Artinian Principal Ideal Rings ... 237
- 8.4 Kolyvagin Systems over Discrete Valuation Rings ... 256
- 8.5 Selmer Groups over Artinian Principal Ideal Rings ... 261
- 8.6 Selmer Groups over Discrete Valuation Rings ... 284
- 8.7 Shafarevich-Tate Groups of Elliptic Curves ... 294

Reference ... 307

Index ... 311

Introduction

In this text we consider the Shafarevich-Tate groups of abelian varieties, and especially elliptic curves, over global fields of any characteristic. The principal aim is to show the finiteness of these groups and determine their structure under suitable hypotheses where the ground field is a global field.

Such finiteness theorems for Shafarevich-Tate groups were already known for elliptic curves over number fields due notably to Rubin and later Kolyvagin [K2]. In this book we are mainly concerned with finiteness results that can be demonstrated using Heegner points in any characteristic, whereas Rubin's results used in effect the different method of the Euler system of elliptic units. Much of this book is an exposition of the methods of Euler systems and Kolyvagin systems adapted to Heegner points and CM points.

In this book the cases of global fields of zero characteristic and those of positive characteristic are treated uniformly and simultaneously. For the positive characteristic case there are Heegner points on Drinfeld modular curves. In the zero characteristic case it is necessary to distinguish between the ground field $\mathbb{Q}$ where one has classical modular curves with their Heegner points and a ground field which is a totally real number field different from $\mathbb{Q}$ where there are CM points on corresponding Shimura curves. In both these latter two cases one has Heegner points or CM points on an elliptic curve parametrized by a classical modular curve or a Shimura curve. Similarly, there are Heegner points on elliptic curves over global fields of positive characteristic when the curve is parametrized by a Drinfeld modular curve.

In more detail the content of the individual chapters is the following.

Chapters 1 and 2 give basics essential for this book. These two chapters cover briefly ring class fields, continuous cochain cohomology, moduli spaces for Drinfeld modules, and Shimura curves. As we require Heegner points on classical modular curves and Drinfeld modular curves and CM points on Shimura curves, Chap. 3 covers basics on complex multiplication of elliptic curves and Drinfeld modules.

Chapter 4 concerns the local theory over local fields. We define local conditions on first cohomology groups and the finite-singular comparison map required for defining Kolyvagin systems. Our definition of the finite-singular comparison map is more general than that of Mazur and Rubin [MR] in that we obtain this map from a normal form for endomorphisms of a module over a coefficient ring (Sect. 4.4). This normal form established in this chapter allows the definition of a finite singular comparison map of [MR] to be given in much wider setting. In the special case considered by [MR], in the terminology of Chap. 4, those authors only consider module endomorphisms with a single block (Remark 4.5.9(iv)). Indeed the definition of the finite-singular comparison map in [MR] is insufficient even for the case of elliptic curves over $\mathbb{Q}$ considered by Howard [H], although the finite-singular comparison map there is the identity on a free module of rank 2, a case that is relatively trivial.

In Chap. 5, we consider global fields and Selmer structures on cohomology groups, which are the global equivalent of local conditions for local fields. These Selmer structures are a generalization of the classical Selmer module of an abelian variety. As the classical Selmer module is strongly related to the existence of rational points of algebraic varieties over local and global fields, this chapter also examines local to global principles associated to such questions. The Shafarevich-Tate group of an abelian variety is a measure of the failure of local to global principles of the associated principal homogeneous spaces of this variety. This chapter ends with a proof of the classical result of the finiteness of the classical Selmer modules of abelian varieties in all ground field characteristics.

In Chap. 6, we define Euler systems, a concept introduced by Kolyvagin [K1] and also in much more generality by Rubin [Ru1]. In particular as Heegner points play a major role in the last two chapters, we consider in detail the Euler system provided by Heegner points on classical modular curves and Drinfeld modular curves and CM points on Shimura curves. We define a Heegner system which is an axiomatic system of points on an elliptic curve where this notion covers both Heegner points and CM points.

In Chap. 7, Kolyvagin systems over arbitrary global fields are introduced. These were first defined by Mazur and Rubin [MR] for Galois representations of $\text{Gal}(\mathbb{Q}^{\text{sep}}/\mathbb{Q})$ and also later by Howard [H], in the case of elliptic curves over the rational numbers. In this chapter we derive the basic properties of Kolyvagin systems obtained from a Heegner system and we obtain properties of the derived cohomology classes.

In the final Chap. 8, we apply the results of Chap. 7 to obtain the main results of this text on Shafarevich-Tate groups of elliptic curves over global fields by means of Heegner point Kolyvagin systems over fields of any characteristic. In particular we show under hypotheses the finiteness of these Shafarevich-Tate groups and determine their structure. The method can be adapted to abelian varieties over global fields of any characteristic but this generalization is not included here.

Preliminaries

Contents

1.1	Rings and Fields	1
1.2	Groups	4
1.3	Continuous Cochain Cohomology	5
1.4	Orders in Imaginary Quadratic Fields	17
1.5	Ring Class Fields	17
1.6	Group Representations over a Local Principal Ideal Ring	20

This chapter covers some basic results notably on continuous cochain cohomology and ring class fields. The foundational notation of this text is detailed in the rest of this chapter and is mainly that of the monograph [4] augmented by that for the number field case.

1.1 Rings and Fields

Fields

(1.1.1) For any field L, we write

L^{sep} for the separable closure of L;
$G_L = \text{Gal}(L^{\text{sep}}/L)$.

(1.1.2) Suppose that F is a global field that is to say either a finite field extension of $\mathbb{Q}$ or a finite field extension of the rational field $\mathbb{F}_p(t)$ where $\mathbb{F}_p$ is the prime field of p elements. Denote by:

Σ_F the set of all places of the field F, archimedean and non-archimedean;

$\kappa(z)$ the residue field at z of F for a non-archimedean place $z \in \Sigma_F$;

F_z the completion of F at the place $z \in \Sigma_F$.

(1.1.3) Suppose that F is a global field of positive characteristic. Let k be the exact field of constants of F, that is, the algebraic closure of the prime subfield $\mathbb{F}_p$ in F. Then k is a finite field and F is the function field of a smooth projective irreducible algebraic curve C/k. The curve C/k is uniquely determined by F up to k-isomorphism.

(1.1.4) Let F be a global field. Denote by $\mathrm{Cyc}(F)$ the set of *cycles* of F that is to say $\mathrm{Cyc}(F)$ contains all formal products $\prod_{\mathfrak{p}} \mathfrak{p}^{n_{\mathfrak{p}}}$ over all places $\mathfrak{p}$ of F, both non-archimedean and archimedean, where $n_{\mathfrak{p}}$ are non-negative integers almost all of which are zero and where $n_{\mathfrak{p}}$ takes only the values 0 or 1 if $\mathfrak{p}$ is archimedean.

The set cycles of F are equipped with the evident partial order $\leq$ where $c \leq c'$ if and only if c'/c belongs to $\mathrm{Cyc}(F)$.

We write $\mathrm{Supp}(c)$ for the *support* of the cycle $c \in \mathrm{Cyc}(F)$ which is the set of prime divisors, archimedean or non-archimedean, with non-zero exponent in c. For a cycle c and a prime divisor z of F we sometimes write $z|c$ to mean that z is in the support of c.

If $\mathfrak{p}$ is a non-archimedean prime of F we write $N(\mathfrak{p})$ for the order of the residue field of F at $\mathfrak{p}$.

(1.1.5) Suppose that F'/F is a finite Galois field extension of a global field F and $\mathfrak{p} \in \Sigma_F$ is a non-archimedean prime divisor of F unramified in F'. For any place $\mathfrak{q}$ of F' lying over $\mathfrak{p}$ we denote by $\mathrm{Fr}_{\mathfrak{q}/\mathfrak{p}}$ the unique element of the decomposition group of $\mathfrak{q}$ in $\mathrm{Gal}(F'/F)$ which induces the endomorphism $x \mapsto x^{N(\mathfrak{p})}$ of the residue field $\kappa(\mathfrak{q})$. This $\mathrm{Fr}_{\mathfrak{q}/\mathfrak{p}}$ is called an *arithmetic Frobenius element* at $\mathfrak{p}$.

As $\mathfrak{q}$ ranges over all primes of F' lying over $\mathfrak{p}$, the elements $\mathrm{Fr}_{\mathfrak{q}/\mathfrak{p}}$ range over a conjugacy class in $\mathrm{Gal}(F'/F)$ denoted $\mathrm{Fr}_{\mathfrak{p}}$ called the conjugacy class of arithmetic Frobenius elements at $\mathfrak{p}$.

The set of *geometric Frobenius elements* associated to $\mathfrak{p}$ is the conjugacy class of $\mathrm{Gal}(F'/F)$ of inverses in $\mathrm{Gal}(F'/F)$ of the arithmetic Frobenius elements in the conjugacy class $\mathrm{Fr}_{\mathfrak{p}}$.

We sometimes write $\mathrm{Fr}_{\mathfrak{p},\mathrm{arith}}$ and $\mathrm{Fr}_{\mathfrak{p},\mathrm{geom}}$ to distinguish the two types of Frobenius element but without any other qualification we always mean an arithmetic Frobenius.

Remarks 1.1.6

(i) For a global field F, the elements of $\mathrm{Cyc}(F)$ are also called *effective divisors* on F but where archimedean places of F are also included so they may be considered as *Arakelov divisors*. We write 1 for the unit cycle or divisor.

1.1 Rings and Fields

(ii) We shall always write cycles or divisors multiplicatively $\prod_{\mathfrak{p}} \mathfrak{p}^{n_{\mathfrak{p}}}$; this is usual for algebraic number fields but although it is normal to write divisors on algebraic function fields additively we shall here write them multiplicatively.

Special set of places; imaginary quadratic field extensions

(1.1.7) Let k be a global field. A set of places place $\infty \subset \Sigma_k$ of k is said to be *special* if one of the following holds:

(a) k is a number field and ∞ consists of all (equivalence classes of) archimedean places of k;
(b) k is an algebraic function field and ∞ is set containing one single place of k.

A special set of places of k will always be denoted by the symbol ∞. If k is a number field then evidently ∞ is uniquely determined by k but this is not the case if k is a global function field. We sometimes call the places in ∞ the *infinite places*.

(1.1.8) Suppose that ∞ is a special set of places of k. Then an extension field F/k is *imaginary quadratic* with respect to ∞ if F is a separable quadratic extension field of k in which every place in ∞ remains inert.

If k has characteristic zero this means that no archimedean place of k splits into two distinct places in F. If k has positive characteristic this means that the unique place in ∞ of k remains inert in F.

Remarks 1.1.9

(i) For the number fields, we are often concerned in this book with the case where k is a totally real number field, ∞ is the set of archimedean places of k, and F is a totally imaginary quadratic field extension of k.
(ii) For the function field case, the single place in ∞ may be any place of k but it is distinguished from the other places. This is because Drinfeld modules and Drinfeld modular curves for the field k are then relative to this place in ∞ and the automorphic representations corresponding to the Drinfeld modular curves are *special* at the place contained in ∞; this is the reason for introducing special sets of places (see [4, Appendix B]).
(iii) Let k be the rational function field $\mathbb{F}_q(t)$ over the finite field $\mathbb{F}_q$, where $q = p^r$ and p is a prime number different from 2. Then k is the function field of the projective line over $\mathbb{F}_q$. Let ∞ be the singleton set containing the place at infinity rational over $\mathbb{F}_q$ of the projective line. Then ∞ is a special set of places of k.

Let $d \in \mathbb{F}_q[t]$ be a polynomial in t over $\mathbb{F}_q$. The extension field $k(\sqrt{d})$ of k is imaginary quadratic with respect to ∞ if and only if either the degree of d is odd or the degree of d is even and the leading coefficient of d is not a square in $\mathbb{F}_q$.

Ring of integers with respect to a special set of places

(1.1.10) Let k be a global field equipped with a special set ∞ of places. The *ring of integers of k with respect to ∞* is defined to be the subring A of k of elements $x \in k$ such that

$$|x|_v \leq 1 \text{ for all } v \notin \infty.$$

That is to say the ring of integers A of k with respect to ∞ is given by:

(a) If k is a number field then A is the integral closure of $\mathbb{Z}$ in k;
(b) If k has positive characteristic then k is the function field of a smooth projective irreducible curve C over a finite field and A is defined to be $\Gamma(C \setminus \{\infty\}, \mathcal{O}_C)$, the affine coordinate ring of the affine curve C without the point ∞.

We shall sometimes write "the ring of integers of k" omitting reference to the special set of places ∞ where this is understood.

1.2 Groups

(1.2.1) If G is an abelian group denote by:

G_m the kernel of multiplication by the integer $m \in \mathbb{N}$ on G;
$|G|$ the order of the group G which is either a positive integer or $+\infty$;
$\exp(G)$ the exponent of G which is maximum order of an element of G.

(1.2.2) Suppose that G is a locally compact topological abelian group. Denote by $\check{G}$ the Pontryagin dual of G, namely, the topological group $\text{Hom}_{\text{cts}}(G, \mathbb{T})$ where $\mathbb{T} = \{z \in \mathbb{C}^* \mid |z| = 1\}$ is the circle group and the group $\text{Hom}_{\text{cts}}(G, \mathbb{T})$ of continuous homomorphisms is equipped with the compact-open topology.

If G is a finite abelian group equipped with the discrete topology then $\check{G}$ is identified with the group of one-dimensional complex characters of G, that is to say the group of homomorphisms $\text{Hom}(G, \mathbb{C}^*)$. We always assume that a character of a finite abelian group is irreducible.

1.3 Continuous Cochain Cohomology

(1.3.1) We shall give a brief account of continuous cochain cohomology. This is required because we have profinite Galois groups acting on modules, such as the Tate module of an elliptic curve, which are themselves equipped with a topology. For the case of discrete modules, continuous cochain cohomology reduces to the "classical" group cohomology and, for the applications we have in view, continuous cochain cohomology groups can normally be treated as though the modules were discrete.

Definition of continuous cochain cohomology

(1.3.2) For the rest of Sect. 1.3, let G be a profinite topological group and M be a topological abelian group equipped with a continuous action by G. For $i \in \mathbb{N}$ let $C^i(G, M)$ be the group of continuous maps from the i-fold product G^i to M. There is an evident coboundary homomorphism

$$d^i : C^i(G, M) \longrightarrow C^{i+1}(G, M)$$

defined by

$$d^i(f)(g_1, \ldots, g_{i+1}) =$$

$$g_1 f(g_2, \ldots, g_{i+1}) + \sum_{j=1}^{i} (-1)^j f(g_1, \ldots, g_{j-1}, g_j g_{j+1}, g_{j+2}, \ldots, g_{i+1})$$

$$+ (-1)^{i+1} f(g_1, \ldots, g_i)$$

and

$$d^0(f)(g) = gf(e) - f(e)$$

where e is the unique element of G^0. These satisfy $d^{i+1}d^i = 0$ and therefore the groups $C^*(G, M)$ form a cochain complex and whose cohomology is denoted $H^i(G, M)$ for $i \in \mathbb{N}$.

The group $H^0(G, M)$ can be identified with M^G, the subgroup of invariants of M under G, by the map $f \mapsto f(e)$ where e is the unique element of G^0.

If M is a discrete G-module then $H^i(G, M)$ evidently coincides with the classical group cohomology of G on M defined as the derived functors of the functor $M \to M^G$ as well as by the above cochain complexes. But for arbitrary topological G-modules M the derived functors of $M \to M^G$ are not defined because of a lack of injective topological modules.

(1.3.3) If $0 \to M' \to M \to M'' \to 0$ is an exact sequence of continuous G-modules such that the topology on M' is that induced by that of M and if there is a continuous set-theoretic section $M'' \to M$, not necessarily a homomorphism, then there is a corresponding exact sequence of cochain complexes

$$0 \longrightarrow C^i(G, M') \longrightarrow C^i(G, M) \longrightarrow C^i(G, M'') \longrightarrow 0$$

where this is exact for all i and also a long exact sequence of cohomology

$$\cdots \longrightarrow H^i(G, M') \longrightarrow H^i(G, M) \longrightarrow H^i(G, M'') \longrightarrow H^{i+1}(G, M') \longrightarrow \cdots$$

If either M'' is discrete or if M' is open in M then a continuous section $M'' \to M$ and hence a long exact sequence of cohomology exist. If M is a finitely generated $\mathbb{Z}_l$-module, for some prime number l, or a finite dimensional $\mathbb{Q}_l$-vector space equipped with its l-adic topology then there is a section $M'' \to M$ and hence a long exact sequence of cohomology.

Cup products, inflation, restriction, and corestriction for continuous cochain cohomology

(1.3.4) Let I be a closed normal subgroup of G and M be a topological G-module. Then M^I is a topological G/I-module and we then obtain the evident homomorphisms of continuous cochain complexes

$$C^i(G/I, M^I) \longrightarrow C^i(G, M)$$

compatible with the coboundary maps. Hence we obtain the functorial inflation homomorphisms of cohomology

$$\inf_{G/H} : H^i(G/H, M^I) \longrightarrow H^i(G, M).$$

(1.3.5) Suppose that I is a subgroup of G with its subspace topology. Then restriction of cochains gives homomorphisms of continuous cochain complexes

$$C^i(G, M) \longrightarrow C^i(I, M)$$

and hence functorial restriction homomorphisms of cohomology

$$\text{res}_{G/H} : H^i(G, M) \longrightarrow H^i(I, M).$$

It is evident that in degree zero the restriction map on cohomology is the inclusion $M^G \to M^I$.

1.3 Continuous Cochain Cohomology

(1.3.6) Suppose now that I is an open subgroup of finite index in G. Let $R \subseteq G$ be a set of coset representatives of I in G. Define a homomorphism of continuous cochain complexes

$$N_{G/I} : C^i(I, M) \longrightarrow C^i(G, M)$$

by the formula

$$(N_{G/I} f)(g_1, \ldots, g_i) = \sum_{r \in R} r_0^{-1} f(r_0 s_1 r_1^{-1}, r_1 g_2 r_2^{-1}, \ldots, r_{i-1} g_i r_i^{-1})$$

where $r_i \in R$ are defined inductively in terms of $r \in R$ and the i-tuple $(g_1, \ldots, g_i)$ by the formulae

$$r_0 = r, \quad r_j \in I r_{j-1} g_j \quad \text{for} \quad j = 1, \ldots, i.$$

It is evident that $N_{G/I} \circ d = d \circ N_{G/I}$, where the d are the coboundary homomorphisms, so that $N_{G/I}$ induces homomorphisms

$$N_{G/I} : H^i(I, M) \longrightarrow H^i(G, M).$$

For $i = 0$ this homomorphism is

$$N_{G/I} : M^I \to M^G, \quad m \mapsto \sum_{r \in R} r^{-1} m.$$

These homomorphisms $N_{G/I}$ are called *corestriction* or *norm*.

(1.3.7) Suppose that

$$[,] : M_1 \times M_2 \longrightarrow M$$

is a continuous pairing of topological G-modules; that is to say $[,]$ is a continuous biadditive map and satisfies $[g m_1, g m_2] = g[m_1, m_2]$ for all $g \in G$. Then $[,]$ induces biadditive maps of cochain groups

$$C^i(G, M_1) \times C^j(G, M_2) \longrightarrow C^{i+j}(M)$$

by the usual formulae

$$(e \cup f)(g_1, \ldots, g_{i+j}) = [e(g_1, \ldots, g_i), g_1 g_2 \cdots g_i f(g_{i+1}, \ldots, g_{i+j})].$$

This satisfies the identity

$$d(e \cup f) = (de) \cup f + (-1)^i f \cup (dg).$$

Therefore this cochain cup product induces cup product pairings on continuous cochain cohomology

$$H^i(G, M_1) \times H^j(G, M_2) \longrightarrow H^{i+j}(G, M).$$

(1.3.8) The inflation, restriction, norm homomorphisms, and cup products on continuous cohomology are functorial and satisfy exactly the same compatibility and change of group properties as for the cohomology of discrete modules. These properties are detailed explicitly in [58, Chapter I, §§4–5].

Comparison of continuous cochain cohomology and cohomology of discrete modules

(1.3.9) We shall see that, under hypotheses, continuous cochain cohomology may be calculated as if the topological G-modules were discrete.

Proposition 1.3.10 *Suppose that $M = \varprojlim M_n$ where each M_n, for $n \in \mathbb{N}$, is a finite discrete G-module and M is equipped with its profinite topology. If $i > 0$ and $H^{i-1}(G, M_n)$ is finite for all n then there is an isomorphism*

$$H^i(G, M) \cong \varprojlim H^i(G, M_n).$$

[For the proof, see [78, Corollary to Proposition 2.2] for an l-adic module M or [58, corollary 2.7.6, p. 142] in general.] □

Proposition 1.3.11 *Suppose that the G-module M is a finitely generated $\mathbb{Z}_l$ module, where l is a prime number, and is equipped with the l-adic topology. Then for all $i \geq 0$ the group $H^i(G, M)$ has no non-zero l-divisible subgroup and the natural map*

$$H^i(G, M) \otimes \mathbb{Q}_l \to H^i(G, M \otimes \mathbb{Q}_l)$$

is an isomorphism.

[A subgroup J of $H^i(G, M)$ is l-divisible if $lJ = J$. For the proof, this proposition follows immediately from either [78, Propositions 2.1 and 2.3] or [58, Propositions 2.7.8 and 2.7.11.] □

1.3 Continuous Cochain Cohomology

Proposition 1.3.12 *Suppose that the G-module M is a finitely generated $\mathbb{Z}_l$-module, where l is a prime number, and is equipped with the l-adic topology. Suppose that the cohomology groups of G with coefficients in finite l-primary modules are finite. Then for all n, $H^n(G, M)$ is a finitely generated $\mathbb{Z}_l$-module and the canonical map*

$$H^n(G, M) \otimes_{\mathbb{Z}_l} \mathbb{Q}_l/\mathbb{Z}_l \longrightarrow H^n(G, M \otimes_{\mathbb{Z}_l} \mathbb{Q}_l/\mathbb{Z}_l)$$

is an isogeny (i.e. it has finite kernel and cokernel) where $\mathbb{Q}_l/\mathbb{Z}_l$ is a discrete module.

[For the proof, see [58, Propositions 2.7.10.]] □

Inflation-restriction sequence

Proposition 1.3.13 *Let H be a closed normal subgroup of G and M be a topological G-module.*

(i) *There is an inflation-restriction sequence*

$$0 \longrightarrow H^1(G/H, M^H) \longrightarrow H^1(G, M) \longrightarrow H^1(H, M).$$

(ii) *Let R be a complete noetherian local ring with finite residue field of characteristic l. Let the G-module M be either a finitely generated R-module or a discrete module. Suppose that for every finite G-module (resp. finite H-module) S of order a power of l the groups $H^i(G, S)$ (resp. $H^i(H, S)$) are finite for $i = 1, 2$. Then there is an exact sequence extending that of (i)*

$$0 \to H^1(G/H, M^H) \to H^1(G, M) \to H^1(H, M)^{G/H} \to H^2(G/H, M^H)$$
$$\to H^2(G, M).$$

Proof If M is a discrete G-module then the exact sequences of (i) and (ii) are standard in the cohomology of profinite groups.

If M is a topological G-module, the exact sequence of (i) is proved exactly as in the case of discrete modules [58, Proposition 1.6.7].

For part (ii), the ring R is isomorphic to $\varprojlim R/\mathfrak{m}^i$ where $\mathfrak{m}$ is the maximal ideal of R. If M is a finitely generated R-module then it follows that M is algebraically and topologically isomorphic to $\varprojlim M/\mathfrak{m}^i M$. Put $M_i = M/\mathfrak{m}^i M$. Therefore from the case of discrete modules we have the exact sequence

$$0 \to H^1(G/H, M_i^H) \to H^1(G, M_i) \to H^1(H, M_i)^{G/H} \to H^2(G/H, M_i^H)$$
$$\to H^2(G, M_i).$$

Here all terms are finite groups from the hypotheses on G and H and hence the inverse systems formed from the groups in this exact sequence all satisfy the Mittag-Leffler condition. Passing to the inverse limit this sequence therefore remains exact and we then obtain the exact sequence of part (ii) from Proposition 1.3.10.

Corollary 1.3.14 *Suppose the G-module M is a finite dimensional $\mathbb{Q}_l$-vector space and that H is a closed normal subgroup of G. Suppose that for every finite G-module (resp. finite H-module) S of order a power of l the groups $H^i(G, S)$ (resp. $H^i(H, S)$) are finite for $i = 1, 2$. Then there is an exact sequence*

$$0 \to H^1(G/H, M^H) \to H^1(G, M) \to H^1(H, M)^{G/H} \to H^2(G/H, M^H) \to H^2(G, M).$$

Proof As G is compact, there is a finitely generated $\mathbb{Z}_l$-submodule N of M which is a G-submodule and for which $N \otimes_{\mathbb{Z}_l} \mathbb{Q}_l \cong M$. The result now follows from Propositions 1.3.11 and 1.3.13(ii).

Continuous cochain cohomology of $\widehat{\mathbb{Z}}$, the profinite completion of $\mathbb{Z}$

(1.3.15) Suppose for this paragraph that G is a finite cyclic group, τ is a generator of G, and M is a G-module and let $\widehat{H}^i(G, M)$ be the Tate modified cohomology groups of G with coefficients in M [58, Chapter 1, §§2, 7]. Then we have for all integers $i \in \mathbb{Z}$ [58, proposition 1.7.1]

$$\widehat{H}^{2i}(G, M) \cong M^G / N_G M, \quad \widehat{H}^{2i-1}(G, M) \cong {}_{N_G} M / I_G M$$

where

$$N_G = \sum_{\sigma \in G} \sigma, \quad I_G = (g - 1, g \in G)$$

and I_G is the augmentation ideal of the group ring $\mathbb{Z}[G]$ of G and so $I_G = (\tau - 1)$ for τ a generator of G. Here we have

$$M^G = \{a \in M \mid ga = a \text{ for all } g \in G\}, \quad N_G M = \{\sum_{g \in G} ga \mid a \in M\}$$

and

$${}_{N_G} M = \{m \in M \mid (\sum_{g \in G} g)m = 0\}, \quad I_G M = \{\tau a - a \mid a \in M\}.$$

Proposition 1.3.16 *Suppose that the topological group G is isomorphic to $\widehat{\mathbb{Z}}$ the profinite completion of $\mathbb{Z}$. Let F be a topological generator of G and M be a topological G-module.*

1.3 Continuous Cochain Cohomology

Suppose that $M = \varprojlim M_n$ where each M_n, for $n \in \mathbb{N}$, is a finite discrete G-module and M is equipped with its profinite topology. Then we have the natural isomorphisms for the continuous cochain cohomology of G

$$H^i(G, M) \cong \begin{cases} M^G & \text{if } i = 0 \\ M/(F-1)M & \text{if } i = 1 \\ 0 & \text{if } i > 1. \end{cases}$$

The isomorphism for $i = 1$ is given by the evaluation of a 1-cocycle at F.

Proof This is well known at least for finite G-modules [66, 68] but we recall the proofs. We have evidently $H^0(G, M) = M^G$.

Write $G = \varprojlim G/G(s)$ where $G(s) = sG$ runs over the open subgroups of finite index of G and s runs over all positive integers. The groups $H^i(G, M_n)$, which are the profinite cohomology groups of G with coefficients in a finite discrete module M_n, are then given as direct limits [58, Proposition 1.2.5]

$$H^i(G, M_n) \cong \varinjlim_s H^i(G/G(s), M_n^{G(s)}) \qquad (i \geq 0).$$

Fix a topological generator F of G and denote by the same symbol F for the image of F in the quotient groups $G/G(s)$. The group $G/G(s)$ is finite cyclic and it therefore follows from computation of the cohomology of a finite cyclic group that we have the isomorphisms for $i = 1$

$$H^1(G/G(s), M_n^{G(s)}) \cong {}_{N_{G/G(s)}}(M_n^{G(s)})/(F-1)(M_n^{G(s)}).$$

where this isomorphism is given by the evaluation of a 1-cocycle at F.

For positive integers s, t, the transition homomorphisms from $G(s)$ to $G(st)$, on both sides of this isomorphism, are given by the inclusion $M_n^{G(s)} \subseteq M_n^{G(st)}$. As M_n is invariant under $G(t)$ for some t, passing to the direct limit we obtain

$$H^1(G, M_n) \cong M'_n/(F-1)M_n$$

where M'_n is the submodule of M_n of elements annihilated by $1 + F + \ldots + F^u$ for some positive integer u. But M_n is torsion so that if the positive integer r annihilates M_n we have for any $m \in M_n$

$$\sum_{i=0}^{tr-1} F^i m = (1+F+\ldots F^{t-1})(1+F^t+F^{2t}+\ldots+F^{(r-1)t})m = (1+F+\ldots F^{t-1})rm = 0$$

and therefore $M'_n = M_n$ and hence

$$H^1(G, M_n) \cong M_n/(F-1)M_n$$

where the isomorphism is given by evaluation of a cocycle at F. In particular, these groups $H^i(G, M_n)$ are finite for $i = 0, 1$.

The exact sequence

$$0 \longrightarrow (F-1)M_n \longrightarrow M_n \longrightarrow M_n/(F-1)M_n \longrightarrow 0 \qquad (1.3.17)$$

remains exact after applying $\varprojlim_n$ because the inverse system $\{(F-1)M_n\}_n$ with its evident transition maps satisfies the Mittag-Leffler condition as the M_n are finite for all n. Therefore from the exact sequence (1.3.17) and Proposition 1.3.10, we obtain the isomorphism

$$H^1(G, M) \cong \varprojlim_n M_n/(F-1)M_n \cong M/(F-1)M.$$

For the second cohomology group, it follows from the cohomology of a finite cyclic group that we have the isomorphisms

$$H^2(G/G(s), M_n^{G(s)}) \cong (M_n^{G(s)})^{G/G(s)}/N_{G/G(s)}(M_n^{G(s)}) \cong M_n^G/N_{G/G(s)}(M_n^{G(s)}).$$

For positive integers s, t, the transition homomorphism

$$M_n^G/N_{G/G(s)}(M_n^{G(s)}) \to M_n^G/N_{G/G(st)}(M_n^{G(st)})$$

from $G(s)$ to $G(st)$ is given by multiplication by t on $M_n^G \to M_n^G$. It follows that as M_n is finite taking the direct limit over all integers s

$$H^2(G, M_n) \cong \varinjlim_s H^2(G/G(s), M_n^{G(s)}) = 0.$$

We have already shown that the groups $H^i(G, M_n)$ are finite for $i \leq 1$ and all n. It follows from Proposition 1.3.10 that there are isomorphisms

$$H^2(G, M) = \varprojlim H^2(G, M_n) = 0.$$

For the higher cohomology groups, it follows by dimension shifting that $H^i(G, N)$ is zero for all $i \geq 3$ and all finite discrete G-modules N; again using Proposition 1.3.10 we obtain that for all $i \geq 3$

1.3 Continuous Cochain Cohomology

$$H^i(G, M) = \varprojlim H^i(G, M_n) = 0.$$ □

Corollary 1.3.18 *Suppose that the topological group G is isomorphic to $\widehat{\mathbb{Z}}$ and that F is a topological generator of G. Suppose also that R is one of the following:*

(a) A complete noetherian local ring with finite residue field;
(b) The field of fractions of a complete discrete valuation ring with finite residue field.

Suppose that M is a continuous G-module which is either a discrete torsion G-module or a finitely generated R-module. Then we have for the continuous cochain cohomology

$$H^i(G, M) \cong \begin{cases} M^G & \text{if } i = 0 \\ M/(F-1)M & \text{if } i = 1 \\ 0 & \text{if } i \geq 2 \end{cases}$$

where the isomorphism for $i = 1$ is the map of evaluation of a 1-cocycle at F. Furthermore G has cohomological dimension 1.

Proof Suppose that M is a discrete torsion G-module. Then it is a direct limit of finite G-modules. As the cohomology $H^i(G, -)$ commutes with direct limits, the corollary follows from the case of finite G-modules given by Proposition 1.3.16.

Suppose that M is a finitely generated R-module where R is as in case (a). Then M is a projective limit of finite G-modules and so the result follows from Proposition 1.3.16.

Suppose finally that M is a finitely generated R-module where R is as in case (b). Then R is the fraction field of a complete discrete valuation ring S with finite residue field and M is a finite dimensional R-vector space.

Suppose that

$$\rho : G \longrightarrow \mathrm{GL}(V)$$

is a representation where V is a finite dimensional R-vector space. Then ρ is continuous if and only if the characteristic polynomial $\det(F - \mathrm{id}.X|V)$ has coefficients in S. It follows that because M is a continuous G-module, the polynomial $\det(F - \mathrm{id}.X|M)$ has coefficients in S and therefore there is a finitely generated S-submodule N of M which is a $S[[G]]$-module so that there is an isomorphism of $R[[G]]$-modules

$$N \otimes_S R \cong M$$

and where N is an S-torsion free module.

We have the exact sequence $0 \to S \to R \to R/S \to 0$. Tensoring this with N gives the short exact sequence

$$0 \to N \to M \to N \otimes_S R/S \to 0.$$

This gives a cohomology sequence

$$(N \otimes R/S)^G \to H^1(G, N) \to H^1(G, M) \to H^1(G, N \otimes R/S).$$

As $(N \otimes R/S)^G$ is an S-torsion module, the image of

$$(N \otimes R/S)^G \to H^1(G, N)$$

is therefore an S-torsion subgroup of $H^1(G, N)$.

But M is an R-vector space and hence $H^1(G, M)$ is an R-vector space. It follows that the S-torsion subgroup of $H^1(G, N)$ is exactly the image of $(N \otimes R/S)^G \to H^1(G, N)$. The image I of the map

$$H^1(G, M) \to H^1(G, N \otimes R/S)$$

is a divisible subgroup of $H^1(G, N \otimes R/S)$ because $H^1(G, M)$ is divisible; as R/S is discrete and G is compact a cochain with values in $N \otimes R/S$ takes only a finite set of values; therefore this image I is an S-torsion group as well as an S-divisible group. Tensoring the above exact cohomology sequence with R we obtain that

$$H^1(G, N) \otimes_S R \to H^1(G, M)$$

is an isomorphism.

As N is a finitely generated S-module, it is a projective limit of finite G-modules. From Proposition 1.3.16, we then have that $H^1(G, N)$ is S-isomorphic to $N/(F-1)N$. From this isomorphism $H^1(G, N) \cong N/(F-1)N$, we obtain that

$$H^1(G, M) \cong (N \otimes_S R)/(F-1)(N \otimes_S R).$$

In the same way we have the cohomology sequence extending that above where $i \geq 2$

$$H^{i-1}(G, N \otimes R/S) \to H^i(G, N) \to H^i(G, M) \to H^i(G, N \otimes R/S) \to H^{i+1}(G, N).$$

If follows from Proposition 1.3.16 that $H^i(G, N) \cong H^{i+1}(G, N) \cong 0$; hence this sequence reduces to an isomorphism $H^i(G, M) \cong H^i(G, N \otimes R/S)$. But $H^i(G, M)$ is an R-vector space and $H^i(G, N \otimes R/S)$ is an S-torsion group, exactly as for the case $i = 1$; hence both these groups are zero for $i \geq 2$.

We have shown that $H^i(G, T) = 0$ for every discrete torsion G-module T and all $i \geq 2$; it can be seen that $H^1(G, T) = T/(F-1)T$ is non-zero for some finite G-module T. Hence by definition G has cohomological dimension 1 (see Remark 1.3.19. or [68, Chapitre 1, §3.1, Proposition 11 p.17]).

1.3 Continuous Cochain Cohomology

Remarks 1.3.19 (i) Let G be a profinite group. Let $\mathcal{C}_G$ be the abelian category of discrete abelian groups on which G acts continuously and let $\mathcal{C}'_G$ be the sub-abelian category of $\mathcal{C}_G$ of torsion discrete abelian groups on which G acts continuously.

We recall that for any prime number p the *p-strict cohomological dimension* $\operatorname{scd}_p(G)$ (resp. *p-cohomological dimension* $\operatorname{cd}_p(G)$) of G is the least integer n such that for all $M \in \mathcal{C}_G$ (resp. $M \in \mathcal{C}'_G$) we have $H^i(G, M)(p) = 0$ for all $i > n$ where $H^i(G, M)(p)$ denotes the p-primary component of $H^i(G, M)$.

The *strict cohomological dimension* (resp. *cohomological dimension*) of G is defined to be

$$\operatorname{scd}(G) = \sup_p \operatorname{scd}_p(G)$$

(resp.

$$\operatorname{cd}(G) = \sup_p \operatorname{cd}_p(G))$$

where the supremum runs over all prime numbers p.

The strict cohomological dimension $\operatorname{scd}(\widehat{\mathbb{Z}})$ of $\widehat{\mathbb{Z}}$ is equal to 2. For we have $H^2(\widehat{\mathbb{Z}}, \mathbb{Z}) \cong H^1(\widehat{\mathbb{Z}}, \mathbb{Q}/\mathbb{Z}) \cong \mathbb{Q}/\mathbb{Z}$ so that $\operatorname{scd}(\widehat{\mathbb{Z}}) \geq 2$ and equality holds here by [68, Chapitre 1, §3.2, proposition 13, p. 19]. This contrasts with Corollary 1.3.18 where the cohomological dimension of $\widehat{\mathbb{Z}}$ is shown to be equal to 1.

(ii) As the Galois group of the maximal separable abelian extension of a non-archimedean local field is isomorphic to $\widehat{\mathbb{Z}}$, the results of this subsection apply to this Galois group.

Galois cohomology

(1.3.20) Let L be a field and let K be a Galois field extension of L, not necessarily finite. Then $\operatorname{Gal}(K/L)$ is a profinite group for which we may consider the corresponding continuous cochain cohomology groups.

(1.3.21) Let M be a continuous $\operatorname{Gal}(K/L)$-module.

(a) We write $H^i(K/L, M)$ in place of $H^i(\operatorname{Gal}(K/L), M)$ where this is cohomology computed via continuous cochains;

(b) When K is the separable closure of L, we write $H^i(G_L, M)$ or $H^i(L, M)$ for the Galois cohomology group $H^i(\operatorname{Gal}(L^{\operatorname{sep}}/L), M)$ where $G_L = \operatorname{Gal}(L^{\operatorname{sep}}/L)$;

(c) If L'/L is a finite Galois field extension and M is a $\operatorname{Gal}(L'/L)$-module, we write $H^i(L'/L, M)$ for $H^i(\operatorname{Gal}(L'/L), M)$;

(d) If the module M is equipped with the discrete topology, then the groups $H^i(K/L, M)$ are "classical" Galois cohomology groups [68].

(1.3.22) The inflation, restriction, and corestriction homomorphisms of continuous cochain cohomology transfer to Galois cohomology. Let $F'' \supseteq F' \supseteq F$ be separable field extensions, not necessarily finite, where F''/F is Galois.

(a) Suppose that F'/F is Galois. Then $\mathrm{Gal}(F''/F')$ is a closed normal subgroup of $\mathrm{Gal}(F''/F)$. Let M be a topological $\mathrm{Gal}(F''/F)$-module. Then $M^{\mathrm{Gal}(F''/F')}$ is a topological $\mathrm{Gal}(F'/F)$-module and we then obtain the functorial inflation homomorphisms of cohomology

$$\inf\nolimits_{F''/F'} : H^i(F'/F, M^{\mathrm{Gal}(F''/F')}) \longrightarrow H^i(F''/F, M);$$

(b) There is a restriction homomorphism

$$\mathrm{res}_{F'/F} : H^i(F''/F, M) \to H^i(F''/F', M)$$

obtained from the inclusion of Galois groups $\mathrm{Gal}(F''/F') \subseteq \mathrm{Gal}(F''/F)$;

(c) If F'/F is finite then $\mathrm{Gal}(F''/F')$ is a closed subgroup of $\mathrm{Gal}(F''/F)$ of finite index. We then have the *norm* or *corestriction* homomorphism

$$N_{F'/F} : H^i(F''/F', M) \to H^i(F''/F, M).$$

Exercise 1.3.23 In order to apply Proposition 1.3.12, it is required to know when the groups $H^i(G, M)$ are finite for a finite G-module M. For this there is the following special case.

Suppose that the profinite group G is one of the following:

(i) F is a global field, F_S is a Galois extension of F, possibly infinite, which is unramified outside a finite set S of places of F, and $G = \mathrm{Gal}(F_S/F)$;
(ii) F is a local field and $G = G_F$;
(iii) F is a non-archimedean local field and G is the inertia subgroup of G_F.

If M is a finite discrete topological G-module, show that for all $i \in \mathbb{N}$ the group $H^i(G, M)$ is finite.

[Hint: It is easier to demonstrate the result when the characteristic of F is coprime to the order of M, but the result remains true even if this simplifying condition is not satisfied.]

1.3.24. Bibliographical Remarks. Section 1.3 only covers basics on continuous cochain cohomology. For more detailed expositions on this, see [78, §2], [31], [58, Chapter 2, §7] and [62, Appendix B]. For comprehensive expositions of the cohomology of profinite groups, see [58, Chapters 1–3] and [68].

1.4 Orders in Imaginary Quadratic Fields

(1.4.1) Let

> k be a global field equipped with a special set of places ∞;
> F/k be an imaginary quadratic field extension with respect to ∞;
> A be the ring of integers of k with respect to ∞;
> B be the integral closure of A in F.

An *order* O in F with respect to A is an A-subalgebra of B whose fraction field is equal to F.

There is a bijection between orders O_c of F with respect to A and cycles c in $\text{Cyc}(k)$ where $\text{Supp}(c)$ contains only finite primes and it is given by

$$c \mapsto A + BI(c)$$

where $I(c)$ is the ideal of A cutting out the cycle c. The cycle c is the *conductor* of the order O_c.

[For more details on orders in imaginary quadratic extensions, see [4, Chapter 2, §2.2.]]

1.5 Ring Class Fields

(1.5.1) Heegner points on modular curves are defined over *ring class fields*. We shall in this section give some brief notes on ring class fields for global fields k equipped with a special set of places ∞.

The hypotheses and notation of the last Sect. 1.4 hold in this section. Let

> $c \in \text{Cyc}(k)$ be a cycle supported only at non-archimedean primes and without places in ∞;
> O_c be the order of F with respect to A and with conductor c (see §1.4);
> A_v, for each place v of A, be the localization of A at v;
> $\widehat{O}_{c,v}$ be the completion of the semi-local ring $O_c \otimes_A A_v$ with respect to the topology defined by its Jacobson radical.

(1.5.2) By definition, the places of ∞ are inert in the imaginary quadratic extension F/k. The completion F_v of F at the place v of k is then a field isomorphic to $F \otimes_k k_v$. The product $\prod_{v \in \infty} F_v^*$ is a well-defined subgroup of the idèle group of F.

For each non-archimedean place v of k, we have the group of units $\widehat{O}_{c,v}^*$ of the complete semi-local ring $\widehat{O}_{c,v}$; again $\widehat{O}_{c,v}^*$ can be considered a subgroup of the idèle group I_F of F.

Let $G_c = \prod_{v \in \infty} F_v^* \prod_{v \notin \infty} \widehat{O}_{c,v}^*$ be the subgroup of the idèle group I_F of the global field F whose components are the units of $\widehat{O}_{c,v}$ for all places $v \notin \infty$ of k and F_v^* for the places $v \in \infty$ and where in the double product v runs over all places of k.

Definition 1.5.3 The subgroup $F^* G_c$ is a subgroup of finite index of the idèle group I_F and the field $F[c]$ defined by the reciprocity isomorphism

$$I_F / F^* G_c \xrightarrow{\cong} \operatorname{Gal}(F[c]/F)$$

is the *ring class field of F/k with conductor c*.

This $F[c]$ is a finite abelian separable extension field of F defined by the subgroup $F^* G_c$ of I_F.

We write $G(c/c')$ for the Galois group of the field extension $F[c]/F[c']$ for cycles $c \geq c'$ of $\operatorname{Cyc}(k)$ supported only at finite primes.

(1.5.4) We have these properties of the decomposition of primes in ring class fields (the proofs are immediate consequences of class field theory; for more details in the function field case, see [4, Chapter 2, §2.3.13]):

(a) The primes ramified in $F[c]/F$ are precisely the primes of F dividing the support of c;
(b) The extension $F[c]/F$ is split completely at every place of F lying above a place of ∞;
(c) If z is a place of k where $z \notin \operatorname{Supp}(c)$ and $z \notin \infty$ then for any positive integer n, the Galois extension $F[cz^n]/F[c]$ is totally ramified at all places of $F[c]$ above z.

Lemma 1.5.5 *Let $F[c], F[c']$ be two ring class fields for $c, c' \in \operatorname{Cyc}(k)$, coprime to ∞. Then in the maximal separable abelian extension of F we have the equality of subfields*

$$F[c] \cap F[c'] = F[\gcd(c, c')]$$

where $\gcd(c, c')$ is the greatest common divisor of the cycles c and c'.

Proof The fields $F[c], F[c']$ are associated by class field theory to the following subgroups of the idèle group I_F of F:

$$G_c = \prod_{v \in \infty} F_v^* \prod_{v \neq \infty} \widehat{O}_{c,v}^*, \quad G_{c'} = \prod_{v \in \infty} F_v^* \prod_{v \neq \infty} \widehat{O}_{c',v}^*.$$

Put $d = \gcd(c, c') = \prod_v v^{n_v}$ where v runs over the non-archimedean prime divisors of k and n_v are natural numbers. We then have the equality of subgroups of the multiplicative group F_v^*, which is the unit group of $F_v = F \otimes_k k_v$,

1.5 Ring Class Fields

$$\widehat{O}^*_{c,v} \cdot \widehat{O}^*_{c',v} = (\widehat{A}_v + \pi_v^{n_v} B \otimes_A \widehat{A}_v)^*$$

where $\widehat{A}_v$ is the completion of A at v and π_v is a local parameter of $\widehat{A}_v$. It then follows that there is an equality of subgroups of I_F

$$F^* G_c G_{c'} = (F^* \prod_{v \in \infty} F_v^* \prod_{v \neq \infty} \widehat{O}^*_{c,v})(F^* \prod_{v \in \infty} F_v^* \prod_{v \neq \infty} \widehat{O}^*_{c',v}) = F^* \prod_{v \in \infty} F_v^* \prod_{v \neq \infty} \widehat{O}^*_{d,v}$$

$$= F^* G_{\gcd(c,c')}$$

whence the result by class field theory.

Proposition 1.5.6 *The finite group $\mathrm{Gal}(F[c]/k)$ is generalized dihedral of the form*

$$\mathrm{Gal}(F[c]/k) \cong \mathrm{Gal}(F[c]/F) \rtimes \mathrm{Gal}(F/k).$$

Proof The reciprocity map rec: $I_F \to \mathrm{Gal}(F^{\mathrm{ab}}/F)$, where I_F is the idèle group of F and F^{ab} is the maximal abelian separable extension of F, is compatible with the action of the group $\mathrm{Gal}(F/k)$ of order 2 generated by the unique non-trivial element σ. Then σ extends uniquely to an automorphism of I_F also denoted by σ. As $a.\sigma(a) \in I_k$ for all $a \in I_F$ where I_k is the idèle group of k, we then have that $F[c]/k$ is Galois and the conjugation action of σ on $\mathrm{Gal}(F[c]/F)$ is given by $\sigma g \sigma^{-1} = g^{-1}$ whence the result.

Remarks 1.5.7

(i) See [4, Chapter 2, §2.3] for more details on ring class fields in the function field case. The results and proofs given there transfer easily to the case of ring class fields of algebraic number fields.
In the case of algebraic number fields, we shall only use ring class fields for totally imaginary quadratic extensions of totally real fields where we use these for Shimura curves and classical modular curves. In §§1.4,1.5 what we have called an imaginary quadratic extension of a totally real field is normally called a *totally imaginary quadratic extension field*.
(ii) We have only defined orders and ring class fields for imaginary quadratic extensions of global fields. But these may be defined more generally.
For example, the construction above may be extended as follows. Let F be an algebraic number field with ring of integers A. An *order* of F is a subring of A whose field of fractions is equal to F.
Let O be an order of F. For a non-zero prime ideal $\mathfrak{p}$ of O let $\widehat{O}^*_\mathfrak{p}$ be the group of units of the completion of O with respect to $\mathfrak{p}$. Let $G = F^* \prod_{v \in \infty} F_\infty^* \prod_\mathfrak{p} \widehat{O}^*_\mathfrak{p}$ where the first product runs over the finite set of archimedean places of F and the second product runs over all non-zero prime ideals $\mathfrak{p}$ of O. Then G can be considered a subgroup of

finite index of the idèle group I_F of the global field F. The quotient group I_F/G is isomorphic, via the reciprocity map, to the Galois group of a finite abelian Galois extension field of F which is the ring class field of F given by the order O.

1.6 Group Representations over a Local Principal Ideal Ring

(1.6.1) In this section we give some basic results on group representations over a local principal ideal ring. For more details, see [12, §30C].

Let R be a local principal ideal ring with local parameter π and maximal ideal m. That is to say R is a homomorphic image of a discrete valuation ring and π generates the maximal ideal of R.

Let G be a finite group and let M be a finite free R-module on which G acts on the left. An $R[G]$-*lattice* is a finitely generated left $R[G]$-module which is a free R-module.

(1.6.2) The following properties are easily verified:

(a) Every short exact sequence of $R[G]$-lattices is split over R;
(b) Every $R[G]$-lattice is an injective R-module;
(c) If $N \subseteq M$ are $R[G]$-lattices then so is M/N (from (b));
(d) Direct summands of $R[G]$-lattices are again $R[G]$-lattices.

(1.6.3) Suppose now that the order of the finite group G is a unit in R. Then we have:

(a) Every $R[G]$-lattice is $R[G]$-projective;
(b) The correspondence

$$M \mapsto M \otimes_R R/\mathfrak{m}$$

is an isomorphism-preserving bijection between the set of $R[G]$-lattices and the set of finitely generated projective $(R/\mathfrak{m})[G]$-modules;
(c) The Krull-Schmidt theorem holds for $R[G]$-lattices and so each $R[G]$-lattice M is uniquely a direct sum of isotypical components M_i where each M_i is a direct sum of copies of the same indecomposable projective $R[G]$-lattice N_i and the N_i are non-isomorphic $R[G]$-lattices;
(d) Suppose that χ is an irreducible character of G with values in $R/\mathfrak{m}$. There is a finitely generated projective irreducible $(R/\mathfrak{m})[G]$-module $\overline{N}$ with character χ. Then there is an indecomposable $R[G]$-lattice N such that $N \otimes_R R/\mathfrak{m} \cong \overline{N}$ is an isomorphism of $(R/\mathfrak{m})[G]$-modules. For an $R[G]$-lattice M we denote by M^χ the isotypical component of M corresponding to N and M^χ is called the χ- *isotypical component* of M (by (1.6.2)(c)).

(1.6.4) Suppose now that $R/\mathfrak{m}$ contains all the $|G|$th roots of unity, the characteristic of the residue field of R if it is non-zero is coprime to $|G|$, the group G is abelian, and R is a complete local principal ideal ring. Then we have:

(a) Every irreducible character of G over $R/\mathfrak{m}$ is one-dimensional and its character takes values in $R/\mathfrak{m}$;
(b) Any irreducible character χ of G over $R/\mathfrak{m}$ lifts to a group homomorphism $\chi': G \to R$ by taking Teichmüller representatives in R of elements of $R/\mathfrak{m}$;
(c) For an $R[G]$-lattice M and any irreducible $R/\mathfrak{m}$ character χ of G, then M^χ is the isotypical component of M and is equal to $\frac{1}{|G|} \sum_{g \in G} \chi'(g^{-1}) g M$.

Elliptic Curves; Drinfeld Modules; Shimura Curves; Elliptic Spaces

Contents

2.1	Elliptic Curves	23
2.2	Drinfeld Modules	26
2.3	Shimura Curves	28
2.4	Elliptic Spaces	36
2.5	Moduli Schemes of Elliptic Spaces	37
2.6	Modularity of Elliptic Curves over Global Fields	41
2.7	Torsion Points on Elliptic Curves	42

2.1 Elliptic Curves

(2.1.1) Let L be a field. Then an elliptic curve E/L is defined to be a one-dimensional abelian variety over L.

Similarly for a scheme S, a family of elliptic curves E/S is a S-group scheme E with identity element $e : S \to E$ where the fibre E_x over any point $x \in S$ of the structure morphism $E \to S$ is an elliptic curve over the residue field $\kappa(x)$ whose zero point is induced by the section s.

[For basics on elliptic curves, see [74, 75].]

The conductor of an elliptic curve over a global field

(2.1.2) Let F be a global field and E/F be an elliptic curve. The conductor of an elliptic curve E/F is a cycle on F supported only at the places of bad reduction of E and whose multiplicities are defined in terms of the Galois representation of $\text{Gal}(F^{\text{sep}}/F)$ given by E. We define the conductor in the next paragraphs:

(2.1.3) Let L be a non-archimedean local field of residue characteristic $p > 0$. For a finite Galois extension of fields J/L, let G be the Galois group of J/L. Let R be the valuation ring of the field J, v_J be the normalized valuation on J, and $\mathfrak{m}$ be the maximal ideal of R. The ith higher ramification group is defined to be

$$G_i = \{\sigma \in G \mid v_J(\alpha^\sigma - \alpha) \geq i + 1 \text{ for all } \alpha \in R\}.$$

Then G_i is the largest normal subgroup of G which acts trivially on $R/\mathfrak{m}^{i+1}$. In particular, G_0 is the inertia subgroup of G. Put $g_i = |G_i|$.

(2.1.4) Let k be a field of characteristic different from the residue characteristic of L. Let W be a finite dimensional representation over k of $\text{Gal}(J/L)$ so that there is a homomorphism

$$\rho : \text{Gal}(J/L) \longrightarrow \text{End}_k(W).$$

Let G_0 be the inertia subgroup of $\text{Gal}(J/L)$. The *tame part of the conductor* of ρ is equal to

$$\epsilon(\rho) = \dim_k \left(\frac{W}{W^{G_0}} \right).$$

The *wild part of the conductor* (or the *Swan conductor*) of ρ is equal to

$$\text{sw}(\rho) = \sum_{i=1}^{\infty} \frac{g_i}{g_0} \dim_k \left(\frac{W}{W^{G_i}} \right).$$

The *exponent of the conductor* of ρ is then equal to

$$f(\rho) = \epsilon(\rho) + \text{sw}(\rho).$$

(2.1.5) Let E be an elliptic curve defined over the non-archimedean local field L of residue characteristic p. Let l be a prime number distinct from the residue characteristic of L and let $T_l(E)$ be the l-adic Tate module of E/L; put $V_l = T_l(E) \otimes_{\mathbb{Z}_l} \mathbb{Q}_l$. Then V_l is equipped with an action by $\text{Gal}(L^{\text{sep}}/L)$

$$\rho : \text{Gal}(L^{\text{sep}}/L) \longrightarrow \text{End}_{\mathbb{Q}_l}(V_l).$$

Let G_0 be the inertia subgroup of $\text{Gal}(L^{\text{sep}}/L)$. The *tame part of the conductor* of E is equal to

$$\epsilon(E) = \dim_{\mathbb{Q}_l} \left(\frac{V_l}{V_l^{G_0}} \right).$$

2.1 Elliptic Curves

Let $E[l]$ be the l-torsion subgroup of E so that $E[l]$ is finite group scheme over L and for this paragraph we do not distinguish between $E[l]$ and the group of points $E[l](L^{\text{sep}})$ which has the structure of a two-dimensional vector space over $\mathbb{Z}/l\mathbb{Z}$. Let $J = L(E[l])$, that is to say J is the smallest field of rationality over L of the l-torsion points of E.

The *wild part of the conductor* of E/L is equal to

$$\text{sw}(E) = \sum_{i=1}^{\infty} \frac{g_i}{g_0} \dim_{\mathbb{Z}/l\mathbb{Z}} \left(\frac{E[l]}{E[l]^{G_i}} \right).$$

The *exponent of the conductor* of E/L is then

$$f(E/L) = \epsilon(E) + \text{sw}(E).$$

The exponent of the conductor is independent of the choice of prime number l (see [75, Chap. IV, Theorem 10.2]).

(2.1.6) Suppose that E is an elliptic curve defined over a global field F. The *conductor* of E/F is then the cycle

$$f(E/F) = \prod_v v^{f(E/F_v)}$$

where the product runs over all non-archimedean places v of F and where $f(E/F_v)$ is the local exponent of the conductor of $E \otimes_F F_v$ over the local field F_v and where F_v denotes the completion of F at v. The conductor is known to be independent of the prime number l used in its definition (see [75, Chap. IV, Theorem 10.2]).

Ogg's formula

Proposition 2.1.7 (Ogg) *Let L be a local field and E/L be an elliptic curve. Let*

$v_L(\mathfrak{D}_{E/L})$ *be the valuation of the minimal discriminant of E/L;*
$f(E/L)$ *be the exponent of the conductor of E/L;*
$m(E/L)$ *be the number of components of the special fibre of the minimal proper regular model of E/L.*

Then we have

$$v_L(\mathfrak{D}_{E/L}) = f(E/L) + m(E/L) - 1.$$

In particular if E/L has good reduction, then $f(E/L) = 0$.

[For the proof, see, for example, [75, Chapter IV, §11].

Remarks 2.1.8

(i) Suppose that F is a global field of positive characteristic. Then F is the function field of a smooth projective irreducible curve C over the finite field k which is the exact field of constants of F.
An elliptic curve E/F then extends to a morphism of k-schemes

$$\mathcal{E} \to C$$

where $\mathcal{E}/k$ is an elliptic surface over k, that is to say $\mathcal{E}$ is a is a smooth projective irreducible surface over k where the morphism $\mathcal{E} \to C$ has a section and where all fibres of this morphism, except a finite number, are elliptic curves.
Similarly if F is a number field with ring of integers A, then an elliptic curve E/F extends to an arithmetic elliptic surface which is a morphism of A-schemes

$$\mathcal{E} \to \mathrm{Spec}\, A$$

whose generic fibre is isomorphic to the elliptic curve E and where all but finitely many fibres are elliptic curves.
This extension of the elliptic curve E/F over a global field to an elliptic surface over a finite field or an arithmetic elliptic surface is not unique. But there is a canonical choice of such an extension, namely, the *Néron model* of an elliptic curve (see [75, Chap. IV]).

(ii) It follows from Ogg's formula that if E/F is an elliptic curve over a global field F, then the local exponent of the conductor $f(E/F_v) = 0$ for all places v except possibly for the places of bad reduction of E/F. Hence the conductor $\prod_v v^{f(E/F_v)}$ is a product over finitely many places of F as the curve E/F has good reduction at all but finitely many places of F.

(iii) The conductor of an abelian variety over a global field is defined in exactly the same way as in (2.1)–(2.1) for an elliptic curve (see [77]).

2.2 Drinfeld Modules

(2.2.1) Let k be a global field of characteristic $p > 0$ equipped with a special set ∞ of places; in fact ∞ contains precisely one place. Let A be the ring of integers of k with respect to ∞ (see Sect. 1.1). Let $a \mapsto |a|$ be the absolute value on k corresponding to the unique place in ∞ and normalised so that $|a|_\infty = \sharp(A/aA)$, the cardinality of A/aA, for all non-zero $a \in A$.

We shall briefly define Drinfeld modules for the ring A.

[For more details and basic results on Drinfeld modules, see [16].]

2.2 Drinfeld Modules

Drinfeld modules over a field

(2.2.2) Let L be a field of characteristic $p > 0$ and $\mathbb{G}_a$ be the additive group scheme over L. The ring of endomorphisms $\text{End}_L(\mathbb{G}_a)$ is isomorphic to the non-commutative ring $L\{\tau\}$ of non-commutative polynomials

$$\sum_{i=0}^{m} a_i \tau^i$$

where $a_i \in L$ and where multiplication is defined to satisfy the commutation rule

$$(a\tau^i)(b\tau^j) = ab^{p^i} \tau^{i+j} \qquad (a, b \in A).$$

The isomorphism is given by, where $\mathbb{G}_a = \text{Spec } L[X]$,

$$L\{\tau\} \to \text{End}_L(\mathbb{G}_a)$$

$$\sum_{i=0}^{m} a_i \tau^i \mapsto \{\mathbb{G}_a \to \mathbb{G}_a, X \mapsto \sum_{i=0}^{m} a_i X^{p^i}\}.$$

Define the degree of the non-commutative polynomial $\sum_{i=0}^{m} a_i \tau^i$ to be p^m where m is the largest integer for which $a_m \neq 0$.

[See [16, §1] for more details on the endomorphism ring $\text{End}(\mathbb{G}_a)$.]

(2.2.3) A *Drinfeld module* over L of rank r, where r is a positive real number, is a ring homomorphism

$$\phi : A \to \text{End}(\mathbb{G}_a)$$

such that

$$\deg(\phi(a)) = |a|_{\infty}^{r}$$

for all non-zero $a \in A$.

It can be shown that the rank r of Drinfeld module is always a positive integer (see [16, §3]).

Given the Drinfeld module ϕ, the constant term of $\phi(a)$ as a non-commutative polynomial, as a runs over all elements of A, provides a ring homomorphism $\iota : A \to L$. The *characteristic* of the Drinfeld module ϕ is defined to be the kernel of ι, which is a prime ideal of A. If ι is injective, then ϕ is said to have *infinite characteristic*; otherwise, it has *finite characteristic*.

(2.2.4) Let $\phi : A \to \mathrm{End}(G_1)$ and $\psi : A \to \mathrm{End}(G_2)$ be two Drinfeld modules over the field L where G_1, G_2 are both isomorphic to $\mathbb{G}_a/L$. Then a morphism $f : \phi \to \psi$ of Drinfeld modules is a L-scheme morphism $f : G_1 \to G_2$ such that

$$\psi_a f = f \phi_a \text{ for all } a \in A.$$

Drinfeld modules over a scheme

(2.2.5) Let S be a locally noetherian A-scheme of characteristic $p > 0$, i.e. there is a morphism $S \to \mathrm{Spec}\, \mathbb{F}_p$. A *Drinfeld module* over S of rank r is an invertible sheaf $\mathcal{L}$ on S and a morphism of rings

$$\phi : A \to \mathrm{End}(\mathcal{L})$$

such that over any affine open subscheme U where $\mathcal{L}$ is trivial we have

$$\phi_a(x) = \sum_{i=0}^{m} b_i x^{p^i}, \quad b_i \in \Gamma(U, \mathcal{O}_U),$$

for $a \in A$ where

$$|a|_\infty^r = p^m$$

and b_m is a unit of the ring $\Gamma(U, \mathcal{O}_U)$.

2.3 Shimura Curves

(2.3.1) Shimura curves are a generalization of classical modular curves and they can parametrise elliptic curves over totally real number fields in the same way as classical modular curves may parametrise rational elliptic curves. We give some brief notes, mostly without proofs, in this section on quaternionic Shimura curves.

For any abelian group, or ring, A let $\widehat{A}$ denote the group, or ring, $A \otimes_\mathbb{Z} \widehat{\mathbb{Z}}$ where $\widehat{\mathbb{Z}}$ is the profinite completion of $\mathbb{Z}$.

Construction

(2.3.2) Let k be a totally real number field of degree d over $\mathbb{Q}$. Let S be a finite set of non-archimedean places of k such that $|S| + d$ is odd. Let $\tau_1, \ldots, \tau_d : k \to \mathbb{R}$ be the real embeddings of k.

Let B be a quaternion algebra over k which ramifies precisely at the places $S \cup \{\tau_2, \ldots, \tau_d\}$,

$$\mathrm{Ram}(B) = S \cup \{\tau_2, \ldots, \tau_d\} \tag{2.3.3}$$

2.3 Shimura Curves

that is to say $B \otimes k_v$, for $v \in S \cup \{\tau_2, \ldots, \tau_d\}$, is a division algebra over k_v and at any other place v of k then $B \otimes k_v$ is isomorphic to the matrix algebra $M_2(k_v)$; note that the place τ_1 is excluded from Ram(B).

The condition that $|S| + d$ be odd is necessary and sufficient for such a B to exist; furthermore, the quaternion algebra B satisfying this ramification condition (2.3.3) is unique up to isomorphism.

(2.3.4) Let G be the group scheme over the rational field $\mathbb{Q}$ determined by the condition that for any $\mathbb{Q}$-algebra A there is a functorial isomorphism

$$G(A) \cong (B \otimes_{\mathbb{Q}} A)^*.$$

Denote by $G_i/\mathbb{R}$ the group scheme

$$G_i = G \times_{k, \tau_i} \mathbb{R}$$

that is to say $G_i/\mathbb{R}$ is the base change of $G/\mathbb{Q}$ via the map $\mathbb{Q} \to k \xrightarrow{\tau_i} \mathbb{R}$. Then we have an isomorphism of $\mathbb{R}$-group schemes

$$G \times_{\mathbb{Q}} \mathbb{R} \cong G_1 \times \ldots \times G_d.$$

(2.3.5) Let $\operatorname{Res}_{\mathbb{C}/\mathbb{R}}(\mathbb{G}_m/\mathbb{C})$ be the $\mathbb{R}$-group scheme obtained from $\mathbb{G}_m/\mathbb{C}$ by restriction of scalars from $\mathbb{C}$ to $\mathbb{R}$. Let h_0 be the morphism

$$h_0 : \operatorname{Res}_{\mathbb{C}/\mathbb{R}}(\mathbb{G}_m/\mathbb{C}) \longrightarrow G \times_{\mathbb{Q}} \mathbb{R}$$

which on $\mathbb{R}$-valued points is given by

$$x + iy \mapsto (\begin{pmatrix} x & y \\ -y & x \end{pmatrix}, 1, \ldots, 1)$$

where $G(\mathbb{R}) \cong G_1(\mathbb{R}) \times \ldots \times G_d(\mathbb{R})$ and $G_1(\mathbb{R}) \cong \operatorname{GL}_2(\mathbb{R})$ and where $G_i(\mathbb{R})$ are groups of units of division algebras over $\mathbb{R}$ for $i = 2, \ldots, d$ corresponding to the embeddings τ_i.

The group $G(\mathbb{R})$ acts on the target group scheme $G \times_{\mathbb{Q}} \mathbb{R}$ of the morphism h_0 by conjugation. Let X be the conjugacy class of the morphism h_0 under this action of $G(\mathbb{R})$.

For $z \in \mathbb{C}$ and $\begin{pmatrix} a & b \\ c & d \end{pmatrix} \in \operatorname{GL}_2(\mathbb{R})$ we write the Möbius transformation as

$$\begin{pmatrix} a & b \\ c & d \end{pmatrix}.z = \frac{az + b}{cz + d}.$$

Then the map

$$g \circ h_0 = g.h_0.g^{-1} \mapsto g.i, \ g \in \mathrm{GL}_2(\mathbb{R}) \longrightarrow \mathbb{C} \setminus \mathbb{R}$$

identifies the conjugacy class X with $\mathbb{C} \setminus \mathbb{R}$, where the latter is the complex plane with the real line removed. In this way, the conjugacy class X has a natural complex manifold structure.

(2.3.6) Let H be an open compact subgroup of $G(\mathbb{A}_f) = \widehat{B}^*$ where $\mathbb{A}_f$ is the finite adèle ring of k (i.e. the adèles without the components at the archimedean places) and $\widehat{B}^* = (B \otimes_{\mathbb{Z}} \widehat{\mathbb{Z}})^*$.

Then there is a smooth algebraic curve M_H/k defined over the field k, called a Shimura curve, whose Riemann surface is

$$M_H^{\mathrm{an}} = B^* \backslash (X \times (\widehat{B}^*/H)) \tag{2.3.7}$$

where B^* acts on $X \cong \mathbb{C} \setminus \mathbb{R}$ via the isomorphism $B_{\tau_1}^* \cong \mathrm{GL}_2(\mathbb{R})$ and the standard action by Möbius transformations of $\mathrm{GL}_2(\mathbb{R})$ on $\mathbb{C} \setminus \mathbb{R}$ and where B^* acts on the second component $\widehat{B}^*$ in the natural way.

(2.3.8) We may then denote by $[z, b]$ the point of the complex Shimura curve M_H^{an} represented by the pair $(z, b) \in X \times \widehat{B}^*$ where X is identified with $\mathbb{C} \setminus \mathbb{R}$.

Compactification of the curves M_H

(2.3.9) In the case where $k = \mathbb{Q}$ and S is the empty set, we have that B is isomorphic to the rational matrix algebra $M_2(\mathbb{Q})$. The Shimura curves $M_H/\mathbb{Q}$ are then classical modular curves.

The smooth curves $M_H/\mathbb{Q}$, when $B \cong M_2(\mathbb{Q})$, are not usually compact but can be compactified to smooth projective curves over $\mathbb{Q}$ by adjoining a finite number of cusps; the corresponding projective curve obtained from M_H is uniquely determined and denoted M_H^*.

In this way are obtained the classical smooth projective modular curves $X_0(N)$, $X_1(N)$, etc..

(2.3.10) If $B \not\cong M_2(\mathbb{Q})$, then the Shimura curves M_H^{an} are compact Riemann surfaces and the corresponding algebraic curves M_H/k are projective smooth algebraic curves over k. For this case, we then put $M_H^* = M_H$.

The systems of curves M_H, M_H^*, N_H, N_H^*

2.3 Shimura Curves

(2.3.11) As H varies over the open compact subgroups of $\widehat{B}^*$, the curves M_H form a projective system where an inclusion of subgroups $H \subseteq H'$ induces $M_H \to M_{H'}$ which is a finite morphism of k-schemes.

(2.3.12) The group $\widehat{B}^*$ acts on the system of curves M_H where $g \in \widehat{B}^*$ induces a k-isomorphism by right multiplication by g on the cosets $\widehat{B}^*/H$

$$[g] : M_H \xrightarrow{\cong} M_{g^{-1}Hg}.$$

On the points of the corresponding complex analytic curves, this morphism is given by

$$[z, b] \mapsto [z, bg], \quad M_H^{\mathrm{an}} \longrightarrow M_{g^{-1}Hg}^{\mathrm{an}}.$$

Via this action, the centre $\widehat{k}^*$ of $\widehat{B}^*$ acts on M_H and on M_H^* through its quotient $\widehat{k}^*/(\widehat{k}^* \cap H)$. The quotient curve of M_H by the group $\widehat{k}^*/(\widehat{k}^* \cap H)$ is denoted N_H; similarly the quotient of the projective curve M_H^* by this group is written N_H^* and both N_H, N_H^* are smooth curves over k. The Riemann surface associated to N_H^{an} associated to N_H may be identified with $B^* \backslash (X \times \widehat{B}^*/H\widehat{k}^*)$. A point on the Riemann surface N_H^{an} can similarly be written as $[z, b]$ where $(z, b) \in X \times \widehat{B}^*$.

(2.3.13) (*Hecke Correspondences*) Let H, H' be compact open subgroups of $\widehat{B}^*$. For $g \in \widehat{B}^*$, we then have the map $[g] : M_{H \cap gH'g^{-1}}^* \to M_{g^{-1}Hg \cap H'}^*$ which gives rise to the diagram where the vertical maps are the evident transition morphisms

$$\begin{array}{ccc} M_{H \cap gH'g^{-1}}^* & \xrightarrow{[g]} & M_{g^{-1}Hg \cap H'}^* \\ \downarrow & & \downarrow \\ M_H^* & \dashrightarrow & M_{H'}^* \end{array}$$

and where the lower arrow is defined by this diagram and is a correspondence $[HgH'] : M_H^* \dashrightarrow M_{H'}^*$, i.e. a "many-valued map". The correspondence $[HgH']$ is a *Hecke correspondence*.

Replacing H, H' by $H\widehat{F}^*, H'\widehat{F}^*$ we obtain similarly the Hecke correspondence $[HgH'] : N_H^* \dashrightarrow N_{H'}^*$. [See also paragraph (2.3) below.]

CM points on Shimura curves

(2.3.14) Let

B be a quaternion algebra over k;
S_B be the finite set of non-archimedean places of k where B ramifies;
H be an open compact subgroup of $G(\mathbb{A}_f) = \widehat{B}^*$;

M_H/k be the Shimura curve corresponding to B and H.

Let F be a totally imaginary quadratic extension field of the totally real field k such that every prime $v \in S_B$ is either ramified or inert in F/k. This hypothesis implies that there is an injective homomorphism of k-algebras $t: F \to B$ [80, Chap. III, Théorème 3.8, p.78]. Let $t_v: F \otimes_k k_v \to B \otimes_k k_v$ and $\widehat{t}: \widehat{F} \to \widehat{B}$ be the homomorphisms induced by t on the completions at a place v of k and on the finite adèles.

(2.3.15) The embedding $\tau_1: k \to \mathbb{R}$ provides an isomorphism $B \otimes_k k_{\tau_1} \cong M_2(\mathbb{R})$. The injection of k-algebras $t: F \to B$ composed with $B \otimes_k k_{\tau_1} \cong M_2(\mathbb{R})$ induces an injective homomorphism of multiplicative groups $F^* \to \mathrm{GL}_2(\mathbb{R})$ where k^* maps to the diagonal subgroup of $\mathrm{GL}_2(\mathbb{R})$.

As $F = k(\sqrt{\theta})$ for some totally negative element $\theta \in k$, and $t(F^*)$ acts on $\mathbb{C}$ by Möbius transformations (2.3), it is straightforward to check that:

- There is a unique fixed point $w \in \mathbb{C}$ with $\mathrm{Im}(w) > 0$ under the action of $t(F^*) \subset \mathrm{GL}_2(\mathbb{R})$;
- The stabilizer $\{b \in B^* | b.w = w\}$ of w is equal to $t(F^*)$.

Definition 2.3.16 The set of *CM points by F/k* on the curve M_H, resp. N_H, is defined to be, where w is the fixed point of the complex upper half-plane $\mathcal{H}$ under the action of $t(F^*)$ as in (2.3),

$$\mathrm{CM}(M_H, F) = \{[w, b] \in M_H(\mathbb{C}) | b \in \widehat{B}^*\}$$

$$\mathrm{CM}(N_H, F) = \{[w, b] \in N_H(\mathbb{C}) | b \in \widehat{B}^*\}.$$

Here $\mathrm{CM}(N_H, F)$ is the image of $\mathrm{CM}(M_H, F)$ under the natural map $M_H \to N_H$.

The Jacobians of M_H^* and N_H^*

(2.3.17) Let $J(M_H^*)/k$ be the abelian variety $\mathrm{Pic}^\circ(M_H^*)$ which is the connected component of the identity of the Picard scheme of M_H^* and similarly $J(N_H^*)/k = \mathrm{Pic}^\circ(N_H^*)$ for N_H^*.

For the compact open subgroup $H \subset G(\mathbb{A}_f) = \widehat{B}^*$, there is a finite set $S(H) \supseteq S$ of non-archimedean primes of k such that $H = H_{S(H)} H^{S(H)}$ where

$$H_{S(H)} \subseteq \prod_{v \in S(H)} B_v^*, \quad H^{S(H)} = {\prod_{v \notin S(H)}}' H_v$$

where $H^{S(H)}$ is a maximal compact subgroup of

2.3 Shimura Curves

$$G(\mathbb{A}_f^{S(H)}) = {\prod_{v \notin S(H)}}' B_v^*$$

where $\prod'$ denotes the restricted product.

Define the Hecke algebra $\mathbf{T}_H^{S(H)}$ to be

$$\mathbf{T}_H^{S(H)} = \mathbb{Z}[H^{S(H)} \backslash G(\mathbb{A}_f^{S(H)}) / H^{S(H)}]$$

which is a subalgebra of $\mathbb{Z}[H \backslash \widehat{B}^* / H]$. Then the commutative algebra $\mathbf{T}_H^{S(H)}$ acts naturally on $J(M_H^*)/k$ [56, §1.16] and there is a homomorphism of commutative rings

$$\theta : \mathbf{T}_H^{S(H)} \longrightarrow \operatorname{End}(J(M_H^*)).$$

Write

$$\mathbf{T} = \theta(\mathbf{T}_H^{S(H)}) \subseteq \operatorname{End}(J(M_H^*)).$$

Then $\mathbf{T} \otimes_{\mathbb{Z}} \mathbb{Q}$ is a finite product of fields

$$\mathbf{T} \otimes_{\mathbb{Z}} \mathbb{Q} \cong \prod_{j \in J} L_j$$

where J is a finite index set and where [56, §1.17]

$$J = J_1 \cup J_2$$

L_j is a totally real field if $j \in J_1$

L_j is a CM field if $j \in J_2$.

Denote by θ_j the composite ring homomorphism

$$\theta_j : \mathbf{T}_H^{S(H)} \longrightarrow \mathbf{T} \longrightarrow L_j.$$

(2.3.18) (*Hecke Operators*) With the factorization $H = H_{S(H)} H^{S(H)}$ as in (2.3), for every prime place $v \notin S(H)$ of k, there is a unique maximal O_v-order $R(v) \subset B_v$ where $H_v = R(v)^*$ and O_v is the localization at v of the ring of integers of k.

For a place $v \notin S(H)$, the *Hecke correspondence* $T(v)$ is defined to be $[H b_v H]$: $M_H^* \dashrightarrow M_H^*$ for any element $b_v \in R(v) \cap B_v^* \subset \widehat{B}^*$ satisfying $\operatorname{ord}_v \operatorname{nr}(b_v)_v = 1$ where nr is the reduced norm on the quaternion algebra $\widehat{B}$ and ord_v denotes the valuation on k_v^* (see (2.3)).

Under a selected isomorphism $H_v \cong GL_2(O_v)$, the element b_v can be taken to be $\begin{pmatrix} \pi & 0 \\ 0 & 1 \end{pmatrix}$ where π is a uniformizing parameter of k_v. This correspondence $T(v)$ is independent of the choice of b_v.

The *Hecke correspondence* $T(v) : N_H^* \dashrightarrow N_H^*$ is similarly defined by taking $H\widehat{F}^*$ instead of H, for every prime place $v \notin S(H)$ of k. The correspondence $T(v)$ on N_H^* is similarly independent of the choice of b_v.

(2.3.19) (*Automorphic Forms*). Let σ_2 be the complex representation of $G_1(\mathbb{R}) \times \ldots \times G_d(\mathbb{R})$ defined by

$$\sigma_2 = \left\{ \begin{array}{c} \text{the weight 2 holomorphic discrete} \\ \text{series representation of } G_1(\mathbb{R}) \end{array} \right\} \otimes \left\{ \begin{array}{c} \text{the trivial representation of} \\ G_1(\mathbb{R}) \times \ldots \times G_d(\mathbb{R}) \end{array} \right\}.$$

Let $\mathbb{A}$ be the adèle ring of k. Let $\pi = \sigma_2 \otimes \pi_f$ be an irreducible unitary cuspidal automorphic representation of $B_{\mathbb{A}}^* = (B \otimes_k \mathbb{A})^*$ whose archimedean component is σ_2 and which satisfies $\pi_f^H \neq 0$. Then $\mathbf{T}_H^{S(H)}$ acts on π_f^H through a ring homomorphism $\lambda_\pi : \mathbf{T}_H^{S(H)} \longrightarrow \mathbb{C}$ which factors as

$$\lambda_\pi : \mathbf{T}_H^{S(H)} \xrightarrow{\theta_j} L_j \xrightarrow{\rho} \mathbb{C}$$

for a unique pair $(j, \rho) \in J \times \text{Hom}(L_j, \mathbb{C})$.

Furthermore, the Jacquet-Langlands correspondence associates to each irreducible cuspidal unitary representation $\pi = \sigma_2 \times \pi_f$ with $\pi_f^H \neq 0$ an irreducible cuspidal representation $JL(\pi)$ of $GL_2(\mathbb{A}_f)$ which corresponds to a Hilbert modular newform of weight $(2, \ldots, 2)$ and central character ω_π. This representation is characterised by

$$JL(\pi)_{\mathfrak{p}} \cong \pi_{\mathfrak{p}} \text{ as representations of } GL_2(k_{\mathfrak{p}}) \cong B_{\mathfrak{p}}^* \text{ for all } \mathfrak{p} \notin \text{Ram}(B).$$

(2.3.20) (*The Representation $V_{\mathcal{L}}(\theta_j)$*). Let l be a prime number and $G_{k,S(H)}$ be the maximal separable extension of k unramified outside $S(H)$. For each prime place $\mathcal{L}|l$ of L_j and for the representation π, denote by $V_{\mathcal{L}}(\theta_j)$ the $\mathcal{L}$-adic representation of $G_{k,S(H)}$ associated to $JL(\pi)$. Then $V_{\mathcal{L}}(\theta_j)$ is a two-dimensional absolutely irreducible representation defined over the local field $L_{j,\mathcal{L}}$ and such that for all places $\mathfrak{p} \notin S(H)$ of k

$$\det(1 - x.\text{Fr}_{\mathfrak{p},\text{geom}} | V_{\mathcal{L}}(\theta_j)) = 1 - \theta_j(T(\mathfrak{p}))x + \omega_\pi(\mathfrak{p})N(\mathfrak{p})x^2$$

where $T(\mathfrak{p})$ is the Hecke operator at $\mathfrak{p}$ [56, equation (1.12.2)], $N(\mathfrak{p})$ is the residue field order at $\mathfrak{p}$, and $\omega_\pi : Z(\mathbb{A}) \to \mathbb{C}^*$ is the central character of π. This representation $V_{\mathcal{L}}(\theta_j)$ with these properties is unique up to isomorphism.

2.3 Shimura Curves

Proposition 2.3.21 *For each pair $(j, \rho) \in J \times \text{Hom}(L_j, \mathbb{C})$ there is a unique irreducible cuspidal unitary representation $\pi = \sigma_2 \otimes \pi_f$ of $B_{\mathbb{A}}^*$ satisfying $\pi_f^H \neq 0$ and $\lambda_\pi = \rho \circ \theta_j$. We write $\pi = \pi(\rho \circ \theta_j)$.*

[For the proof, see [56, Proposition 1.18].] □

Proposition 2.3.22 (Decomposition of $J(M_H^*)$)

(i) *For every prime number l, there is an isomorphism of $\mathbf{T} \otimes \mathbb{Q}_l[G_k]$-modules*

$$H^1_{\text{ét}}(M_H^* \otimes_k k^{\text{sep}}, \mathbb{Q}_l) \cong \bigoplus_{j \in J} \left(\bigoplus_{\mathcal{L}|l} V_{\mathcal{L}}(\theta_j) \right)^{a_j}$$

where

$$a_j = \dim_{\mathbb{C}} (\pi(\sigma \circ \theta_j))_f^H \geq 1 \tag{2.3.23}$$

$\mathcal{L}$ in the summand $\bigoplus_{\mathcal{L}|l} V_{\mathcal{L}}(\theta_j)$ runs through all primes of L_j over l, and $\mathbf{T}$ acts on $V_{\mathcal{L}}(\theta_j)$ via θ_j.

(ii) *There is a $\mathbf{T}$-linear isogeny defined over k*

$$J(M_H^*) \longrightarrow \prod_{j \in J} A_j^{a_j}$$

where A_j is an abelian variety of dimension $\dim(A_j) = [L_j : \mathbb{Q}]$ where $\text{End}_k(A_j) = O_{L_j}$, the ring of integers of L_j, and on which $\mathbf{T}$ acts via θ_j, and where we have an isomorphism of $\mathbf{T} \otimes \mathbb{Q}_l[G_k]$-modules

$$H^1_{\text{ét}}(A_j \times_k k^{\text{sep}}, \mathbb{Q}_l) \cong \bigoplus_{\mathcal{L}|l} V_{\mathcal{L}}(\theta_j)$$

where $\mathcal{L}$ runs over the primes of k dividing l, and a_j is as in (2.3.23).

(iii) *If j_1, j_2 are distinct elements of J, then A_{j_1} and A_{j_2} are not isogenous abelian varieties over k.*

[For the proof see [56, Proposition 1.18].] □

Proposition 2.3.24 (Decomposition of $J(N_H^*)$) *With the notation of proposition 2.3.23, there is a $\mathbf{T}$-linear isogeny $J(N_H^*) \to \prod_{j \in J_1} A_j^{a_j}$ defined over k, where the product runs over the subset J_1 of J, and which completes the commutative diagram where the lower horizontal arrow is the evident projection and the upper horizontal arrow is obtained by the quotient map $M_H^* \to N_H^*$*

$$\begin{array}{ccc} J(M_H^*) & \longrightarrow & J(N_H^*) \\ \downarrow & & \downarrow \\ \prod_{j \in J} A_j^{a_j} & \longrightarrow & \prod_{j \in J_1} A_j^{a_j} \end{array}$$

[For the proof see [56, Proposition 1.18].] □

2.3.25. Bibliographical Remarks. There are many expositions of Shimura curves and Shimura varieties. For example, for Shimura varieties, see [49, 50]. For Shimura curves, see [14, 56] and [83, §1].

For an exposition of the arithmetic of quaternion algebras, see [80].

2.4 Elliptic Spaces

(2.4.1) Let L be a field. An *elliptic space* $E = (E_0, A, \phi)/L$ consists of a group scheme E_0/L and a Dedekind domain A such that L is an A-field (i.e. there is a ring homomorphism $A \to L$) where E_0/L is equipped with a ring homomorphism $\phi : A \to \mathrm{End}_L(E_0)$ of one of the following kinds:

(1) E_0 is an elliptic curve over L and $A \cong \mathbb{Z}$ is the ring of endomorphisms of E_0/L defined over L of multiplication by integers, $k = \mathbb{Q}$ is the fraction field of A, and k is equipped with the special set of places ∞ whose only element is the unique archimedean place of k and so A is the ring of ∞-integers of k;
(2) E_0, A, ϕ is a Drinfeld module of rank 2 over L for a given global field k of positive characteristic where E_0 is isomorphic to $\mathbb{G}_a/L$, k is equipped with a special set of places ∞, A is the ring of integers of k with respect to ∞, and $\phi : A \to \mathrm{End}_L(\mathbb{G}_a)$ is the ring homomorphism defining the Drinfeld module structure.

(2.4.2) In the same way, an *elliptic space* $E = (E_0, A, \phi)/S$ over a locally noetherian scheme S is one of the following:

(1) An elliptic curve E_0 over S, $A \cong \mathbb{Z}$ and $\phi : A \to \mathrm{End}_S(E_0)$ gives the ring of endomorphisms of E_0/S of multiplication by integers, $k = \mathbb{Q}$ is the fraction field of A and k is equipped with the special set of places ∞ whose only element is the unique archimedean place of k and so A is the ring of ∞-integers of k;
(2) A rank 2 Drinfeld module over S for a given global field k of positive characteristic where k is equipped with a special set of places ∞, A is the ring of integers of k with respect to ∞, E_0 is the additive S-group scheme $\mathbf{Spec}_{O_S} \bigoplus_{n=0}^{\infty} \mathcal{L}^{\otimes n}$ where $\mathcal{L}$ is a line bundle on S, and $\phi : A \to \mathrm{End}_S(E_0)$ is a k_0-algebra homomorphism where k_0 is the finite field of constants of the global field k.

2.5 Moduli Schemes of Elliptic Spaces

Remarks 2.4.3

(i) Evidently if L has characteristic zero, then an elliptic space over L must be an elliptic curve, but this is no longer the case if L has positive characteristic.

It is necessary to specify the ring A in saying that (E_0, A, ϕ) is an elliptic space because for Drinfeld modules changing the ring A may change the rank of the Drinfeld module and we are here only interested in Drinfeld modules of rank 2 (see Sect. 3.1 on complex multiplication of elliptic spaces).

Drinfeld himself originally called his modules "elliptic modules", so with "elliptic spaces" we have simply extended his terminology to include elliptic curves.

(ii) One would like "elliptic spaces" corresponding to points on Shimura curves, but a modular interpretation is not known for general Shimura curves.

One exception is the following. Let B be a quaternion algebra over $\mathbb{Q}$ which becomes isomorphic to $M_2(\mathbb{R})$ at the unique archimedean place ∞. Let O be a maximal order of B. Then associated to O is a Shimura curve which is a coarse moduli scheme of pairs (A, i) where A is a two-dimensional abelian variety over $\mathbb{C}$ and $i : O \to \mathrm{End}(A)$ is a ring embedding.

2.5 Moduli Schemes of Elliptic Spaces

(2.5.1) Let $E = (E_0, A, \phi)/S$ be an elliptic space over a locally noetherian scheme S. In particular, we then have that E_0/S is a group scheme and A is a Dedekind domain, of the type in (2.4), with a ring homomorphism $A \to \mathrm{End}_S(E_0)$.

The moduli scheme $\mathbf{Y}_0^{\mathrm{Ell.Sp.}}(I)$

(2.5.2) Let I be a non-zero ideal of the Dedekind domain A. Let S be a locally noetherian A-scheme and $E = (E_0, A, \phi)/S$ be an elliptic space. In the case where E is a Drinfeld module, we may assume that E is a standard Drinfeld module given by the pair $(G_\mathcal{L}, \phi)$, where $\mathcal{L}$ is a line bundle on S, and $G_\mathcal{L}$ is the additive S-group scheme $\mathbf{Spec}_{O_S} \bigoplus_{n=0}^\infty \mathcal{L}^{\otimes n}$, and where $\phi : A \to \mathrm{End}(G_\mathcal{L})$ is a k_0-algebra homomorphism where k_0 is the finite field of constants of the global field $k = \mathrm{fract}(A)$.

Definition 2.5.3 An I-cyclic subgroup Z of the elliptic space $(E_0, A, \phi)/S$ is a finite flat subgroup scheme Z/S of E_0/S and a homomorphism of A-modules

$$\psi : A/I \to E_0(S)$$

such that there is an equality of relative Cartier divisors of E_0/S

$$\sum_{m \in A/I} \psi(m) = Z.$$

(2.5.4) With the ring A and ideal I of A fixed, let $\mathbf{Y}_0^{\text{Ell.Sp.}}(I)$ denote the coarse moduli scheme of elliptic spaces equipped with a cyclic I-structure; that is to say, $\mathbf{Y}_0^{\text{Ell.Sp.}}(I)$ is a coarse moduli scheme for the functor on the category $A - \mathcal{S}ch$ of locally noetherian A-schemes given by

$$A - \mathcal{S}ch \to \mathcal{S}ets$$

$$S \mapsto \left\{ \begin{array}{c} S - \text{isomorphism classes of pairs } (E, Z) \text{ where} \\ E = (E_0, A, \phi)/S \text{ is an elliptic space and } Z/S \text{ is an} \\ I - \text{cyclic subgroup of } E_0 \end{array} \right\}$$

(2.5.5) If L is an algebraically closed field, then the L-valued points of $\mathbf{Y}_0^{\text{Ell.Sp.}}(I)$ are represented by pairs (E, Z) where $E = (E_0, A, \phi)/L$ is an elliptic space and Z/L is an I-cyclic subgroup of E_0.

(2.5.6) It is known in the cases of both elliptic curves and rank 2 Drinfeld modules that the moduli schemes $\mathbf{Y}_0^{\text{Ell.Sp.}}(I)$ exist. If we need to distinguish between the moduli scheme $\mathbf{Y}_0^{\text{Ell.Sp.}}(I)$ for elliptic curves and for Drinfeld modules, we write $\mathbf{Y}_0(N)$ for the case of elliptic curves, where N is a positive integer generator of the ideal I, and $\mathbf{Y}_0^{\text{Drin}}(I)$ for the case of Drinfeld modules and similarly for any variants of the moduli scheme $\mathbf{Y}_0^{\text{Ell.Sp.}}(I)$.

The moduli schemes $\mathbf{X}_0^{\text{Ell.Sp.}}(I), Y_0^{\text{Ell.Sp.}}(I), X_0^{\text{Ell.Sp.}}(I)$

(2.5.7) We summarise here basic results on the moduli scheme $\mathbf{X}_0^{\text{Ell.Sp.}}(I)$ and some of its variants obtained by concatenating the cases of elliptic curves and Drinfeld modules.

(2.5.8) The scheme $\mathbf{Y}_0^{\text{Ell.Sp.}}(I)$ is normal and two-dimensional. Furthermore, $\mathbf{Y}_0^{\text{Ell.Sp.}}(I)$ is an A-scheme of finite type and hence is either a geometric surface over the finite field of constants of k or is an arithmetic surface over $\mathbb{Z}$.

(2.5.9) Let $k[0]$ be the Hilbert class field of k, that is to say $k[0]$ is the maximal unramified abelian extension of k which is split completely at ∞; this latter condition, for the case where k is an algebraic number field and ∞ consists of the archimedean places of k, means that the real places of k extend to only real places of $k[0]$ and not to complex places.

We write $Y_0^{\text{Ell.Sp.}}(I)/k$ for the generic fibre of the A-scheme $\mathbf{Y}_0^{\text{Ell.Sp.}}(I)/A$. Then the exact field of constants of the curve $Y_0^{\text{Ell.Sp.}}(I)/k$ is $k[0]$, that is to say $k[0]$ is the algebraic closure of k in the function field of $Y_0^{\text{Ell.Sp.}}(I)$ (see [28, §8.3] for the Drinfeld case).

(2.5.10) As the generic fibre of $\mathbf{Y}_0^{\text{Ell.Sp.}}(I)/A$ is a smooth curve defined over the field $k[0]$, we have that the normal surface $\mathbf{Y}_0^{\text{Ell. Sp.}}(I)$ is fibred over $\text{Spec } A[0]$, where $A[0]$ is the integral closure of A in the Hilbert class field $k[0]$ of k. Furthermore, $\mathbf{Y}_0^{\text{Ell.Sp.}}(I)$ has

2.5 Moduli Schemes of Elliptic Spaces

only a finite number of isolated singular points. These singular points only occur at points corresponding to supersingular elliptic curves or supersingular Drinfeld modules of finite characteristic dividing I. By blowing up these singular points, we obtain a smooth surface, arithmetic or geometric, which is the minimal smooth desingularization of $\mathbf{Y}_0^{\text{Ell.Sp.}}(I)$.

(2.5.11)   The A-scheme $\mathbf{Y}_0^{\text{Ell.Sp.}}(I)$ may be compactified to a scheme $\mathbf{X}_0^{\text{Ell. Sp.}}(I)/A$ by adding the cusps to the geometric or arithmetic surface $\mathbf{Y}_0^{\text{Ell.Sp.}}(I)$. The generic fibre of $\mathbf{X}_0^{\text{Ell.Sp.}}(I)/A$ is a smooth projective curve defined over $k[0]$ and written $X_0^{\text{Ell.Sp.}}(I)/k[0]$.

(2.5.12)   Suppose that $k = \mathbb{Q}$. Then I is an ideal of $\mathbb{Z}$ and is therefore generated by a uniquely determined positive integer N. Then $X_0^{\text{Ell.Sp.}}(I)/k[0]$ becomes the classical modular curve $X_0(N)/\mathbb{Q}$ which is the coarse moduli scheme of elliptic curves equipped with a cyclic subgroup of order N where this curve is compactified by a finite number of cusps which correspond to "degenerate" elliptic curves.

(2.5.13)   Suppose now that k is a global field of positive characteristic. Suppose that ∞ is a special set of places of k, A is the ring of integers of k relative to ∞, and I is a non-zero ideal of A.

Then $X_0^{\text{Ell.Sp.}}(I)/k[0]$ becomes the Drinfeld modular curve $X_0^{\text{Drin}}(I)/k(0]$ which is the coarse moduli scheme of Drinfeld modules of rank 2 for A equipped with an I-cyclic structure; this curve is compactified by a finite number of cusps which correspond to "degenerate" Drinfeld modules. Here $k[0]$ is the Hilbert classs field of k, that is to say $k[0]$ is the maximal unramified abelian extension of k which is split completely at ∞.

Theorem 2.5.14   *Let $J(I)/k[0]$ be the Jacobian of the smooth projective curve $X_0^{\text{Ell. Sp.}}(I)/k[0]$. The abelian variety $J(I)/k[0]$ has good reduction at all places of $k[0]$ prime to I and ∞. The cusps of $X_0^{\text{Ell.Sp.}}(I)/k[0]$ generate a torsion subgroup of $J(I)(\bar{k})$, where $\bar{k}$ is the algebraic closure of k.*

Proof   That $J(I)/k[0]$ has good reduction at all places prime to Supp I and ∞ follows from the modular interpretation of the reduction of the scheme $\mathbf{Y}_0^{\text{Ell.Sp.}}(I)$ at such places. That the cusps generate a torsion subgroup is a consequence of [24, Chapter VI, Corollary 5.12] for the function field case and is the Manin-Drinfeld theorem when $k = \mathbb{Q}$.   □

Hecke operators

(2.5.15)   Let

  z be a closed point of Spec A with support disjoint from Spec A/I;
  $\mathfrak{m}$ be the maximal ideal of A defining z.

The *Hecke correspondence* T_z on the moduli space of elliptic spaces $\mathbf{Y}_0^{\text{Ell.Sp.}}(I)$ may be defined via modular data as follows. If L is an algebraically closed field and x is an L-rational point of $\mathbf{Y}_0^{\text{Ell.Sp.}}(I)/A$ given by the pair (E, Z) where $E = (E_0, A, \phi)/L$ is an elliptic space and Z is an I-cyclic subgroup of the group scheme E_0/L, then

$$T_z(x) = \sum_H (E_0/H, (Z+H)/H)$$

where H runs over all m-cyclic subgroups of the group scheme E_0 and where the right-hand side here is a cycle on $\mathbf{Y}_0^{\text{Ell.Sp.}}(I)$ of degree $|\kappa(z)| + 1$ over the field L.

(2.5.16) It can be checked that the Hecke correspondences T_z extend to correspondences denoted by the same symbol T_z on the compactification $\mathbf{X}_0^{\text{Ell.}\tilde{\text{Sp}}.}(I)$ of $\mathbf{Y}_0^{\text{Ell.Sp.}}(I)$ and also on the generic fibres $X_0^{\text{Ell.Sp.}}(I)$, $Y_0^{\text{Ell.Sp.}}(I)$ of these moduli schemes.

Analytic form of the moduli scheme $\mathbf{Y}_0^{\text{Ell.Sp.}}(I)$

(2.5.17) As in (2.4), we have the global field k with a special set of places ∞ and A is the ring of ∞-integers of k. In characteristic zero, $k = \mathbb{Q}$ and ∞ is the unique archimedean place. Let

k_∞ be the completion at ∞ of k;
$\overline{k}_\infty$ be the algebraic closure of k_∞;
$\widehat{\overline{k}}_\infty$ be the completion at a place above ∞ of $\overline{k}_\infty$;
$\Omega = \widehat{\overline{k}}_\infty - k_\infty$.

In the case where $k = \mathbb{Q}$, then we have $k_\infty \cong \mathbb{R}$ and $\overline{k}_\infty \cong \widehat{\overline{k}}_\infty \cong \mathbb{C}$; in this case we then have $\Omega = \mathbb{C} \setminus \mathbb{R}$ which is the union of the upper and lower complex half-planes. If k has positive characteristic, then Ω is called the "Drinfeld upper half-plane".

(2.5.18) The group $\text{GL}(2, A)$, of invertible 2×2-matrices with coefficients in A, acts on Ω via Möbius transformations; that is to say, if $g = \begin{pmatrix} a & b \\ c & d \end{pmatrix} \in \text{GL}(2, A)$ and $\omega \in \Omega$, then we have

$$g\omega = \frac{a\omega + b}{c\omega + d}.$$

Let Γ be a discrete subgroup of $\text{GL}(2, A)$. Then the quotient space $\Gamma \backslash \Omega$ may be equipped with the structure of a rigid analytic space over k_∞ in the case where k has positive characteristic (see [17, 28]) and with the structure of a Riemann surface in the case where $k = \mathbb{Q}$.

Let $\Gamma_0(I)$ be the congruence subgroup of $\mathrm{GL}(2, A)$ given by

$$\Gamma_0(I) = \{ \begin{pmatrix} a & b \\ c & d \end{pmatrix} \in \mathrm{GL}(2, A) \mid c \equiv 0 \text{ (modulo } I)\}.$$

Let $k[0]$ be the Hilbert class field of k with respect to ∞, that is to say $k[0]$ is the maximal unramified abelian extension of k which is split completely at ∞ where for the case $k = \mathbb{Q}$ this means that $k[0] = \mathbb{Q}$.

Fix an embedding $\sigma : k[0] \hookrightarrow \widehat{\overline{k}}_\infty$. Then the set $Y_0^{\mathrm{Ell.Sp.}}(I)(\widehat{\overline{k}}_\infty)$ of $\widehat{\overline{k}}_\infty$-valued points of the curve $Y_0^{\mathrm{Ell.Sp.}}(I)/k[0]$ is also a rigid analytic space over k_∞, when k has positive characteristic, and is a Riemann surface, when $k = \mathbb{Q}$, and there is an isomorphism of rigid analytic spaces or Riemann surfaces

$$Y_0^{\mathrm{Ell.Sp.}}(I)(\widehat{\overline{k}}_\infty) \cong \Gamma_0(I) \backslash \Omega.$$

2.5.19. Bibliographical Remarks. For a comprehensive presentation of the moduli schemes of elliptic curves, see [38]. Another standard reference on this is [73].

For more on Drinfeld moduli schemes, see [4, §2.4, Appendices A and B] and [16]. For properties of the generic fibres of these moduli schemes, which are curves over global fields of positive characteristic, see [25].

2.6 Modularity of Elliptic Curves over Global Fields

(2.6.1) Let

k be a global field equipped with a special set ∞ of places;

A be the ring of integers of k with respect to ∞;

E/k be an elliptic curve where if k has positive characteristic then E/k is assumed to have split (Tate) multiplicative reduction at the place in ∞ and if k has characteristic zero then it is assumed that $k = \mathbb{Q}$;

I be the non-zero ideal of the ring A which is the conductor of the elliptic curve E/k, with any place at ∞ removed.

Theorem 2.6.2 (Wiles et al., Drinfeld) *There is a finite surjective morphism of k-schemes*

$$\psi : X_0^{\mathrm{Ell.Sp.}}(I) \to E.$$

$\square$

Remarks 2.6.3

(i) In the case where $k = \mathbb{Q}$ and $E/\mathbb{Q}$ is an elliptic curve over the rational numbers, this theorem was formerly known as the conjecture of Shimura-Taniyama-Weil.
If k is a global field of characteristic $p > 0$, then this theorem is the analogue of the Shimura-Taniyama-Weil conjecture in positive characteristic.

(ii) A conspicuous absence from theorem 2.6.2 is the modularity of elliptic curves over any number field. The latter is described by the Langlands conjectures, and some special cases of this are known notably the modularity of elliptic curves over real quadratic fields due to Freitas, Le Hung, and Siksek [21].

Another lacuna in the above theorem is the case of an elliptic curve E over a global field k of positive characteristic where E/k does not have split multiplicative reduction at any place. The modularity of these elliptic curves in some form would follow from Lafforgue's proof of the Langlands conjecture for $GL(n)$ over k.

2.6.4. Bibliographical Remarks. When $k = \mathbb{Q}$, Theorem 2.6.2 is due to Wiles, Breuil, Conrad, and Taylor. Wiles first proved the theorem for semi-stable elliptic curves over $\mathbb{Q}$, and this was extended, based on Wiles' method, to all elliptic curves over $\mathbb{Q}$ by Breuil, Conrad, and Taylor. For more explanations and further references, see [18] and [74, Theorem 13.6.] as well as the original papers [81] and [6].

The function field case of theorem 2.6.2 is a consequence of Drinfeld's work on the Langlands conjecture for $GL(2)$ over a function field of positive characteristic. For proofs of theorem 2.6.2 in this case, see [28] or [4, theorem B.11.17, p. 502].

2.7 Torsion Points on Elliptic Curves

This section is a summary of the results of Serre and Igusa for the Galois action on torsion points of elliptic curves over global fields. The Galois action takes different forms depending on the nature of the elliptic curve and its ground field.

Elliptic curves over algebraic number fields

(2.7.1) Let E be an elliptic curve defined over a number field L which is a finite extension of the rational field. Let E_∞ be the torsion subgroup of $E(L^{\text{sep}})$, where L^{sep} denotes the separable closure of L. Then the nature of the Galois group $\text{Gal}(L(E_\infty)/L)$ is as follows.

Let P be the set of all prime numbers. For any prime number l, let $T_l(E)$ be the Tate module of E and let $\mathbb{Z}_l$ be the l-adic completion of $\mathbb{Z}$. Then the Galois action of $G = \text{Gal}(L^{\text{sep}}/L)$ provides a continuous homomorphism $\rho_l : G \to \text{GL}(T_l(E))$.

2.7 Torsion Points on Elliptic Curves

Theorem 2.7.2 *Suppose that E has complex multiplication and $l \in P$. Put $R = \text{End}_{L^{\text{sep}}}(E)$, which is an order of an imaginary quadratic extension of $\mathbb{Q}$. Assume that the elements of R are defined over L. Put $R_l = R \otimes_{\mathbb{Z}} \mathbb{Z}_l$. Then $T_l(E)$ is a free R_l-module of rank 1. The image of $G = \text{Gal}(L^{\text{sep}}/L)$ under $\rho_l : G \to \text{GL}(T_l(E))$ commutes with the elements of R_l and is therefore contained in R_l^*. The image of G under $\rho = \prod_{l \in P} \rho_l$ is an open subgroup of the product $\prod_{l \in P} R_l^*$.* □

[See [67, §4.5] for more details.]

Theorem 2.7.3 (Serre [67]) *If E does not have complex multiplication, then the image of $G = \text{Gal}(L^{\text{sep}}/L)$ under $\rho = \prod_{l \in P} \rho_l$ is an open subgroup of $\text{GL}_2(\widehat{\mathbb{Z}})$ where $\widehat{\mathbb{Z}}$ is the profinite completion of $\mathbb{Z}$.* □

[See [67] and [67, §4.5] for both the cases of elliptic curves with complex multiplication and without complex multiplication.]

Elliptic curves over global fields of positive characteristic

(2.7.4) Let F be a global field of positive characteristic p where k is the exact field of constants of F and let E/F be an elliptic curve. Let

$G = \text{Gal}(F^{\text{sep}}/F)$, where F^{sep} is the separable closure of F;

n be an integer prime to p;

E_n be the finite F-group scheme of n-torsion points of E;

E_∞ be the torsion subgroup of $E(F^{\text{sep}})$ of points of order prime to p.

The elliptic curve E/F is said to be *isotrivial* if there is a finite Galois extension field F' of F such that the curve $E \times_F F'$ is definable over a finite subfield of F'.

(2.7.5) The action of the Galois group G on E_n provides a homomorphism

$$\rho_n : G \to \text{Aut}(E_n) \cong \text{GL}_2(\mathbb{Z}/n\mathbb{Z}).$$

The determinant

$$\det : \text{Aut}(E_n) \to (\mathbb{Z}/n\mathbb{Z})^*$$

induces a homomorphism

$$G \to (\mathbb{Z}/n\mathbb{Z})^*.$$

Let H_n be the subgroup of $(\mathbb{Z}/n\mathbb{Z})^*$ generated by the powers of $q = |k|$ modulo n. Then H_n is naturally isomorphic to the Galois group of the field of nth roots of unity over k. Let Γ_n be the subgroup of $\mathrm{GL}_2(\mathbb{Z}/n\mathbb{Z})$ defined by the exact sequence of finite groups, where det is the restriction to Γ_n of the determinant homomorphism on $\mathrm{GL}_2(\mathbb{Z}/n\mathbb{Z})$,

$$0 \to \mathrm{SL}_2(\mathbb{Z}/n\mathbb{Z}) \to \Gamma_n \xrightarrow{\det} H_n \to 0. \tag{2.7.6}$$

(2.7.7) Passing to the projective limit of the previous exact sequence over all integers n prime to p, we obtain the exact sequence of profinite groups

$$0 \longrightarrow \mathrm{SL}_2(\widehat{\mathbb{Z}}^{(p)}) \longrightarrow \widehat{\Gamma} \longrightarrow \widehat{H} \longrightarrow 0 \tag{2.7.8}$$

where $\widehat{H}$ is the subgroup of $\widehat{\mathbb{Z}}^{(p)*}$ topologically generated by q, where

$$\widehat{\mathbb{Z}}^{(p)} = \prod_{l \neq p} \mathbb{Z}_l$$

is the profinite prime-to-p completion of $\mathbb{Z}$, and $\widehat{\Gamma}$ is a closed subgroup of $\mathrm{GL}_2(\widehat{\mathbb{Z}}^{(p)})$.

(2.7.9) Passing to the projective limit of the exact sequence (2.7.6) where n runs over all powers of a prime number l where $l \neq p$, we obtain the exact sequence

$$0 \longrightarrow \mathrm{SL}_2(\mathbb{Z}_l) \longrightarrow \widehat{\Gamma}_l \longrightarrow \widehat{H}_l \longrightarrow 0.$$

Theorem 2.7.10 (Igusa [I]) *Suppose that E/F is not isotrivial. Then the profinite group $\mathrm{Gal}(F(E_\infty)/F)$ is an open subgroup of $\widehat{\Gamma}$.* □

Theorem 2.7.11 *Suppose that E/F is not isotrivial. Then for all prime numbers $l \neq p$ the profinite group $\mathrm{Gal}(F(E_{l^\infty})/F)$ is an open subgroup of $\widehat{\Gamma}_l$ and is equal to $\widehat{\Gamma}_l$ for all but finitely many l.*

[This is a restatement of theorem 2.7.10.] □

(2.7.12) For a given prime number l different from p, a selected basis of the Tate module $T_l(E)$ over $\mathbb{Z}_l$, the l-adic completion of $\mathbb{Z}$ provides an isomorphism of $\mathrm{Gal}(F(E_{l^\infty})/F)$ with a subgroup of $\mathrm{GL}_2(\mathbb{Z}_l)$.

Proposition 2.7.13 *Let E/F be an elliptic curve which is not isotrivial. Let L be a finite extension field of F in which k is algebraically closed. For each prime number l different*

2.7 Torsion Points on Elliptic Curves

from p, select a basis of the Tate module $T_l(E)$ over $\mathbb{Z}_l$, the l-adic completion of $\mathbb{Z}$. Then there is an infinite set S of prime numbers of positive Dirichlet density such that for all $l \in S$ we have

(a) *The fields $F(E_{l^\infty})$ and L are linearly disjoint over F;*
(b) $E(L)_{l^\infty} = 0$;
(c) $\begin{pmatrix} 1 & 0 \\ 0 & -1 \end{pmatrix} \in \mathrm{Gal}(F(E_{l^\infty})/F)$.

[For the proof, see [4, Proposition 7.3.10].] □

Exercise 2.7.14 Let F be a global field of positive characteristic and E/F be an isotrivial elliptic curve. Show that the group $\mathrm{Gal}(F(E_\infty)/F)$ is an extension of a finite group by the abelian profinite group $\widehat{\mathbb{Z}}$.

Complex Multiplication, Heegner Points, CM Points

Contents

- 3.1 Complex Multiplication of Elliptic Spaces 47
- 3.2 Heegner Points 48
- 3.3 CM Points on Shimura Curves 51
- 3.4 Heegner Points and CM Points on Elliptic Curves 55

3.1 Complex Multiplication of Elliptic Spaces

(3.1.1) Let

k be a global field equipped with a special set of places ∞;
A be the ring of ∞ integers of k;
$E = (E_0, A, \phi)/L$ be an elliptic space over a field L (see (2.4)).

In particular, we then have that E_0/L is a one-dimensional commutative group scheme and A is a Dedekind domain, of the type in (2.4), with a ring homomorphism $A \to \mathrm{End}_L(E_0)$ and L is equipped with a ring homomorphism $A \to L$ where k, A are as in (2.4) so that either k has positive characteristic or $k = \mathbb{Q}$.

(3.1.2) The elliptic space E is said to have *complex multiplication*:

- If E_0 is an elliptic curve, L has characteristic zero, and

$$\dim_k \mathrm{End}(E) \otimes_A k = 2$$

- If E is a rank 2 Drinfeld module, E has infinite characteristic, the ring $\mathrm{End}(E)$ is commutative, and

$$\dim_k \mathrm{End}(E) \otimes_A k = 2$$

(3.1.3) If E/L is an elliptic space with complex multiplication, then $\mathrm{End}(E) \otimes_A k$ is an imaginary quadratic field extension of k with respect to ∞ and $\mathrm{End}(E)$ is A-isomorphic to an A-order of F. It follows that there is an ideal $I(c)$ of A such that

$$\mathrm{End}_L(E) \cong A + BI(c)$$

where B is the integral closure of A in $I(c)$ is the ideal of A cutting out a cycle c. The cycle c is the *conductor* of the order O_c (see Sect. 1.4).

(3.1.4) The complex multiplication of elliptic curves and Drinfeld modules are parallel theories and in both cases there is a "main theorem of complex multiplication" which describes the Galois action on these elliptic spaces with complex multiplication.

Let E/L be an elliptic space with complex multiplication by the order O_c with conductor c of the imaginary quadratic extension field F/k with respect to ∞. Then E/L can be defined over the ring class field $F[c]$ (Sect. 1.5). Furthermore, the finite set of elliptic spaces with complex multiplication by O_c is permuted simply transitively by the Galois group $\mathrm{Gal}(F[c]/F) \cong \mathrm{Pic}(O_c)$ by the main theorem of complex multiplication.

3.1.5. Bibliographical Remarks. For space reasons, it is not possible to present here a fuller account of the theory of complex multiplication. For more detailed expositions of the complex complex multiplication of elliptic curves, see [10, Chap. XIII, pp. 292-296], [75, Chap.2, pp. 95-186], and [73]. For complex multiplication of Drinfeld modules of rank 2, see [16] or [4, §2.5]. See [29] for explicit class field theory for function fields. For complex multiplication of abelian varieties, see [43] or the original monograph of Shimura and Taniyama.

3.2 Heegner Points

(3.2.1) In this section we explain the construction of Heegner points on the modular curves $X^{\mathrm{Ell.Sp.}}(\mathfrak{J})$ of elliptic spaces.

Let

k be a global field equipped with a special set of places ∞;
A be the ring of integers of k with respect to ∞;
F be an imaginary quadratic extension field of k with respect to ∞ (see Sect. 1.1);

3.2 Heegner Points

B be the integral closure of A in F;
c be a cycle of k with support disjoint from the places in ∞;
O_c denote the order of F relative to A and with conductor c (Sect. 1.4).

The Heegner Condition

(3.2.2) Let $\mathfrak{J}$ be a non-zero ideal of A which satisfies the

Heegner condition: All prime ideal components of $\mathfrak{J}$ split completely in the imaginary quadratic field extension F/k.

This Heegner condition on $\mathfrak{J}$ and F is a hypothesis required for the construction of Heegner points.

We then have a factorization, in general not uniquely determined,

$$\mathfrak{J}B = \mathfrak{J}_1 \mathfrak{J}_2 \tag{3.2.3}$$

where $\mathfrak{J}_1, \mathfrak{J}_2$ are ideals of B such that $\mathfrak{J}_1^\tau = \mathfrak{J}_2$ and where τ is the non-trivial element of $\mathrm{Gal}(F/k)$.

(3.2.4) The order O_c equals $A + I(c)B$ where $I(c)$ is the ideal of A cutting out the cycle c. Put

$$\mathfrak{J}_j(O_c) = \mathfrak{J} + I(c)\mathfrak{J}_j \subset O_c \text{ for } j = 1, 2.$$

Then $\mathfrak{J}_j(O_c)$, $j = 1, 2$, are ideals of O_c and we have

$$\mathfrak{J}_j(O_c) = \mathfrak{J}_j \cap O_c \text{ if } \mathfrak{J} \text{ and } c \text{ are coprime} \qquad (j = 1, 2).$$

It follows that if c and $\mathfrak{J}$ are coprime then $\mathfrak{J}_j(O_c)$ is an invertible ideal of O_c, i.e. it is a locally free O_c-module of rank 1, for $j = 1, 2$; furthermore, if c and $\mathfrak{J}$ are coprime we have isomorphisms of A-algebras

$$O_c/\mathfrak{J}_j(O_c) \cong B/\mathfrak{J}_j \cong A/\mathfrak{J} \text{ for } j = 1, 2.$$

Heegner points on $\mathbf{Y}_0^{\mathrm{Ell.Sp.}}(\mathfrak{J})$

(3.2.5) Let $E = (E_0, A, \phi)/L$ be an elliptic space over a field L as in (2.4). Suppose that E has complex multiplication by the order O_c of F with conductor c, that is to say $\mathrm{End}(E) \cong O_c$ (see Sect. 1.4). Then E/L is a group scheme which is either an elliptic curve over L or isomorphic to $\mathbb{G}_a/L$.

Suppose that the ideal $\mathfrak{J} \neq 0$ of A and the field F satisfy the Heegner condition and we write $\mathfrak{J} = \mathfrak{J}_1 \mathfrak{J}_2$ as in (3.2.3). Then

$$Z = \{x \in E | \, ax = 0 \text{ for all } a \in \mathfrak{J}_1\}$$

is a finite closed subgroup scheme of E and defined over L. Suppose that the support of c and $\mathfrak{J}$ are coprime. Because of the isomorphism of A-algebras $O_c/\mathfrak{J}_1 \cong A/\mathfrak{J}$ (cf. (3.2.4)) the subscheme Z is an $\mathfrak{J}$-cyclic subgroup scheme of E.

Definition 3.2.6 The pair (E, Z) defines an L-rational point on the moduli space $\mathbf{Y}_0^{\text{Ell.Sp.}}(\mathfrak{J})$ called a *Heegner point*.

The Heegner point (E, Z) is defined over the ring class field $F[c]$ by the main theorem of complex multiplication so that $(E, Z) \in \mathbf{Y}_0^{\text{Ell.Sp.}}(\mathfrak{J})(F[c])$ (see (3.1.4)).

Analytic construction of Heegner points

(3.2.7) For the rest of this Sect. 3.2, let the field L be either $\mathbb{C}$, if k has characteristic zero, or the field $\mathbb{C}_\infty = \hat{\bar{k}}_\infty$, if k has positive characteristic, where $\hat{\bar{k}}_\infty$ is the completion of the algebraic closure of k with respect to the unique place in ∞.

Elliptic curves over the complex numbers are given by tori $\mathbb{C}/\Lambda$ where $\Lambda \subset \mathbb{C}$ is a rank 2 lattice over $\mathbb{Z}$.

Similarly if k has positive characteristic, a Drinfeld module of rank 2 for k of infinite characteristic is given by an A-lattice Λ of rank 2 in $\mathbb{C}_\infty$ and the Drinfeld module of infinite characteristic can be considered to be the quotient $\mathbb{C}_\infty/\Lambda$ [16, Chapter 2, §§1-2]. We simply call an A-lattice a lattice.

Then we have that an elliptic space over L can be represented as a torus of the form $\mathbb{C}/\Lambda$ or $\mathbb{C}_\infty/\Lambda$ depending on the characteristic of k where Λ is a lattice.

(3.2.8) If k has characteristic zero assume that $k \cong \mathbb{Q}$ and ∞ is the archimedean place of $\mathbb{Q}$. If k has positive characteristic then no other restriction is placed on k.

Let $\mathfrak{J}$ be an ideal of A that satisfies the Heegner condition so that there is a factorization $\mathfrak{J}B = \mathfrak{J}_1\mathfrak{J}_2$. Let a be a divisor class in the Picard group $\text{Pic}(O_c)$ of the A-order O_c of F. Assume that c and the ideal $\mathfrak{J}$ are coprime.

Fix an embedding $F \to L$. Select Λ to be a projective O_c-module of rank 1 in the class a and contained as an A-lattice in L. Then $\Lambda' = \mathfrak{J}_1(O_c)^{-1}\Lambda$ is a projective O_c-module of rank 1 contained as a lattice in L. Let E and E' be the elliptic spaces for A over the field L corresponding to the lattices Λ and Λ', respectively. Then E, E' have complex multiplication by O_c.

The inclusion of O_c-modules $\Lambda \subset \Lambda'$ corresponds to an $\mathfrak{J}$-cyclic isogeny $f : E \to E'$ of the elliptic spaces E, E', as the kernel of this isogeny is isomorphic as an A-module to $O_c/\mathfrak{J}_1(O_c) \cong A/\mathfrak{J}$.

Definition 3.2.9 The pair $(E, \ker(f))$ defines a point in $\mathbf{Y}_0^{\text{Ell.Sp.}}(\mathfrak{J})(L)$ which is the *Heegner point* written $(a, \mathfrak{J}_1, c)$ associated to F with conductor c and class a. It is assumed here that $\mathfrak{J}$ satisfies the Heegner condition and that c and $\mathfrak{J}$ are coprime.

3.3 CM Points on Shimura Curves

(3.2.10) We also write $(a, \mathfrak{J}_1, c)$ for the corresponding point to $(a, \mathfrak{J}_1, c)$ of the generic fibre $Y_0^{\text{Ell.Sp.}}(\mathfrak{J}) = Y_0^{\text{Ell.Sp.}}(\mathfrak{J}) \times_A k$.

By the main theorem of complex multiplication, the elliptic spaces E, E' are defined over the ring class field field $F[c]$ (see §3.1); hence we have

$$(a, \mathfrak{J}_1, c) \in \mathbf{Y}_0^{\text{Ell.Sp.}}(\mathfrak{J})(F[c]).$$

3.3 CM Points on Shimura Curves

(3.3.1) The notation of Sect. 2.3 holds in this section and especially that on CM points (2.3)–(2.3.16).

The Shimura curve N_H/k is associated to H which is an open compact subgroup of $\widehat{B}^*$ (see (2.3)). Let $x = [w, b]$ be a CM point on N_H where $b \in \widehat{F}^*$ and where $w \in \mathbb{C}$ is the unique point with $\text{Im}(w) > 0$ fixed by $t(F^*)$ under the action of B^* ((2.3) and Definition 2.3.16).

(3.3.2) Class field theory implies that the CM points of N_H are defined over abelian extension fields, that is to say we have $\text{CM}(N_H, F) \subseteq N_H(F^{\text{ab}})$ where F^{ab} is the maximal abelian extension of F. The Galois action on $\text{CM}(N_H, F)$ is via the reciprocity map $\text{rec}: \widehat{F}^* \to \text{Gal}(F^{\text{ab}}/F)$, where $\widehat{F}^*$ is the group of finite idèles of F, by this formula

$$\text{rec}(a)[w, b] = [w, \widehat{t}(a)b] \qquad (a \in \widehat{F}^*)$$

where $\widehat{t}: \widehat{F} \to \widehat{B}$ is the map of paragraph (2.3). [For the proof, see [49, II, 5.1] and the sign correction to this noted in that text.]

Proposition 3.3.3 *Let $F(x) \subseteq F^{\text{ab}}$ be the field of definition of the CM point $x = [w, b] \in N_H$. Then the reciprocity map induces an isomorphism*

$$\text{rec}: F^* \backslash \widehat{F}^* / \widehat{t}^{-1}(bH\widehat{k}^* b^{-1}) \cong \text{Gal}(F(x)/F).$$

Proof We have that $N_H = M_H/(\widehat{k}^*/(\widehat{k}^* \cap H))$ (2.3). So by (3.3.2) we have if $[w, b] \in \text{CM}(N_H, F)$ and $a \in \widehat{F}^*$

$$\text{rec}(a)[w, b] = [w, b]$$

if and only if there is $\lambda \in B^*$, $h \in H$ and $g \in \widehat{k}^*$ such that

$$(w, \widehat{t}(a)b) = (\lambda w, \lambda bgh) \in (\mathbb{C} \setminus \mathbb{R}) \times \widehat{B}^*.$$

As the stabilizer of $w \in \mathbb{C}$ under the action of B^* is $t(F^*)$, this latter equality holds if and only if there is $\mu \in F^*, h \in H$ and $g \in \widehat{k}^*$ such that $\widehat{t}(a)b = t(\mu)bgh$ and this last holds if and only if $a \in F^*\widehat{t}^{-1}(bH\widehat{k}^*b^{-1})$. □

Definition 3.3.4 Let O_F be the ring of integers of F and O_k be that of k. If Z is an open compact subgroup of $\widehat{O}_F^*$ which contains $\widehat{O}_k^*$, then the reciprocity map provides an isomorphism

$$\text{rec}: \widehat{F}^*/F^*\widehat{k}^*Z \cong \text{Gal}(F_Z/F)$$

where F_Z is an abelian extension of F, uniquely determined by Z.

If $x = [w, b] \in N_H$ is a CM point, then associated to x is the open compact subgroup of $\widehat{O}_F^*$ given by $Z(x) = \widehat{t}^{-1}(bH\widehat{O}_k^*b^{-1})$ which contains $\widehat{O}_k^*$ and satisfies

$$\text{rec}: \widehat{F}^*/F^*\widehat{k}^*Z(x) \cong \text{Gal}(F(x)/F).$$

The subgroup $Z(x)$ is uniquely determined by x. In this case, we have $F_{Z(x)} = F(x)$.

Definition 3.3.5 The curve N_H^* depends only on the subgroup $H\widehat{k}^* \subseteq \widehat{B}^*$ and not on H. For any open compact subgroup $H \subseteq \widehat{B}^*$ there is an O_F-order $R \subset B$ such that $\widehat{R}^* \subseteq H\widehat{O}_F^*$ and in particular that $\widehat{R}^*\widehat{k}^* \subseteq H\widehat{k}^*$.

Let $x = [w, b] \in N_H$ be a CM point on N_H. It then follows, from Proposition 3.3.3 and Definition 3.3.4, that for any CM point x on N_H there is a ring class field $F[c]$, for some cycle c on k, such that $F(x) \subseteq F[c]$.

There is then a smallest ring class field $F[c]$ for some cycle c on k such that $F(x) \subseteq F[c]$. The cycle $c(x)$ is the *modulator* of the CM point x (see also Exercise 3.3.10 below).

Proposition 3.3.6 *Let Z be an open compact subgroup of $\widehat{O}_F^*$ which contains $\widehat{O}_k^*$. Then the finite group $\text{Gal}(F_Z/k)$ is generalized dihedral of the form*

$$\text{Gal}(F_Z/k) \cong \text{Gal}(F_Z/F) \rtimes \text{Gal}(F/k).$$

Proof The reciprocity map $\text{rec}: \widehat{F}^* \to \text{Gal}(F^{\text{ab}}/F)$, where F^{ab} is the maximal abelian separable extension of F, is compatible with the action of the group $\text{Gal}(F/k)$ of order 2 generated by the unique non-trivial element σ. Then σ extends uniquely to an automorphism of $\widehat{F}^*$ also denoted by σ. As $a.\sigma(a) \in \widehat{k}^*$ for all $a \in \widehat{F}^*$ where $\widehat{k}^*$ is the the idèle group of k, we then have that F_Z/k is Galois and the conjugation action of σ on $\text{Gal}(F_Z/F)$ is given by $\sigma g \sigma^{-1} = g^{-1}$ whence the result. [This proposition is an extension of Proposition 1.5.6.] □

Proposition 3.3.7 *Let $\widehat{O}_k^* \subseteq Z' \subseteq Z \subseteq \widehat{O}_F^*$ be open compact subgroups of $\widehat{O}_F^*$. Then there is an exact sequence of groups*

3.3 CM Points on Shimura Curves

$$0 \longrightarrow \frac{F^* \cap \widehat{k}^* Z}{F^* \cap \widehat{k}^* Z'} \longrightarrow \frac{Z}{Z'} \longrightarrow \mathrm{Gal}(F_{Z'}/F_Z) \longrightarrow 0.$$

Proof From (3.3.4) we have $\widehat{F}^*/F^*\widehat{k}^*W \cong \mathrm{Gal}(F_W/F)$ for $W = Z$ or Z'; hence we have $\mathrm{Gal}(F_{Z'}/F_Z) \cong (F^*\widehat{k}^*Z)/(F^*\widehat{k}^*Z')$ and the exact sequence follows from basic group theory where we note that the natural map $Z/Z' \to \widehat{k}^*Z/\widehat{k}^*Z'$ is an isomorphism. □

Proposition 3.3.8

(i) The kernel of the norm homomorphism $N_{F/k} : O_F^* \to O_k^*$ is the group of roots of unity of F.

(ii) Let $\widehat{O}_k^* \subseteq Z' \subseteq Z \subseteq \widehat{O}_F^*$ be open compact subgroups of $\widehat{O}_F^*$ and let $G = \mathrm{Gal}(F/k)$. Then the natural maps give an exact sequence of groups, which defines the subgroup H_0 of $H^1(G, O_F^*)$,

$$0 \longrightarrow \frac{O_F^* \cap Z}{O_F^* \cap Z'} \xrightarrow{f_{Z,Z'}} \frac{F^* \cap \widehat{k}^* Z}{F^* \cap \widehat{k}^* Z'} \longrightarrow H^1(G, O_F^*)/H_0 \longrightarrow 0$$

where $H^1(G, O_F^*)$ has order at most 2.

Proof

(i) As K is a totally imaginary quadratic extension of the totally real field k, the unit groups of F and k have the same rank. Therefore the kernel of the homomorphism $N_{F/k} : O_F^* \to O_k^*$ is a finite group and therefore consists of the roots of unity of F and it follows that this kernel is the whole group of roots of unity of F.

(ii) The injectivity of the first map $f_{Z,Z'}$ follows from the equality $O_F^* \cap \widehat{k}^* Z' = O_F^* \cap Z'$. The morphism of affine schemes $\mathrm{Spec}\, O_F \to \mathrm{Spec}\, O_k$ is Galois [51, p.44] with Galois group $G \cong \mathrm{Gal}(F/k)$. Hence there is a spectral sequence of étale cohomology

$$H^i(G, H^j_{\text{ét}}(\mathrm{Spec}\, O_F, \mathbb{G}_m)) \Longrightarrow H^{i+j}_{\text{ét}}(\mathrm{Spec}\, O_k, \mathbb{G}_m)$$

where $\mathbb{G}_m$ is the multiplicative group scheme [51, Chap. III, Theorem 2.20, p.105]. The short exact sequence of low degree terms associated to this spectral sequence then gives the exact sequence

$$0 \to H^1(G, O_F^*) \to \mathrm{Pic}(O_k) \to \mathrm{Pic}(O_F)^G.$$

Replacing the group Z' by a larger group Z'' such that $Z' \subseteq Z'' \subseteq Z$ decreases the cokernel of $f_{Z,Z'}$ in that there is an evident surjective homomorphism $\mathrm{cok}(f_{Z,Z'}) \to \mathrm{cok}(f_{Z,Z''})$. We may therefore reduce to proving the exact sequence for the case when $Z' = \widehat{O}_k^*$.

Again if Z is replaced by a smaller subgroup Z'' such that $Z' \subseteq Z'' \subseteq Z$, then the natural homomorphism $\text{cok}(f_{Z'',Z'}) \to \text{cok}(f_{Z,Z'})$ is an injection; this follows from a diagram chase and again the equality $O_F^* \cap \widehat{k}^* Z'' = O_F^* \cap Z''$. We may then reduce to proving the exact sequence for the case when $Z = \widehat{O}_F^*$.

Suppose then that $Z = \widehat{O}_F^*$, $Z' = \widehat{O}_k^*$. Then the part of the exact sequence that has been constructed gives

$$\text{cok}(f_{Z,Z'}) \cong \frac{F^* \cap \widehat{k}^* \widehat{O}_F^*}{F^* \cap \widehat{k}^* \widehat{O}_k^*} / O_F^* / O_k^* \cong \frac{F^* \cap \widehat{k}^* \widehat{O}_F^*}{k^* O_F^*}.$$

But this last group $(F^* \cap \widehat{k}^* \widehat{O}_F^*)/(k^* O_F^*)$ is isomorphic to the kernel Ξ of the natural homomorphism $\text{Pic}(O_k) \to \text{Pic}(O_F)^G$, where this isomorphism is given as follows. If $\beta \in F^* \cap \widehat{k}^* \widehat{O}_F^*$ then $\beta = \lambda \mu$ where $\lambda \in \widehat{k}^*, \mu \in \widehat{O}_F^*$. The ideal I of O_k defined by the idèle $\lambda \in \widehat{k}^*$ depends only on β, and furthermore, the divisor class of I in $\text{Pic}(O_k)$ lies in Ξ as $IO_F = \beta O_F$ is principal. The map $\beta \mapsto$ (class of I) $\in \text{Pic}(O_k)$ defines the isomorphism $(F^* \cap \widehat{k}^* \widehat{O}_F^*)/(k^* O_F^*) \cong \Xi$. Now Ξ is isomorphic to $H^1(G, O_F^*)$ as already shown which gives the exact sequence of the proposition.

The group $H^1(G, O_F^*)$ is a quotient of $\ker : N_{F/k} : O_F^* \to O_k^*$ where this last is isomorphic to the group of roots of unity of F by part (i) and hence is a cyclic group. As G has order 2 it follows that $H^1(G, O_F^*)$ has order at most 2. □

Proposition 3.3.9 *Let $I_0 \subseteq O_F$ be the ideal of O_F given by $\bigcap_{u \neq 1}(u - 1)O_F$ where u runs over the roots of unity, different from 1, contained in F. Let $\widehat{O}_k^* \subseteq Z' \subseteq Z \subseteq \widehat{O}_F^*$ be open compact subgroups of $\widehat{O}_F^*$.*

Suppose that $Z \subseteq \widehat{O}_c^*$ for some cycle c of k, without archimedean components, such that $I(c)O_F \not\supseteq I_0$, where $I(c)$ is the ideal of O_k defined by c. Then we have

- *$F^* \cap \widehat{k}^* Z = k^*$;*
- *$O_F^* \cap Z = O_k^*$;*
- *There is an isomorphism $\text{Gal}(F_{Z'}/F_Z) \cong Z/Z'$.*

Proof First, we assert that if c is a cycle of k without archimedean components with corresponding ideal $I(c)$ of O_k such that $I(c)O_F \not\supseteq I_0$ and if $Z \subseteq \widehat{O}_c^*$ is a subgroup then $F^* \cap \widehat{k}^* Z = k^*$ and $O_F^* \cap Z = O_k^*$.

To see this, let $\beta \in F^* \cap \widehat{k}^* Z \subseteq F^* \cap \widehat{k}^* \widehat{O}_c^*$. Then we may write $\beta = ab, a \in \widehat{k}^*, b \in \widehat{O}_c^*$. We then have, where σ is complex conjugation,

$$v = \beta/\beta^\sigma = b/b^\sigma \in O_F^*.$$

We then have $N_{F/k}(v) = 1$. By Proposition 3.3.8(i), from $N_{F/k}(v) = 1$ it follows that v is a root of unity in F. Furthermore, as v is in the kernel of the homomorphism

$O_F^* \to (O_F/I(c)O_F)^*$ and the ideal $I(c)O_F$ does not contain I_0, it follows that $v = 1$ and hence $\beta \in k^*$.

We have then shown that

$$F^* \cap \widehat{k}^* Z = k^*.$$

It follows that

$$O_k^* \subseteq O_F^* \cap Z \subseteq O_c^* \subseteq O_F^* \cap \widehat{k}^* \widehat{O}_c^* \subseteq O_F^* \cap k^* = O_k^*$$

so that $O_F^* \cap Z = O_k^*$.

The isomorphism $\mathrm{Gal}(F_{Z'}/F_Z) \cong Z/Z'$ now follows from Proposition 3.3.7. □

Exercise 3.3.10 The Shimura curve N_H^*/k is said to be of *classical type* if $H = \widehat{R}^*$ where $R \subset B$ is an O_k-order of the quaternion algebra B and O_k is the ring of integers of k.

(i) The curve N_H depends only on the subgroup $H\widehat{k}^*$ of $\widehat{B}^*$. Show that for each open compact subgroup H of $\widehat{B}^*$ there is an O_k-order $R \subseteq B$ such that $\widehat{R}^* \subseteq H\widehat{O}_k^*$ so that we have $\widehat{R}^*\widehat{k}^* \subseteq H\widehat{k}^*$. Conclude that the Shimura curves of classical type $N_{\widehat{R}^*}$, where R runs over all O_k-orders of B, are cofinal in the system of curves $\{N_H\}_H$.
(ii) Let N_H^*/k be a Shimura curve of classical type. Show that the field of definition $F(x) \subseteq F^{\mathrm{ab}}$ of a CM point $x = [w, b] \in N_H$ is a ring class field of F of the form $F[c]$ for some cycle c of k (see Sect. 1.5).

3.4 Heegner Points and CM Points on Elliptic Curves

(3.4.1) In this section we consider Heegner points on elliptic curves parametrized by modular curves. We first consider elliptic curves parametrized by classical modular curves or Drinfeld modular curves and in the last part we have elliptic curves parametrized by Shimura curves.

Heegner points on elliptic curves parametrized by classical modular curves or Drinfeld modular curves

(3.4.2) Let

k be a global field equipped with a special set of places ∞;
A be the ring of integers of k with respect to ∞;
F be an imaginary quadratic extension field of k with respect to ∞ (see Sect. 1.1);
B be the integral closure of A in F;

c be a cycle of k with support disjoint from the places in ∞;
O_c denote the order of F relative to A and with conductor c (Sect. 1.4).

Assume that if k has characteristic zero then $k = \mathbb{Q}$.

(3.4.3) Let E/k be an elliptic curve that is to say a one-dimensional abelian variety over k.

We assume that if k has positive characteristic then $E \times_k k_\infty/k_\infty$ is a Tate curve (i.e. E has split multiplicative reduction at the unique place in ∞), where k_∞ is the completion of k at the unique place in ∞.

Let $\mathfrak{J}$, an ideal of A, be the conductor of E/k without the components contained in ∞. We further assume that the support of the cycle c is coprime to $\mathfrak{J}$ and that the Heegner condition holds (see (3.2.2)) so that we may write $\mathfrak{J}B$ as a product $IB = \mathfrak{J}_1\mathfrak{J}_2$ of ideals $\mathfrak{J}_1, \mathfrak{J}_2$ of B where $\mathfrak{J}_1, \mathfrak{J}_2$ are permuted transitively by $\mathrm{Gal}(F/k)$.

(3.4.4) *(Rigidification by ξ for Classical Modular Curves and Drinfeld Modular Curves.)* The modularity theorem 2.6.2 shows that there is a finite surjective morphism of projective k-schemes

$$\pi : X_0^{\mathrm{Ell.Sp.}}(\mathfrak{J}) \to E$$

where $X_0^{\mathrm{Ell.Sp.}}(\mathfrak{J})/k$ is the generic fibre of the moduli scheme $\mathbf{X}_0^{\mathrm{Ell.Sp.}}(\mathfrak{J})/A$. The map π is not uniquely determined and maybe translated in the group scheme E. But π may be rigidified as follows.

Let Ξ be the divisor $\sum x_i$ on $X_0^{\mathrm{Ell.Sp.}}(\mathfrak{J})/k$ consisting of the cusps x_i all with multiplicity 1 and let Γ be the cuspidal subgroup of $\mathrm{Pic}(X_0^{\mathrm{Ell.Sp.}}(\mathfrak{J}))(k^{\mathrm{sep}})$ where Γ is the subgroup generated by the linear equivalence classes of the differences of the cusps $x_i - x_j$. This group Γ is finite by the Manin-Drinfeld theorem.

The curve $X_0^{\mathrm{Ell.Sp.}}(\mathfrak{J}) \otimes_k k^{\mathrm{sep}}$ is a union $\bigcup_i X_i$ of irreducible components X_i which are connected projective curves.

Then the pullback of Ξ to each irreducible component of $X_0^{\mathrm{Ell.Sp.}}(\mathfrak{J}) \times_k k^{\mathrm{sep}}$ has degree 1. Let $m \in \mathbb{N}$ be any integer which annihilates the subgroup Γ of $\mathrm{Pic}(X_0^{\mathrm{Ell.Sp.}}(\mathfrak{J}))$.

Then each Hecke operator acts on the class of $m\Xi$ by multiplication by its degree. Let $m\xi_i \in \mathrm{Pic}(X_i)$ be the pullback to X_i of the class of $m\Xi$ for each i. We may then define the surjective finite k-morphism

$$\pi : X_0^{\mathrm{Ell.Sp.}}(\mathfrak{J}) \to E$$

by

$$P \in X_i(k^{\mathrm{sep}}) \mapsto mP - m\xi_i \quad (\text{for all } i).$$

Then with this morphism, we say that π *is rigidified by* ξ.

3.4 Heegner Points and CM Points on Elliptic Curves

[See paragraph (3.4.11) for a corresponding rigidification by ξ for compact Shimura curves.]

Definition 3.4.5 Let c be a cycle on k, with support coprime to Supp $\mathfrak{J}$ and to the places in ∞. Let a be a class in the Picard group $\mathrm{Pic}(O_c)$ of the order O_c of F with conductor c. Let $(a, \mathfrak{J}_1, c)$ be the Heegner point in $X_0^{\mathrm{Ell.Sp.}}(\mathfrak{J})(F[c])$ given by Definition 3.2.9. Then the point

$$\pi(a, \mathfrak{J}_1, c) \in E(F[c])$$

written $(a, \mathfrak{J}_1, c, \pi)$ is a $F[c]$-rational point of the elliptic curve E called a *Heegner point of the elliptic curve E with class a and conductor c*.

(3.4.6) Let l be any prime number distinct from the characteristic of k. Let $T_l(E)$ be the l-adic Tate module of E, that is to say $T_l(E)$ is the dual of $H^1_{\mathrm{ét}}(E \times_k k^{\mathrm{sep}}, \mathbb{Q}_l)$ where k^{sep} denotes the separable closure of k. For z a closed point of Spec A which is prime to $\mathrm{Supp}(\mathfrak{J})$, we put

$$a_z = \mathrm{Tr}(\mathrm{Fr}_z | T_l(E))$$

which is the trace of a Frobenius element $\mathrm{Fr}_z \in \mathrm{Gal}(k^{\mathrm{sep}}/k)$ at z acting on $T_l(E)$. We have that a_z is an integer in $\mathbb{Z}$. Put, where this is an equality of divisors on the curve E,

$$\Delta(a, c, z, \pi) = a_z(a, \mathfrak{J}_1, c, \pi) - \frac{|O_c^*|}{|A^*|} \mathrm{Tr}_{F[zc]/F[c]}(a^\sharp, \mathfrak{J}_1, cz, \pi)$$

where $a^\sharp$ is any element of $\mathrm{Pic}(O_{cz})$ lifting a under the natural homomorphism $t_{cz,c} : \mathrm{Pic}(O_{cz}) \longrightarrow \mathrm{Pic}(O_c)$ induced by the inclusion $O_{cz} \subseteq O_c$.

(3.4.7) Suppose that z is a closed point of Spec A. Assume that z and the cycle $c \in \mathrm{Cyc}(k)$ are coprime to $\mathrm{Supp}(\mathfrak{J})$ and to all places in ∞. We may then tabulate the values of $\Delta(a, c, z, \pi)$, where

$$t_{c,c/z} : \mathrm{Pic}(O_c) \longrightarrow \mathrm{Pic}(O_{c/z})$$

is the homomorphism induced from the inclusion of orders $O_{c/z} \to O_c$ and $[[b]]$ for an invertible ideal b of O_c denotes the class in $\mathrm{Pic}(O_c)$ given by b:

Proof of Table 3.4.8 In the case where k has positive characteristic, this is a restatement of [4, table 4.8.5] which in turn follows immediately from [4, table 4.6.9], obtained using Bruhat-Tits trees with complex multiplication [4, Chapter 3] and the positive characteristic analogue [4, theorem 4.5.7] of the Eichler-Shimura congruence, namely, we have

$$T_z(a, \mathfrak{J}_1, c, \pi) = a_z(a, \mathfrak{J}_1, c, \pi)$$

Table 3.4.8 Values of $\Delta(a, c, z, \pi)$

(1) 0	If z remains prime in F/k and is prime to c
(2) $(a[[\mathfrak{m}'_z]]^{-1}, \mathfrak{J}_1, c, \pi)$	If z is ramified in F/k and is prime to c where $\mathfrak{m}'_z$ is the prime ideal of O_c lying above the ideal $\mathfrak{m}_z$ of A defining z
(3) $(a[[\mathfrak{p}_1]]^{-1}, \mathfrak{J}_1, c, \pi) + (a[[\mathfrak{p}_2]]^{-1}, \mathfrak{J}_1, c, \pi)$	If z is split completely in F/k and is prime to c where $\mathfrak{m}_z O_c = \mathfrak{p}_1 \mathfrak{p}_2$ and $\mathfrak{p}_1, \mathfrak{p}_2$ are two distinct prime ideals of O_c
(4) $(a^\flat, \mathfrak{J}_1, c/z, \pi)$	If $z \in \mathrm{Supp}(c)$, where $a^\flat = t_{c,c/z}(a)$

where a_z is the trace of the Frobenius at z on $T_l(E)$ and T_z is the Hecke operator at z acting on the Drinfeld modular curve $X_0^{\mathrm{Drin}}(\mathfrak{J})$.

In the case where k has zero characteristic and hence $k = \mathbb{Q}$, Table 3.4.8 follows in exactly the same way using the Bruhat-Tits trees with complex multiplication [5, Chapter 3] to derive the analogue of the table [5, table 4.6.9] for this case where $k = \mathbb{Q}$ and then applying the classical Eichler-Shimura congruence for the Hecke operator T_z on the classical modular curve $X_0(\mathfrak{J})$. The case of this table where k is the rational number field $\mathbb{Q}$ is essentially well known and for this case $k = \mathbb{Q}$ part of the table is stated in [34, §3, p470]; the case (1) of Table 3.4.8 where z remains prime in F/k and is coprime to c is proved in [22, Prop. 3.7]. □

Theorem 3.4.9 *Let $c, z \in \mathrm{Cyc}(k)$ be cycles, coprime to ∞, where z is a prime divisor such that $z \notin \mathrm{Supp}(c)$. Assume that z and c are both primes to $\mathrm{Supp}(\mathfrak{J})$. Let Z be a place of $F[cz]$ lying over z. Let $a^\sharp \in \mathrm{Pic}(O_{cz})$ be any class lifting $a \in \mathrm{Pic}(O_c)$. In the group $E_z(\kappa(Z))$, where E_z denotes the reduction modulo z of the Néron model of E, we then have the relation*

$$\mathrm{Fr}_z(a^\sharp, \mathfrak{J}_1, cz, \pi) \equiv (a, \mathfrak{J}_1, c, \pi) \pmod{Z}.$$

Proof For the case where k has positive characteristic, this is an immediate consequence of [5, theorem 4.6.19(i)].

If k has characteristic zero and $k = \mathbb{Q}$, this congruence relation is proved in [22, Prop. 3.7]. □

CM points on elliptic curves parametrized by Shimura curves

(3.4.10) The notation of Sects. 2.3 and 3.3 holds in this subsection. Suppose that

k is a totally real field with ring of integers O_k;
∞ is the set of distinct archimedean places of k;
F/k is a totally imaginary quadratic extension field of k and O_F is the ring
  of integers of F;

3.4 Heegner Points and CM Points on Elliptic Curves

E/k is an elliptic curve;
R is a coefficient ring of residue characteristic l;
H is an open compact subgroup of $\widehat{B}^*$, where B is a quaternion algebra over
$\quad k$ as in Sect. 2.3;
N_H^*/k is the projective Shimura curve defined by H (see (2.3)).

(3.4.11) (*Rigidification by ξ for Compact Shimura Curves.*) Suppose that there is a finite surjective morphism of k-schemes

$$\pi : N_H^* \longrightarrow E.$$

The map π is defined only up to translation in the group scheme E.

A rigidification of π, when the Shimura curve N_H is not compact, is given in paragraph (3.4.4) (i.e. the case of classical modular curves).

Suppose then that N_H is compact (i.e. $B \neq M_2(\mathbb{Q})$). A canonical map from N_H^* to the Jacobian $\text{Jac}(N_H^*)$ can be constructed as follows. Let $\tau : k \to \mathbb{C}$ be a fixed archimedean place of k. Then $N_H^* \otimes_{k,\tau} \mathbb{C}$ is a finite disjoint union $\bigcup_i X_i$ of irreducible components X_i which are connected compact Riemann surfaces.

There is a canonical integral divisor $\Xi \in \text{Pic}(N_H^*/k)$ and a positive integer m such that Ξ has degree m on each irreducible component X_i and Ξ is constructed in [14, §3.5].

The rational divisor class $\xi \in \text{Pic}(X) \otimes_{\mathbb{Z}} \mathbb{Q}$ is then defined to be $\Xi \otimes (1/m)$. The canonical rational divisor class ξ of degree 1 in $\text{Pic}(X_i) \otimes_{\mathbb{Z}} \mathbb{Q}$ (called a "Hodge class" in [56, (1.19)] and [14, §3.5]) has the property that the pullback of ξ to each irreducible component $X_i/\mathbb{C}$ has degree 1 and each Hecke operator $[HbH]$ acts on ξ by multiplication by its degree [14, Remark 3.7].

We then define a map $\psi : N_H^* \to \text{Jac}(X) \otimes_{\mathbb{Z}} \mathbb{Q}$ where $p \in N_H^*$ maps to the class of $p - \xi_i$. It can be shown that the positive integral multiple $m\psi$ of ψ is k-morphism of k-schemes and we then define the map

$$\phi : N_H^* \to \text{Jac}(N_H^*)$$

by $\phi(p) = m(p) - m\xi_i$ for all $p \in X_i(k^{\text{sep}})$ and all i.

The morphism of k-schemes $\pi : N_H^* \to E$ factors as $N_H^* \xrightarrow{\phi} \text{Jac}(N_H^*) \xrightarrow{\widehat{\pi}} E$ where $\widehat{\pi} : \text{Jac}(N_H^*) \to E$ is a k-morphism. We say that π is *rigidified by ξ* if the induced map $\widehat{\pi}$ is a homomorphism of abelian varieties.

Evidently any surjective k-morphism $\pi : N_H^* \to E$ may be rigidified by ξ by translating π in E. We assume for the rest of this chapter that $\pi : N_H^* \to E$ is rigidified by ξ both when N_H is compact and non-compact (by (3.4.4)).

Definition 3.4.12 Let $x = [w, b]$ be a CM point in N_H^* where $b \in \widehat{F}^*$ and where $w \in \mathbb{C}$ is the unique point with $\text{Im}(w) > 0$ fixed by $t(F^*)$ under the action of B^* ((2.3) and Definition 2.3.16).

The field of definition of $x = [w, b]$ is by definition the field $F(x)$ so that $[w, b] \in N_H^*(F(x))$ (see Proposition 3.3.3).

Then the point

$$\pi[w, b] \in E(F(x))$$

written $[w, b, \pi]$ is a $F(x)$-rational point of the elliptic curve E called a *CM point of the elliptic curve E*.

(3.4.13) Let l be any prime number and $T_l(E)$ be the l-adic Tate module of E. For z a closed point of Spec A which is prime to Supp($\mathfrak{J}$), we put

$$a_z = \text{Tr}(\text{Fr}_z | T_l(E))$$

which is the trace of a Frobenius element $\text{Fr}_z \in \text{Gal}(k^{\text{sep}}/k)$ at z acting on $T_l(E)$ and where a_z is an integer in $\mathbb{Z}$.

(3.4.14) Let S be the finite set of non-archimedean places where the quaternion algebra B is ramified (see (2.3)). Fix $U \supseteq S$ a finite set of non-archimedean places of k such that H can be written as $H = H_U H^U$ where

$$H_U \subseteq \prod_{v \in U} B_v^*, \quad H^U = \prod_{v \notin U} H_v$$

and each H_v, for $v \notin U$, is a maximal compact subgroup of $B_v^* \cong \text{GL}_2(k_v)$. The group H^U is then a maximal compact subgroup of the restricted product $\prod_{v \notin U}' B_v^*$.

Definition 3.4.15 Let $x = [w, b]$ be a CM point and let $H = H_U H^U$ be the factorization of H defined in the previous paragraph (3.4.14). Let $\mathcal{S} = \bigcup_{r \geq 0} \mathcal{S}_r$ be the following set of square-free cycles of O_k depending on x. We let $\mathcal{S}_1$ be the following set of prime ideals of O_k:

$$\mathcal{S}_1 = \left\{ \mathfrak{p} \,\middle|\, \begin{array}{c} \text{is a non-zero prime ideal of } O_k \\ \text{which is inert in } F/k \\ \text{and } \mathfrak{p} \text{ is coprime to } U, l, c(x), \text{ and } I_0 \not\subseteq \mathfrak{p} O_F \end{array} \right\}.$$

Here I_0 is the ideal defined in Proposition 3.3.9, $c(x)$ is the modulator of the CM point x (Definition 3.3.5) and l is the prime number which is the residue characteristic of the coefficient ring R.

We define $\mathcal{S}_r$ to be the square-free cycles of k formed from the elements of $\mathcal{S}_1$

$$\mathcal{S}_r = \{\mathfrak{p}_1 \ldots \mathfrak{p}_r \mid \mathfrak{p}_i \in \mathcal{S}_1 \text{ are distinct}\} \quad (r > 1).$$

3.4 Heegner Points and CM Points on Elliptic Curves

We put formally $\mathcal{S}_0 = \{1\}$ where 1 denotes the ideal O_k of O_k.

Definition 3.4.16 Let $x = [w, b] \in \mathrm{CM}(N_H, F)$ be a CM point on N_H (see (2.3.15)). Define the CM points $x(\mathfrak{n}) \in \mathrm{CM}(N_H, F)$ for all $\mathfrak{n} \in \mathcal{S}$ as follows:
Let h be any map of sets $h : \mathcal{S} \to \widehat{B}^*$ such that

- For a prime cycle $\mathfrak{p} \in \mathcal{S}_1$ then $H_{\mathfrak{p}} = M_{\mathfrak{p}}^*$ for a maximal $O_{k,\mathfrak{p}}$-order $M_{\mathfrak{p}} \subseteq B_{\mathfrak{p}}$, let $h(\mathfrak{p}) \in M_{\mathfrak{p}} \cap B_{\mathfrak{p}}^*$ be any element such that $\mathrm{ord}_{\mathfrak{p}} \mathrm{nr}(h(\mathfrak{p})) = 1$ where nr is the reduced norm of the quaternion algebra $B_{\mathfrak{p}}$ [80, p.1] and $\mathrm{ord}_{\mathfrak{p}}$ is the normalized absolute value on the local field $k_{\mathfrak{p}}$;
- For $\mathfrak{n} \in \mathcal{S}_r$ put $h(\mathfrak{n}) = \prod_i h(\mathfrak{p}_i) \in \widehat{B}^*$ where $\mathfrak{n} = \mathfrak{p}_1 \ldots \mathfrak{p}_r$;
- $h(1) = 1$.

The CM point $x(\mathfrak{n})$ for $\mathfrak{n} \in \mathcal{S}$ is then defined by

$$x(\mathfrak{n}) = [w, bh(\mathfrak{n})] \in \mathrm{CM}(N_H, F).$$

The points $x(\mathfrak{n}) \in \mathrm{CM}(N_H, F)$ are independent of the choice of map h (see Remark 3.4.17).

Remark 3.4.17 For each prime cycle $\mathfrak{p} \in \mathcal{S}_1$, the element $h(\mathfrak{p}) \in \widehat{B}^*$ is roughly speaking a "minimal modification at $\mathfrak{p}$" of the identity element of $\widehat{B}^*$.

The points $x(\mathfrak{n})$ are independent of the choice of map h, where this is shown as follows. Let $h, h' : \mathcal{S} \to \widehat{B}^*$ be two maps satisfying the above conditions. Then for $\mathfrak{p} \in \mathcal{S}_1$ we have $h(\mathfrak{p}), h'(\mathfrak{p}) \in M_{\mathfrak{p}} \cap B_{\mathfrak{p}}^*$ such that $\mathrm{ord}_{\mathfrak{p}} \mathrm{nr}(h(\mathfrak{p})) = 1 = \mathrm{ord}_{\mathfrak{p}} \mathrm{nr}(h'(\mathfrak{p}))$; then we have $h(\mathfrak{p}).h'(\mathfrak{p})^{-1} \in M_{\mathfrak{p}}^*$ because $\mathrm{ord}_{\mathfrak{p}} \mathrm{nr}(h(\mathfrak{p}).h'(\mathfrak{p})^{-1}) = 0$; hence $h(\mathfrak{p}).h'(\mathfrak{p})^{-1} \in H_{\mathfrak{p}}$. Therefore $bh(\mathfrak{p})H = bh'(\mathfrak{p})H$ so that the CM point $[w, bh(\mathfrak{n})] = [w, bh'(\mathfrak{n})]$ is independent of the choice of h.

Lemma 3.4.18 *Let E be a mixed characteristic local field and let E'/E be the quadratic unramified field extension. Let O_E, $O_{E'}$ be their respective rings of valuation integers. Let π be a local parameter of E. Let $R \subseteq M_2(E)$ be a maximal O_E-order and $i : E' \to M_2(E)$ an E-embedding.*

If $i^{-1}(R^*) = O_{E'}^*$, and if $g \in R$ satisfies $\mathrm{ord}(\det(g)) = 1$, where ord is the normalized valuation on E, then we have

$$i^{-1}(gR^*g^{-1}) = i^{-1}(R^* \cap gR^*g^{-1}) = (O_E + \pi O_{E'})^*$$

and there is a group isomorphism, where $\kappa_{E'}, \kappa_E$ are the residue fields of E', E,

$$\frac{i^{-1}(R^*)}{i^{-1}(gR^*g^{-1})} \cong \kappa_{E'}^*/\kappa_E^*.$$

Furthermore, the cardinality of $R^*/(R^* \cap gR^*g^{-1})$ is given by

$$\left|\frac{R^*}{R^* \cap gR^*g^{-1}}\right| = |\kappa_E| + 1.$$

Proof By choosing a suitable basis, we may assume that $R = M_2(O_E)$ and $g = \begin{pmatrix} 1 & 0 \\ 0 & \pi \end{pmatrix}$. Then R^* is a maximal compact subgroup of $M_2(R)^* = GL_2(R)$ and for any $\gamma \in GL_2(E)$ then $i^{-1}(\gamma R^* \gamma^{-1})$ is a compact subgroup of $(E')^*$, as the map i is proper, and hence $i^{-1}(\gamma R^* \gamma^{-1}) \subseteq O_{E'}^*$.

By assumption $i^{-1}(R^*) = O_{E'}^*$ and as we evidently have $i^{-1}(gR^*g^{-1}) = O_{E'}^* \cap i^{-1}(gR^*g^{-1})$, we obtain

$$i^{-1}(gR^*g^{-1}) = i^{-1}(R^* \cap gR^*g^{-1}).$$

Now $O = R \cap gRg^{-1}$ is an Eichler order of level 1 of $M_2(E)$ and equals

$$O = \left\{\begin{pmatrix} a & b \\ \pi c & d \end{pmatrix} \Big| a, b, c, d \in O_E\right\}.$$

Hence $i^{-1}(O)$ is an O_E-order of $O_{E'}$ and therefore $i^{-1}(O) = O_E + \pi^c O_{E'}$ for some natural number c. As $i^{-1}(R^*) = O_{E'}^*$, we must have $i^{-1}(R) = O_{E'}$. The O_E-module $i^{-1}(R)/i^{-1}(O)$ is O_E-isomorphic to $O_E/\pi^c O_E$ but which is also isomorphic via i to a submodule of $R/O \cong O_E/\pi O_E$. Therefore we have $c \leq 1$. If $c < 1$ then $i(O_{E'}) \subseteq O$. The 2-sided ideal $I = \left\{\begin{pmatrix} \pi a & b \\ \pi c & \pi d \end{pmatrix} \Big| a, b, c, d \in O_E\right\}$ of O is topologically nilpotent for the π-adic topology; hence we have $i^{-1}(I) \subseteq \pi O_{E'}$ and an injective homomorphism $\kappa_{E'} \to \kappa_E \times \kappa_E \cong O/I$ which is impossible. Therefore we must have $c = 1$. Hence we have

$$i^{-1}(gR^*g^{-1}) = i^{-1}(R^* \cap gR^*g^{-1}) = i^{-1}(O^*) = (O_E + \pi O_{E'})^*.$$

That the cardinality of $R^*/(R^* \cap gR^*g^{-1})$ equals $|\kappa_E| + 1$ now follows (see Exercise 3.4.24). □

Proposition 3.4.19 *The field of definition $F(x(\mathfrak{n}))$ of the CM point $x(\mathfrak{n})$ is contained in the ring class field $F[c(x)\mathfrak{n}]$ where $c(x)$ is the modulator of x (Definition 3.3.5).*

Let $Z(x)$, $Z(x(\mathfrak{n}))$ be the open compact subgroups of $\widehat{O}_F^*$ which contain $\widehat{O}_k^*$ associated to the CM points $x, x(\mathfrak{n})$ by Definition 3.3.4. The quotient $Z(x)/Z(x(\mathfrak{n}))$ is given by the natural isomorphisms

3.4 Heegner Points and CM Points on Elliptic Curves

$$Z(x)/Z(x(\mathfrak{n})) \cong \prod_{\mathfrak{p}|\mathfrak{n}} Z(x)_\mathfrak{p}/Z(x(\mathfrak{n}))_\mathfrak{p}, \quad Z(x)_\mathfrak{p}/Z(x(\mathfrak{n}))_\mathfrak{p} \cong \frac{(O_F/\mathfrak{p}O_F)^*}{(O_k/\mathfrak{p}O_k)^*} \quad (for\ \mathfrak{p}|\mathfrak{n}).$$

Proof We have the reciprocity isomorphism from Definition 3.3.4

$$\widehat{F}^*/F^*\widehat{k}^*Z(x(\mathfrak{n})) \cong \mathrm{Gal}(F(x(\mathfrak{n}))/F)$$

and similarly for $Z(x)$ and $\mathrm{Gal}(F(x)/F)$.

The point x is $[w,b]$ and $x(\mathfrak{n}) = [w, bh(\mathfrak{n})]$ (see Definition 3.4.16). We have from Definition 3.3.4 that the compact subgroup $Z(x(\mathfrak{n}))$ is given by

$$Z(x(\mathfrak{n})) = \hat{\imath}^{-1}(bh(\mathfrak{n})H\widehat{O}_k^* h(\mathfrak{n})^{-1}b^{-1})$$

and similarly $Z(x) = Z(x(1)) = \hat{\imath}^{-1}(bH\widehat{O}_k^* b^{-1})$ where $h(1) = 1$ by definition.

Let U be a finite set of non-archimedean places of k such that $H = H_U H^U$ satisfying the conditions in (3.4.14). Then the product $H = H_U H^U$ shows that $Z(x)$ can be decomposed as

$$Z(x) = Z(x)_U Z(x)^U, \quad Z(x)_U \subseteq \prod_{v \in U} O_{F,v}^*, \quad Z(x)_v = t_v^{-1}(b_v H_v O_{k,v}^* b_v^{-1}) \quad (v \notin U)$$

$$Z(x)^U = \prod_{v \notin U} Z(x)_v.$$

Put

$$U(\mathfrak{n}) = U \cup \{\mathfrak{p} \in \mathcal{S}_1 \text{ where } \mathfrak{p}|\mathfrak{n}\}.$$

We then have, as $h(\mathfrak{n}) = \prod_{\mathfrak{p}|\mathfrak{n}} h(\mathfrak{p})$,

$$Z(x(\mathfrak{n})) = Z(x)_U \prod_{\mathfrak{p}|\mathfrak{n}} Z(x(\mathfrak{n}))_\mathfrak{p} \prod_{\mathfrak{p} \notin U(\mathfrak{n})} Z(x)_\mathfrak{p}, \quad Z(x(\mathfrak{n}))_v = Z(x)_v \quad (\text{for all } v \notin U(\mathfrak{n})).$$

Then the group $Z(x(\mathfrak{n}))$ decomposes as $Z(x)/Z(x(\mathfrak{n})) \cong \prod_{\mathfrak{p}|\mathfrak{n}} Z(x)_\mathfrak{p}/Z(x(\mathfrak{n}))_\mathfrak{p}$.

Recall from the definition of $h(\mathfrak{n})$ (Definition 3.4.16) that for a prime place $\mathfrak{p} \in \mathcal{S}_1$ then $H_\mathfrak{p} = M_\mathfrak{p}^*$ for a maximal $O_{k,\mathfrak{p}}$-order $M_\mathfrak{p} \subseteq B_\mathfrak{p}$. We then apply Lemma 3.4.18 for any prime $\mathfrak{p}$ of k dividing $\mathfrak{n}$, where we recall that the primes dividing $\mathfrak{n}$ are inert in F/k by Definition 3.4.15. In the notation of that lemma, we put $E = k_\mathfrak{p}$, $E' = F_\mathfrak{q}$ where $\mathfrak{q}$ is the unique prime of F over k, $R = M_\mathfrak{p} \subseteq B_\mathfrak{p} \cong M_2(k_\mathfrak{p})$ where $R^* = H_\mathfrak{p}$, $i = \mathrm{Ad}(b_\mathfrak{p}^{-1}) \circ t_\mathfrak{p} : F \otimes k_\mathfrak{p} \to B_\mathfrak{p}$, $g = h(\mathfrak{p})$.

We have

$$i^{-1}(R^*) = Z(x)_{\mathfrak{p}}, \quad i^{-1}(gR^*g^{-1})) = Z(x(\mathfrak{n}))_{\mathfrak{p}}.$$

We then have from Lemma 3.4.18 that $i^{-1}(gR^*g^{-1}) = i^{-1}(R^* \cap gR^*g^{-1}) = \widehat{Q}_{\mathfrak{p}}^*$ where we put $Q = O_k + \mathfrak{p} O_F$ which is an O_k-order of F of conductor $\mathfrak{p}$. Hence we obtain, where $\widehat{O}_{\mathfrak{n}}^*$ is obtained from the unit group of the O_k-order $O_{\mathfrak{n}}$ of F with conductor $\mathfrak{n}$,

$$(Z(x) \cap \widehat{O}_{\mathfrak{n}}^*)_{\mathfrak{p}} = Z(x)_{\mathfrak{p}} \cap (\widehat{O}_{\mathfrak{n}}^*)_{\mathfrak{p}} =$$

$$Z(x)_{\mathfrak{p}} \cap \widehat{Q}_{\mathfrak{p}}^* = i^{-1}(R^*) \cap \widehat{Q}_{\mathfrak{p}}^* =$$

$$i^{-1}(R^* \cap gR^*g^{-1}) = i^{-1}(gR^*g^{-1}) = Z(x(\mathfrak{n}))_{\mathfrak{p}}.$$

On the other hand, if $\mathfrak{p}$ does not divide $\mathfrak{n}$ then we have

$$Z(x(\mathfrak{n}))_{\mathfrak{p}} = Z(x)_{\mathfrak{p}} = Z(x)_{\mathfrak{p}} \cap (\widehat{O}_F^*)_{\mathfrak{p}} = Z(x)_{\mathfrak{p}} \cap (\widehat{O}_{\mathfrak{n}}^*)_{\mathfrak{p}} = (Z(x) \cap \widehat{O}_{\mathfrak{n}}^*)_{\mathfrak{p}}.$$

Hence we have shown that $Z(x(\mathfrak{n}))_{\mathfrak{p}} = (Z(x) \cap \widehat{O}_{\mathfrak{n}}^*)_{\mathfrak{p}}$ for all non-archimedean prime places of k. Therefore we obtain

$$Z(x(\mathfrak{n})) = Z(x) \cap \widehat{O}_{\mathfrak{n}}^*.$$

We then have from the isomorphism of Lemma 3.4.18 that for a non-archimedean prime $\mathfrak{p}$ of k which divides $\mathfrak{n}$, where $\mathfrak{p}$ is by definition inert in F/k,

$$Z(x)_{\mathfrak{p}} / Z(x(\mathfrak{n}))_{\mathfrak{p}} \cong \frac{(O_F / \mathfrak{p} O_F)^*}{(O_k / \mathfrak{p} O_k)^*}.$$

Finally, it follows from $Z(x) \supseteq \widehat{O}_{c(x)}^*$ that $Z(x(\mathfrak{n})) \supseteq \widehat{O}_{c(x)}^* \cap \widehat{O}_{\mathfrak{n}}^* = \widehat{O}_{c(x)\mathfrak{n}}^*$ hence we have $F(x(\mathfrak{n})) \subseteq F[c(x)\mathfrak{n}]$. □

Proposition 3.4.20 (Splitting of Primes in $F(x(\mathfrak{n}))/F$.) *Let $\mathfrak{p}$ be a non-archimedean prime of k which does not divide the modulator $c(x)$ of the CM point $x \in \mathrm{CM}(N_H, F)$. Then we have:*

(i) *If $\mathfrak{n} \in \mathcal{S}$ and $\mathfrak{p} \nmid \mathfrak{n}$, then each prime of F over $\mathfrak{p}$ is unramified in $F(x(\mathfrak{n}))/F$;*
(ii) *If $\mathfrak{n} \in \mathcal{S}$ and $\mathfrak{p} \nmid \mathfrak{n}$ and $\mathfrak{p}$ is inert in F/k, then the prime place $\mathfrak{p} O_F$ of F splits completely in $F(x(\mathfrak{n}))/F$;*
(iii) *If $\mathfrak{n}\mathfrak{p} \in \mathcal{S}$ where $\mathfrak{p}$ is the given prime, then each prime of $F(x(\mathfrak{n}))$ above $\mathfrak{p}$ is totally ramified in $F(x(\mathfrak{n}\mathfrak{p}))/F(x(\mathfrak{n}))$.*

3.4 Heegner Points and CM Points on Elliptic Curves

Proof The proofs are similar to those of the corresponding results for ring class fields Sect. 1.5.

(i) This holds as $F(x) \subseteq F[c(x)]$, by Proposition 3.4.19 where $c(x)$ is the modulator of x, and the ring class field extension $F[c(x)]/F$ is unramified outside the support of $c(x)$.

(ii) The decomposition group of $\mathfrak{p}O_F$ in the extension $F(x(\mathfrak{n}))/F$ is the image of the composite map, where $Z(x(\mathfrak{n}))$ is the open compact subgroup associated to x as in Definition 3.3.4,

$$(F \otimes_k k_\mathfrak{p})^* \longrightarrow \widehat{F}^*/F^*\widehat{k}^*Z(x(\mathfrak{n})).$$

This map factors through $(F \otimes_k k_\mathfrak{p})^*/(O_F \otimes_{O_k} O_{k,\mathfrak{p}})^* k_\mathfrak{p}^* \cong 0$ as $\mathfrak{p}O_F$ is unramified in $F(x(\mathfrak{n}))/F$.

(iii) The inertia group of $F(x(\mathfrak{np}))/F(x(\mathfrak{n}))$ at any prime $\mathfrak{P}$ over $\mathfrak{p}$ is equal to the image of the composite map, where $Z(\mathfrak{np}), Z(x(\mathfrak{n}))$ are as in Definition 3.3.4 and the middle isomorphism is from Proposition 3.4.19,

$$f: O_{F,\mathfrak{p}}^* \xrightarrow{f_1} \frac{(O_F/\mathfrak{p}O_F)^*}{(O_k/\mathfrak{p}O_k)^*} \cong \frac{Z(\mathfrak{n})}{Z(\mathfrak{np})} \xrightarrow{f_2} \mathrm{Gal}(F(x(\mathfrak{np}))/F(x(\mathfrak{n}))).$$

The homomorphisms f_1, f_2 are the evident canonical maps (see Proposition 3.3.7 for f_2). It follows that the composite map f is surjective and hence $\mathfrak{P}$ is totally ramified and the ramification is tame as $[F(x(\mathfrak{np})) : F(x(\mathfrak{n}))]$ divides $N\mathfrak{p} + 1$ which is coprime to $N\mathfrak{P}$. □

Definition 3.4.21 (Definition of $u(r, x)$.) Let $x \in N_H$ be a CM point. Let $Z(x)$ be the open compact subgroup of $\widehat{O}_F^*$ which contains $\widehat{O}_k^*$ given from x by defintion 3.3.4. Then $Z(x)$ satisfies the reciprocity isomorphism, where $F(x)$ is the field of definition of x,

$$\widehat{F}^*/F^*\widehat{k}^*Z(x) \cong \mathrm{Gal}(F(x)/F).$$

Define, for the CM point x and any non-negative integer r, the integer $u(r, x)$ by

$$u(r, x) = \begin{cases} [F^* \cap \widehat{k}^*Z(x) : k^*] & \text{if } r = 0 \\ 1 & \text{if } r \geq 1. \end{cases}$$

Here $u(0, x)$ equals either $[O_F^* \cap Z(x) : O_k^*]$ or equals $2[O_F^* \cap Z(x) : O_k^*]$ by Propositions 3.3.8 and 3.3.9.

Proposition 3.4.22 (Norm Relations) *Let $\mathfrak{np} \in \mathcal{S}_{r+1}$ where $\mathfrak{n} \in \mathcal{S}_r, \mathfrak{p} \in \mathcal{S}_1$. Let $T(\mathfrak{p})$ be the Hecke correspondence at $\mathfrak{p}$ (see (2.3)):*

(i) *The extension $F(x(\mathfrak{n}\mathfrak{p}))/F(x(\mathfrak{n}))$ is cyclic of degree $(N\mathfrak{p}+1)/u(r,x)$;*
(ii) *There is an equality of divisors on $N_H^* \times_k \overline{k}$*

$$T(\mathfrak{p})x(\mathfrak{n}) = u(r,x) \sum_\sigma \sigma x(\mathfrak{n}\mathfrak{p})$$

where σ runs over the elements of $\mathrm{Gal}(F(x(\mathfrak{n}\mathfrak{p}))/F(x(\mathfrak{n})))$;
(iii) *There is the equality in the elliptic curve E*

$$T(\mathfrak{p})\pi(x(\mathfrak{n})) = u(r,x)\mathrm{Tr}_{F(x(\mathfrak{n}\mathfrak{p}))/F(x(\mathfrak{n}))}\pi(x(\mathfrak{n}\mathfrak{p}))$$

where $\pi : N_H^* \longrightarrow E$ is the parametrization of (3.4.11).

Proof

(i) Let $Z(x(\mathfrak{n}))$, $Z(x(\mathfrak{n}\mathfrak{p}))$ be the open compact subgroups of $\widehat{O}_F^*$ which contain $\widehat{O}_k^*$ associated to the CM points $x(\mathfrak{n})$, $x(\mathfrak{n}\mathfrak{p})$ by Definition 3.3.4.

Proposition 3.4.19 shows that $Z(x(\mathfrak{n}))/Z(x(\mathfrak{n}\mathfrak{p})) \cong (O_F/\mathfrak{p}O_F)^*/(O_k/\mathfrak{p}O_k)^*$, and this with Proposition 3.3.7 provides an exact sequence, where this sequence defines the map f,

$$0 \longrightarrow \frac{F^* \cap \widehat{k}^* Z(x(\mathfrak{n}))}{F^* \cap \widehat{k}^* Z(x(\mathfrak{n}\mathfrak{p}))} \longrightarrow \frac{(O_F/\mathfrak{p}O_F)^*}{(O_k/\mathfrak{p}O_k)^*} \xrightarrow{f} \mathrm{Gal}(F(x(\mathfrak{n}\mathfrak{p}))/F(x(\mathfrak{n}))) \longrightarrow 0.$$

As the ideal $\mathfrak{n}\mathfrak{p}O_F$ does not contain I_0, Proposition 3.3.9 implies that $F^* \cap \widehat{k}^* Z(x(\mathfrak{n}\mathfrak{p})) = k^*$.

If $r > 0$ then again $F^* \cap \widehat{k}^* Z(x(\mathfrak{n})) = k^*$ and this above exact sequence shows that the map f is an isomorphism.

If $r = 0$, which occurs if and only if $\mathfrak{n} = 1$, then this exact sequence shows that the kernel of f has order equal to $u(0,x) = [F^* \cap \widehat{k}^* Z(x) : k^*]$. Part (i) now follows.

(ii) Let $T(\mathfrak{p})$ be the Hecke operator at $\mathfrak{p}$ ((2.3) and (2.3)). By definition the divisor $T(\mathfrak{p})x(\mathfrak{n})$ contains $x(\mathfrak{n}\mathfrak{p})$. By Lemma 3.4.18, and with the notation and hypotheses of that lemma, we have that $|R^*/(R^* \cap gR^*g^{-1})| = |\kappa_E| + 1$. We apply this to the operator $T(\mathfrak{p})$ where g is taken to be the element $b_\mathfrak{p} \in \widehat{B}^*$ defining $T(\mathfrak{p})$ and $E = k_\mathfrak{p}$, $E' = F \otimes_k k_\mathfrak{p}$ and we obtain that the number of points, counted with multiplicity, in the divisor $T(\mathfrak{p})x(\mathfrak{n})$ on N_H is equal to $|\kappa_E| + 1$.

Part of the exact sequence of part (i) can be written (by Proposition 3.4.19)

$$\frac{Z(x(\mathfrak{n}))}{Z(x(\mathfrak{n}\mathfrak{p}))} \cong \frac{(O_F/\mathfrak{p}O_F)^*}{(O_k/\mathfrak{p}O_k)^*} \xrightarrow{f} \mathrm{Gal}(F(x(\mathfrak{n}\mathfrak{p}))/F(x(\mathfrak{n}))) \longrightarrow 0$$

3.4 Heegner Points and CM Points on Elliptic Curves

where the kernel of f has order $u(r, x)$, as shown in the proof of part (i) and where r is the the number of prime components in $\mathfrak{n}$.

That the number of points in the divisor $T(\mathfrak{p})x(\mathfrak{n})$ is equal to $|\kappa_E| + 1$, together with that the kernel of the homomorphism f has order equal to $u(r, x)$ and the Galois action on CM points given by class field theory (see (3.3.2)), implies that $T(\mathfrak{p})x(\mathfrak{n})$ coincides with the orbit of $x(\mathfrak{np})$ under the action of $\mathrm{Gal}(F(x(\mathfrak{np}))/F(x(\mathfrak{n})))$, with each point counted with multiplicity $u(r, x)$.

(iii) The rigidification of the parametrization π is made via a divisor $m\Xi$, in (3.4.4) for a non-compact Shimura curve N_H (i.e. the case of classical modular curves where $B \neq M_2(\mathbb{Q})$), and via a class $m\xi$ given in (3.4.11), for the case of a compact Shimura curve N_H (i.e. $B \neq M_2(\mathbb{Q})$). This norm relation of part (iii) is a consequence of part (ii) and that $T(\mathfrak{p})$ acts on the class of $m\Xi$, resp. on $m\xi$, by multiplication by $\deg T(\mathfrak{p}) = N\mathfrak{p} + 1$. □

□

Proposition 3.4.23 *Let* $\mathfrak{np} \in \mathcal{S}_{r+1}$ *where* $\mathfrak{n} \in \mathcal{S}_r, \mathfrak{p} \in \mathcal{S}_1$. *Let* $\pi : N_H^* \longrightarrow E$ *be the parametrization of (3.4.11). Then we have:*

(i) *(Norm relation.)* $u(r, x)\mathrm{Tr}_{F(x(\mathfrak{np}))/F(x(\mathfrak{n}))} \pi x(\mathfrak{np}) = a_\mathfrak{p} \pi x(\mathfrak{n})$

   where $a_\mathfrak{p} \in \mathbb{Z}$ *is the trace of the Frobenius* $\mathrm{Fr}_\mathfrak{p}$ *on the Tate module of E.*

(ii) *(Congruence relation.) For each prime* $\mathfrak{q}|\mathfrak{p}$ *above* $\mathfrak{p}$ *in* $F(x(\mathfrak{np}))$ *we have the congruences, considered as an equality in* $\mathbf{N}_H^*(\kappa(\mathfrak{q}))$ *where* $\mathbf{N}_H^*/O_{k,\mathfrak{p}}$ *denotes a proper smooth model over* $O_{k,\mathfrak{p}}$ *of* N_H^*/k,

$$\pi x(\mathfrak{np}) \equiv \mathrm{Fr}_{\mathfrak{p},\mathrm{arith}} \, \pi x(\mathfrak{n}) \equiv \mathrm{Fr}_{\mathfrak{p},\mathrm{geom}} \, \pi x(\mathfrak{n}) \quad \mathrm{mod}\ \mathfrak{q}.$$

Proof

(i) This follows from Proposition 3.4.22(iii) and that $T(\mathfrak{p})$ acts on E by multiplication by $a_\mathfrak{p}$ by the Eichler-Shimura congruence [56, (1.14.3)].
(ii) The field extension $\kappa(\mathfrak{q})/\kappa(\mathfrak{p})$ is quadratic, by Proposition 3.4.20(ii) and (iii). Hence the actions of the Frobenius elements $\mathrm{Fr}_{\mathfrak{p},\mathrm{arith}}$ and $\mathrm{Fr}_{\mathfrak{p},\mathrm{geom}}$ on $\kappa(\mathfrak{q})$ coincide. Combining this remark with the Eichler-Shimura congruence relation [56, (1.14.3)], we conclude that the reduction mod $\mathfrak{q}$ of each point of the support of the divisor $T(\mathfrak{p})x(\mathfrak{n})$, and in particular the point $x(\mathfrak{np})$, is equal to the reduction mod $\mathfrak{q}$ of $\mathrm{Fr}_{\mathfrak{p},\mathrm{arith}}x(\mathfrak{n})$. □

Exercise 3.4.24 With the notation and hypotheses of Lemma 3.4.18, show the final part of the lemma that $|\frac{R^*}{R^* \cap gR^*g^{-1}}| = |\kappa_E| + 1$.

Local Fields and Local Conditions

Contents

4.1	R-Representations, l-adic Representations	69
4.2	Local Duality	72
4.3	Local Conditions	79
4.4	A Normal Form for Endomorphisms	83
4.5	The Finite-Singular Comparison Map	97

4.1 R-Representations, l-adic Representations

(4.1.1) In this section, let k be a local field or a global field of any characteristic.

(4.1.2) A *coefficient ring* is a complete noetherian local ring R with finite residue field of characteristic l and maximal ideal $\mathfrak{m}$.

In this section, let R be a coefficient ring. We are mainly interested in the cases when $R = \mathbb{F}_l$ or $\mathbb{Z}/l^n\mathbb{Z}$ or $\mathbb{Z}_l$, the l-adic completion of $\mathbb{Z}$, or an Iwasawa algebra that is to say a formal power series ring in a finite number of variables such as $\mathbb{Z}_l[[X_1, \ldots, X_n]]$.

(4.1.3) Let k be a global field or a local field of any characteristic. An *R-module scheme over k* is a commutative group scheme T of finite type over k where T is equipped with an action by the ring R compatible with the group scheme structure i.e. for any k-scheme X the abelian group $T(X)$ is an R-module functorial in X.

R-representations

(4.1.4) We define an *R-representation of G_k* in the following way, where there are two separate cases depending on whether l is equal or not equal to the characteristic of k.

- Suppose first that the residue characteristic l of R is different from the characteristic of k. An *R-representation* of G_k is a finitely generated R-module T which is a continuous G_k-module and is unramified outside a finite set of primes of k.

  A finitely generated R-module T inherits a topology from the $\mathfrak{m}$-adic topology on R, and for T to be a G_k-representation, the G_k action is required to be continuous with respect to this topology and the profinite topology on G_k.

  A morphism of R-representations $M \to N$, when l is different from the characteristic of k, is then a continuous homomorphism of $R[[G_k]]$-modules.

- Suppose now that l is equal to the characteristic of k, where this is necessarily non-zero. An *R-representation* of G_k is a finite R-module scheme T/k.

  A morphism of R-representations $M \to N$, when l is equal to the characteristic of k, is a homomorphism of R-module schemes over k.

(4.1.5) We write Mod_{R,G_k} for the category of R-representations of G_k whatever the characteristic of k.

The category Mod_{R,G_k} is abelian, where in the case that l is equal to the characteristic of k this holds because the category of finite commutative group schemes over a field is abelian.

If R is the ring of integers of a finite extension field of $\mathbb{Q}_l$, where l is different from the characteristic of k, then an R-representation of G_k which is a free R-module is also called an *l-adic representation* of G_k with coefficients in R.

(4.1.6) Suppose that ρ is the cyclotomic character, where l is different from the characteristic of k, so that

$$\rho : G_k \to \mathrm{Aut}(\mu_{l^\infty})$$

is given by G_k acting on the l^nth roots of unity in k^{sep} for all n. Then

$$\varprojlim \mu_{l^n}$$

is an l-adic representation of G_k written $\mathbb{Z}_l(1)$ which is a free $\mathbb{Z}_l$-module of rank 1.

Suppose that R is the ring of integers of a finite extension field of $\mathbb{Q}_l$, where l is different from the characteristic of k. Then for a R-representation T of G_k, we write

$$T(1) = T \otimes_{\mathbb{Z}_l} \mathbb{Z}_l(1)$$

for the Tate twist of T.

4.1 R-Representations, l-adic Representations

(4.1.7) Let T be an R-representation of G_k, where R is a coefficient ring. The *quotient category of* T denoted Quot(T) is the category whose objects are the G_k-representations T/IT where I runs over the ideals of R and the set of morphisms Mor($T/IT, T/JT$), for ideals I, J of R, is the ideal $(J : I)$ of R where this is the ideal of scalar multiplications $r : T/IT \to T/JT, r \in (J : I) \subseteq R$.

Dual R-representations: k is a Local Field

(4.1.8) Let k be a local field. Suppose that R is a coefficient ring where the residue field characteristic l of R may or may not be different from the characteristic of k. Let T be an R-representation of G_k.

The *dual representation* T^* is defined in Sect. 4.2 via local duality, and it is defined separately for the two cases where l is equal or not equal to the characteristic of k. In all cases, the dual can be defined by the formula (see (4.2.11), when k has characteristic different from l, and (4.2.18), when k has positive characteristic which is equal to l)

$$T^* = \text{Hom}(T, \mu_{l^\infty}).$$

When k has characteristic different from l, Hom here denotes the group of homomorphisms of $\mathbb{Z}[[G_k]]$-modules. This G_k-module T^* is equipped with an $R[[G_k]]$-module structure (see (4.2.11) *et seq.* for more details).

When k has characteristic equal to l then T and T^* are both finite commutative group schemes which have compatible structures of R-module schemes; T^* is then isomorphic to the Cartier dual group scheme of T (see (4.2.18) for more details).

Dual R-representations: k is a Global Field

(4.1.9) Suppose now that k is a global field. Let R be a coefficient ring where the residue field characteristic l of R is different from the characteristic of k. Let T be an R-representation of G_k. We again define the dual of T to be the R-representation

$$T^* = \text{Hom}(T, \mu_{l^\infty}).$$

Note that in all cases for local or global fields, T^* is not the dual of T in the sense of group representations but it arises naturally via Poitou-Tate local and global dualities or Shatz local duality (see, e.g. (4.3.5)).

Remark 4.1.10 For the case of an R-representation where R is an Iwasawa algebra, then the Galois group G_k acts non-trivially on R. But in all other cases that we consider in this text, G_k acts trivially on R. We shall not consider further in this text the Iwasawa theory of Selmer groups and Shafarevich-Tate groups.

4.2 Local Duality

(4.2.1) Let

k be a local field, archimedean or non-archimedean;
R be a coefficient ring where R has residue characteristic l;
T be an R-representation of G_k.

The local duality of the cohomology of G_k-modules with coefficients in R takes different forms depending on whether the field k has characteristic different from or equal to l.

In the case when k has characteristic equal to l, the cohomology groups may be infinite and the local duality is stated for cohomology groups equipped with a topology for which they are topological groups. The duality theorem is then a statement that one topologized cohomology group is the Pontryagin dual of another. Furthermore, the duality is stated only for G_k representations provided by finite commutative group schemes over k.

In the case when k has characteristic different from l, the higher-dimensional cohomology groups are the usual continuous cochain cohomology and are all equipped with the discrete topology.

We state the local duality theorems for all characteristics in this section.

The Standard Topology on Cohomology Groups

(4.2.2) Let k be a local field of any characteristic. When the characteristic of k coincides with l, the duality theorem is stated for cohomology groups of the form $H^i_{\text{fppf}}(k, G)$ for the fppf topology, where G/k is a commutative k-group scheme and fppf stands for *fidèlement plat et de présentation finie*; the coverings of a given scheme for this Grothendieck topology are surjective families of morphisms which are faithfully flat and locally of finite presentation. It is beyond the scope of this book to describe further the fppf topology and the reader is referred to [79, §34.7 The fppf topology, https://stacks.math.columbia.edu/tag/021L].

For the case where k has characteristic zero or if G is a finite étale k-group scheme, then the groups $H^i(k, G)$ are the usual continuous cochain cohomology groups and for $i \geq 1$ they are equipped with the discrete topology.

We shall usually omit the label fppf from the symbol $H^i_{\text{fppf}}(k, G)$ where it is understood that if the ground field has characteristic p and G is a commutative group scheme, then fppf cohomology is meant.

(4.2.3) Let L/k be a finite extension field of k. Then L is equipped with a unique valuation extending the valuation on k. Hence L is equipped with its valuation topology and k has its own valuation topology which makes k a subspace of L for their respective topologies. The L-vector space $L^{\otimes n}$, where $L^{\otimes n}$ denotes $L \otimes_k L \otimes_k \ldots \otimes_k L$ with n factors, is then equipped

4.2 Local Duality

with its natural topology extending that on L and this topology is uniquely determined by that on k.

Let X/k be an affine scheme of finite type over k. Then a closed immersion of X in an affine space $\mathbb{A}^m/k$ provides a topology on $X(L^{\otimes n}) \subseteq \mathbb{A}^m(L^{\otimes n})$ by restriction of the topology on $\mathbb{A}^m(L^{\otimes n})$ to a subspace. This topology on $X(L^{\otimes n})$ is independent of the immersion chosen and so this topology is uniquely determined by that on k.

(4.2.4) Let G/k be a commutative group scheme of finite type over k. Then G/k is a quasi-projective k-scheme [79, Lemma 39.8.7]. Therefore G/k is a finite union of open affine k-schemes of finite type, and hence by the previous paragraph $G(L^{\otimes n})$ is also equipped with a natural topology whose restrictions to the given open affine subschemes agree with those of the previous paragraph. Again this topology on $G(L^{\otimes n})$ is uniquely determined by that on k and makes $G(L^{\otimes n})$ a topological group.

For $r > 0$, there are $r+1$ homomorphisms of k-algebras

$$L^{\otimes r} \to L^{\otimes r+1}$$

given by

$$\epsilon_s : b_0 \otimes \ldots \otimes b_{r-1} \mapsto b_0 \otimes \ldots \otimes b_{s-1} \otimes 1 \otimes b_s \otimes \ldots b_{r-1} \qquad (s = 0, \ldots, r).$$

These provide continuous homomorphisms of topological groups

$$G(\epsilon_s) : G(L^{\otimes r}) \to G(L^{\otimes r+1}) \qquad (s = 0, \ldots, r).$$

We put

$$\delta^r = \sum_{i=0}^{r} (-1)^i G(\epsilon_i).$$

We may then form the sequence of topological abelian groups that can be checked to be a complex, where the coboundary homomorphisms δ^i are continuous because they are obtained from polynomials,

$$0 \xrightarrow{\delta^0} G(L) \xrightarrow{\delta^1} G(L^{\otimes 2}) \xrightarrow{\delta^2} G(L^{\otimes 3}) \xrightarrow{\delta^3} \cdots$$

Therefore $\mathrm{Ker}(\delta^{r+1}) \subseteq G(L^{\otimes r+1})$ is equipped with its subspace topology. Write

$$\check{H}^r(L/k, G) = \mathrm{Ker}(\delta^{r+1})/\mathrm{Im}(\delta^r)$$

for the rth cohomology group of this complex, where this quotient group is equipped with its quotient topology. Note that $H^0(L/k, G) \cong G(k)$ is a topological isomorphism.

(4.2.5) Define the Čech cohomology groups

$$\check{H}^r(k, G) = \varinjlim \check{H}^r(L/k, G)$$

where L runs over all finite field extensions of k inside an algebraic closure of k. Then $\check{H}^r(k, G) = \varinjlim \check{H}^r(L/k, G)$ may be equipped with its direct limit topology, that is to say a map $f : \check{H}^r(k, G) \to E$, for any topological space E, is continuous if and only if its induced maps $\check{H}^r(L/k, G) \to E$ are continuous for all finite field extensions L/k.

By construction of the groups $H^r_{\mathrm{fppf}}(k, G)$, there are canonical group homomorphisms $\check{H}^r(k, G) \to H^r_{\mathrm{fppf}}(k, G)$.

Proposition 4.2.6 *Let G/k be a commutative group scheme of finite type. Then the canonical group homomorphism $\check{H}^r(k, G) \to H^r_{\mathrm{fppf}}(k, G)$ is an isomorphism.*

Proof For the proof see [Mi, Chap. III, Propn. 6.1, p.338]. □

(4.2.7) Via this isomorphism with the Čech topological cohomology groups $\check{H}^r(k, G)$, the groups $H^i_{\mathrm{fppf}}(k, G)$ are equipped with a standard topology defined above.

These groups $H^i_{\mathrm{fppf}}(k, G)$ equipped with this topology have the following properties (for proofs see [48, Chap. III, lemma 6.5] and [9]):

- The groups $H^i_{\mathrm{fppf}}(k, G)$ are topological groups which are Hausdorff, locally compact, and σ-compact (i.e. a countable union of compact subsets);
- Morphisms induced between cohomology groups by a short exact sequence of commutative group schemes of finite type over k are continuous;
- Restriction homomorphisms and cup products are continuous maps between the groups $H^i_{\mathrm{fppf}}(k, G)$.

Cartier Duality

(4.2.8) Recall that an R-module scheme G is a commutative group scheme where G is equipped with an action by the ring R compatible with the group scheme structure.

Definition 4.2.9 (Cartier Dual) Suppose that N/F is a finite commutative group scheme over a field F of any characteristic. Then $N = \mathrm{Spec}\, A$ for some finite F-algebra A which is also a Hopf algebra. Let $A^* = \mathrm{Hom}_F(A, F)$ be the dual vector space to A. Then A^* is also a Hopf algebra over F and hence $N^* = \mathrm{Spec}\, A^*$ is a finite commutative F-group scheme. The group scheme N^* is the *Cartier dual* group scheme of N/F.

4.2 Local Duality

If N/F is also an R-module scheme then N^* is also an R-module scheme.

(4.2.10) The Cartier dual N^*/F has the following properties:

(a) The Cartier dual N^*/F represents the functor

$$Y \mapsto \mathrm{Hom}(N \times_F Y, \mathbb{G}_m \times_F Y)$$

$$F-\mathit{Schemes} \longrightarrow \mathit{Abelian\ Groups}$$

where $\mathrm{Hom}(N \times_F Y, \mathbb{G}_m \times_F Y)$ is the abelian group of Y-group scheme homomorphisms from $N \times_F Y$ to $\mathbb{G}_m \times_F Y$ [53, Theorem 11.30];

(b) The group scheme $(N^*)^*/F$ is canonically F-isomorphic to N/F;

(c) The functor $N \mapsto N^*$ is a contravariant equivalence from the category of finite commutative F-group schemes to itself.

Local Duality When k has Characteristic Different from l

(4.2.11) If k has characteristic different from l then all cohomology groups $H^i(k, -)$ are equipped with the discrete topology.

Define *the dual* of T to be

$$T^* = \mathrm{Hom}(T, \mu_{l^\infty}).$$

Here Hom denotes the group of homomorphisms of $\mathbb{Z}[[G_k]]$-modules. This G_k-module T^* is equipped with an R-module structure by defining $rf \in T^*$ for $f \in T^*, r \in R$, to be $(rf)(t) = f(rt)$ for all $t \in T$. This gives T^* an $R[[G_k]]$-module structure.

Theorem 4.2.12 (Poitou-Tate Local Duality). *Suppose that k is a local field of characteristic different from l, archimedean or non-archimedean. Let T be an R-representation of G_k. Then the cup product provides a perfect pairing of abelian groups, where inv is the invariant map of class field theory,*

$$H^r(k, T) \times H^{2-r}(k, T^*) \longrightarrow H^2(k, \mu_{l^\infty}) \stackrel{\mathrm{inv}}{\cong} \mathbb{Q}_l/\mathbb{Z}_l. \qquad (4.2.13)$$

Proof The R-representation T can be written as $\varprojlim T_n$ where the T_n are finite R-representations of G_k. That the theorem holds for T_n, for all n, follows from [48, Corollary I.2.3] or [68, §II.5.2] or [58, Theorem 7.2.6], for non-archimedean k, and [58, Theorem 7.2.17] when k is archimedean. The Proposition 1.3.10 then shows that the theorem holds for T on passing to the inverse limit. □

Local Duality When k has Characteristic Equal to l

(4.2.14) Let k be a non-archimedean local field and N/k be a finite commutative group scheme and let N^*/k be its Cartier dual. We emphasize that in the following subsection the residue characteristic l of R may well be equal to the characteristic of k.

Theorem 4.2.15 (Shatz) *Let k be a non-archimedean local field. The cup product pairing*

$$H^r_{\text{fppf}}(k, N) \times H^{2-r}_{\text{fppf}}(k, N^*) \longrightarrow H^2(k, \mathbb{G}_m) \cong \mathbb{Q}/\mathbb{Z}$$

identifies each topological group with the Pontryagin dual of the other.

[For the proof, see [48, Chap. III, Theorem 6.10].□]

(4.2.16) Suppose for this paragraph that N/k is a finite R-module scheme. Then N is a $\mathbb{Z}/l^m\mathbb{Z}$-module scheme for some positive integer m and N has order a power of l.

The finite commutative k-group schemes $\mu_{l^n} = \text{Spec } k[x]/(x^{l^n} - 1)$ form a direct system

$$\mu_1 \to \ldots \to \mu_{l^n} \to \mu_{l^{n+1}} \to \ldots$$

For any $n \in \mathbb{N}$, we then have $\text{Hom}(N, \mu_{l^n})$ which denotes the finite commutative k-group scheme that represents the functor

$$Y \mapsto \text{Hom}(N \times_k Y, \mu_{l^n} \times_k Y)$$

$$k\text{-}\mathscr{S}\!\mathit{chemes} \longrightarrow \mathscr{A}\!\mathit{belian}\ \mathscr{G}\!\mathit{roups}$$

where $\text{Hom}(N \times_k Y, \mu_{l^n} \times_k Y)$ is the abelian group of Y-group scheme homomorphisms from $N \times_k Y$ to $\mu_{l^n} \times_k Y$.

Lemma 4.2.17 *Suppose that N/k is a finite R-module scheme. Then the Cartier dual of N/k is naturally isomorphic to the group scheme $\text{Hom}(N, \mu_{l^n})/k$ for sufficiently large $n \in \mathbb{N}$.*

Proof The group scheme N/k is a finite $\mathbb{Z}/l^n\mathbb{Z}$-module scheme for some n. Then the l^nth power map on N, that is to say $[l^n] : N \to N$, is the group scheme morphism mapping N to the closed point (more precisely, the closed subgroup scheme) 0 of N/k. Let Y be a k-scheme. Then we have a homomorphism of abelian groups obtained from the closed immersion $\mu_{l^n} \to \mathbb{G}_m$

$$\Phi : \text{Hom}(N \times_k Y, \mu_{l^n} \times_k Y) \longrightarrow \text{Hom}(N \times_k Y, \mathbb{G}_m \times_k Y).$$

4.2 Local Duality

Let $\phi : N \times_k Y \to \mathbb{G}_m \times_k Y$ be a homomorphism of Y-group schemes. Then we have the commutative diagram of Y-group schemes

$$\begin{array}{ccc} N \times_k Y & \xrightarrow{\phi} & \mathbb{G}_m \times_k Y \\ {\scriptstyle [l^n]}\downarrow & & \downarrow{\scriptstyle [l^n]} \\ 0 \times_k Y & \longrightarrow & \mathbb{G}_m \times_k Y \end{array}.$$

It follows that the image of ϕ lies in the subgroup scheme $\ker([l^n] : \mathbb{G}_m \times_k Y \to \mathbb{G}_m \times_k Y$, that is to say ϕ factors as $N \times_k Y \to \mu_{l^n} \times_k Y \to \mathbb{G}_m \times_k Y$. Hence Φ is surjective and it is obviously injective; hence it is an isomorphism of abelian groups which is functorial in Y.

As the functors $Y \mapsto \operatorname{Hom}(N \times_k Y, \mu_{l^n} \times_k Y)$, $Y \mapsto \operatorname{Hom}(N \times_k Y, \mathbb{G}_m \times_k Y)$ are represented by the k-group schemes $\operatorname{Hom}(N, \mu_{l^n})$, $\operatorname{Hom}(N, \mathbb{G}_m)$, respectively, we conclude that the natural morphism $\operatorname{Hom}(N, \mu_{l^n}) \to \operatorname{Hom}(N, \mathbb{G}_m)$ is an isomorphism of k-schemes and that the Cartier dual of N/k is naturally isomorphic to $\operatorname{Hom}(N, \mu_{l^n})$. $\square$

(4.2.18) From the previous Lemma 4.2.17, the group schemes $\operatorname{Hom}(N, \mu_{l^n})$ stabilize for sufficiently large n and are isomorphic to the Cartier dual N^*/k for large n. We then define the group scheme $\operatorname{Hom}(N, \mu_{l^\infty})$ to be

$$\operatorname{Hom}(N, \mu_{l^\infty}) = \varinjlim_n \operatorname{Hom}(N, \mu_{l^n}),$$

where this is k-isomorphic to N^*/k and is the *dual R-representation* of N/k.

Theorem 4.2.19 *Suppose that N/k is a finite R-module scheme and $N^* = \operatorname{Hom}(N, \mu_{l^\infty})$ is its Cartier dual. The cup product pairing*

$$H^r_{\mathrm{fppf}}(k, N) \times H^{2-r}_{\mathrm{fppf}}(k, N^*) \longrightarrow H^2(k, \mu_{l^\infty}) \cong \mathbb{Q}_l/\mathbb{Z}_l$$

identifies each topological group with the Pontryagin dual of the other.

Proof The group $H^2(k, \mu_{l^\infty})$ is here defined to be $\varinjlim H^2_{\mathrm{fppf}}(k, \mu_{l^n}) \cong \mathbb{Q}_l/\mathbb{Z}_l$ where $H^2_{\mathrm{fppf}}(k, \mu_{l^n}) \cong \mathbb{Z}/l^n\mathbb{Z}$ (by, e.g. [Sh, Corollary p.421] or [58, Corollary 7.1.6]). The theorem to be demonstrated is then a corollary of Theorem 4.2.15. $\square$

Combined Statement of Local Duality

Theorem 4.2.20 *Let k be a local field, archimedean or non-archimedean, of any characteristic and let T be an R-representation of G_k (4.1.4). Let $T^* = \operatorname{Hom}(T, \mu_{l^\infty})$ be the* *dual of T ((4.2.11) and (4.2.18)). Then the cup product pairing*

$$H^r(k,T) \times H^{2-r}(k,T^*) \longrightarrow H^2(k, \mu_{l^\infty}) \cong \mathbb{Q}_l/\mathbb{Z}_l$$

identifies each topological group with the Pontryagin dual of the other. Here the cohomology groups are equipped with their standard topologies.

Proof This is a restatement of Theorems 4.2.12 and 4.2.19. □

Remarks 4.2.21

(i) If k is a local field of any characteristic and N/k is a finite étale group scheme then $H^1(k, N)$ is finite and is equipped with the discrete topology.
[Here $H^1(k, N)$ denotes Galois cohomology or fppf cohomology where both here coincide for a finite étale group scheme over a field. There is a finite separable extension field k'/k over which N becomes a constant group scheme isomorphic to A'/k' where A' is the sheaf on Spec k' associated to a finite abelian group A. Then we have

$$H^1(k', A') = \text{Hom}(G_{k'}, A) \cong \text{Hom}(G_{k'}^{ab}, A)$$

where $G_{k'}^{ab}$ is the Galois group over k' of the maximal separable abelian extension of k'. It follows from class field theory that $\text{Hom}(G_{k'}^{ab}, A)$ is finite. The Hochschild-Serre spectral sequence

$$H^i(k'/k, H^j(k', N)) \Longrightarrow H^{i+j}(k, N)$$

now shows that $H^1(k, N)$ is finite.]

(ii) If A/k is an abelian variety and n is a positive integer coprime to the characteristic of k then the n-torsion subgroup scheme $A[n]/k$ is étale and hence by part (i) $H^1_{\text{fppf}}(k, A[n])$ is finite and discrete and is even isomorphic to $H^1(k, A[n])$, where the latter group is Galois cohomology.

(iii) If k has characteristic > 0 then for a general finite commutative group scheme N/k the group $H^1(k, N)$ may be infinite, in contrast to the case of étale group schemes considered in (i).

In the following examples k is a local field of characteristic $p > 0$ and with ring of integers A.

- The group $H^1(k, \alpha_p) = k/k^p$ is locally compact, the subgroup A/A^p of k/k^p is compact and open, and the quotient k/Ak^p is an infinite discrete group.

4.3 Local Conditions

- The group $H^1(k, \mathbb{Z}/p\mathbb{Z}) \cong k/L_p(k)$ where $L_p : k \to K$ is the Lang isogeny $x \mapsto x^p - x$ and $H^1(k, \mathbb{Z}/p\mathbb{Z})$ is infinite and discrete;
- The group

$$H^1(k, \mu_p) \cong k^*/k^{*p}$$

is locally compact, the subgroup A^*/A^{*p} of k^*/k^{*p} is compact and open, and the quotient k^*/A^*k^{*p} is an infinite discrete group.

These topological cohomology groups are in particular required in the proof below of Proposition 5.6.3, Steps 3 to 7.

4.2.22. Bibliographical Remarks. The topology on cohomology groups given in paragraphs (4.2.2)–(4.2.7) is that defined in Milne [48, Chap. III] and originally due to Shatz [71]. The topological properties of these cohomology groups stated in paragraph (4.2.7) may possibly not be proved in all circumstances in [48, Chap. III]. But these properties stated in (4.2.7) have been completely established in [9].

The local duality Theorem 4.2.15 in positive characteristic is due to Shatz [71].

The local duality Theorem 4.2.12 for a local field of characteristic zero can be stated in slightly different ways depending on which category $\mathscr{C}$ of Galois representations is considered and where the dualizing module depends on $\mathscr{C}$. See, for example, the local duality theorems for different $\mathscr{C}$ given in [58, Theorem 7.2.6], [58, Theorem 7.2.9], [62, Chapter 1, theorem 4.1]. There is a discussion of dualizing modules for profinite groups in [68, Chap. I, §3.5].

Further results on Cartier duality are given in [53, Chapter 11]; for general results on arithmetic duality, see the monograph [48].

4.3 Local Conditions

(4.3.1) In this section let k be a local field and let T be an R-representation of G_k where R is a coefficient ring with residue field of characteristic l.

By a *local condition* on T is meant a choice of R-submodule $H^1_{\mathcal{F}}(k, T)$ of the cohomology group $H^1(k, T)$. We denote a local condition by $\mathcal{F}$ and its corresponding submodule by $H^1_{\mathcal{F}}(k, T)$.

In the case where the residue characteristic l of R is equal to the characteristic of k, then T/k is a finite R-module scheme and the previous paragraph simply means that $H^1_{\mathcal{F}}(k, T)$ is choice of submodule of the fppf-cohomology group $H^1_{\text{fppf}}(k, T)$.

(4.3.2) If S is an R-representation of G_k which is a sub-representation of T then the *propagated condition* is defined to be the pre-image of $H^1_{\mathcal{F}}(k, T)$ under the induced map

$$H^1(k, S) \to H^1(k, T).$$

Similarly if S is an R-representation of G_k which is a quotient representation of T then the *propagated condition* is defined to be the image of $H^1_{\mathcal{F}}(k, T)$ under the induced map

$$H^1(k, T) \to H^1(k, S).$$

If $\mathcal{C}$ is a category of R-representations of G_k then a *local condition $\mathcal{F}$ functorial over $\mathcal{C}$* means a subfunctor

$$M \mapsto H^1_{\mathcal{F}}(k, M)$$

of $M \mapsto H^1(k, M)$ on $\mathcal{C}$.

(4.3.3) A local condition functorial over a category $\mathcal{C}$ of R-representations of G_k is *cartesian* if for any injective morphism $f : S \to T$ (see (4.1.5)) in the category $\mathcal{C}$ of R-representations then the local condition $\mathcal{F}$ on S is equal to the local condition obtained by propagating $\mathcal{F}$ from T to S.

Examples of Local Conditions

(4.3.4) These are some local conditions that we shall use for k and T.

(a) The *relaxed* condition

$$H^1_{\text{rel}}(k, T) = H^1(k, T).$$

(b) The *strict* condition

$$H^1_{\text{str}}(k, T) = 0.$$

(c) The *unramified* condition when k is non-archimedean and has characteristic different from l

$$H^1_{\text{unr}}(k, T) = \ker(H^1(k, T) \to H^1(k^{\text{unr}}, T))$$

where k^{unr} is the maximal unramified extension of k in k^{sep}. This subgroup $H^1_{\text{unr}}(k, T)$, when $l \neq \text{char. } k$, is isomorphic to $H^1(k^{\text{unr}}/k, T^{G_{k^{\text{unr}}}})$ by the inflation restriction sequence (Proposition 1.3.13(i)).

(d) The *L-transverse* condition when k is non-archimedean and has characteristic different from l

$$H^1_{L-\text{tr}}(k, T) = \ker(H^1(k, T) \to H^1(L, T))$$

4.3 Local Conditions

where L is a totally ramified abelian extension of k of degree $|\kappa^*|$ where κ is the residue field of k.

Dual Local Condition

(4.3.5) The dual of T is $T^* = \text{Hom}(T, \mu_{l^\infty})$ ((4.2.11) and (4.2.18)). The module T^* is then an R-representation of G_k and in particular, when the characteristic of k is equal to l, T^* is a finite R-module scheme.

Then by Poitou-Tate local duality (Theorem 4.2.12) when k has characteristic different from l, and Shatz duality (Theorem 4.2.19) when k has characteristic equal to l, there is a perfect bilinear pairing for the cohomology for the local field k

$$<,>_v: H^r(k, T) \times H^{2-r}(k, T^*) \longrightarrow H^2(k, \mu_{l^\infty}) \overset{\text{inv}}{\cong} \mathbb{Q}_l/\mathbb{Z}_l$$

where the last isomorphism is the invariant map of local class field theory and in positive characteristic $H^i(k, -)$ denotes fppf cohomology. The pairing identifies each component with the Pontryagin dual of the other, where these cohomology groups are equipped with their standard topologies (see Sect. 4.2).

Definition 4.3.6 The *dual local condition* $\mathcal{F}^*$ on T^* is the orthogonal complement of $\mathcal{F}$ under the local pairing.

Finite Condition

(4.3.7) If k is non-archimedean and has characteristic different from l and T is an unramified R-representation of G_k then the unramified local condition on T is also called the *finite condition* and is written $H^1_f(k, T)$ (see also Remark 4.3.11(i)).

Proposition 4.3.8 *Let k be a non-archimedean local field where the residue field l of R is different from the characteristic of k. Then the finite condition is cartesian on any category of unramified G_k-representations with coefficients in R.*

Proof If k has positive characteristic then an R-representation of G_k which is a finite R-module scheme M/k is then a finite étale k-group scheme because $l \neq \text{char. } k$. Therefore in that case the fppf cohomology group $H^1(k, M)$ is isomorphic to the Galois cohomology group $H^1(k^{\text{sep}}/k, M(k^{\text{sep}}))$ so that for this proposition all cohomology groups can be taken to be Galois cohomology.

Suppose now that k is any non-archimedean local field where $l \neq \text{char. } k$. Let M be a G_k-representation with coefficients in R. Let $\mathcal{I}$ be the inertia subgroup of G_k and let κ be the residue field of k. As κ is finite, we have that $H^2(\kappa, M^\mathcal{I}) = 0$ and hence from the inflation-restriction sequence we have the exact sequence

$$0 \longrightarrow H^1(\kappa, M^\mathcal{I}) \longrightarrow H^1(k, M) \longrightarrow H^1(\mathcal{I}, M)^{G_\kappa} \longrightarrow 0.$$

Suppose that $S \to T$ is a homomorphism of unramified G_k-representations with coefficients in R, where in positive characteristic this a homomorphism of k-group schemes. From the above exact sequence we have the commutative diagram with exact rows

$$\begin{array}{ccccccc}
0 & \longrightarrow & H^1_f(k, S) & \longrightarrow & H^1(k, S) & \longrightarrow & \mathrm{Hom}(\mathcal{I}, S) \\
& & \downarrow & & \downarrow & & \downarrow \\
0 & \longrightarrow & H^1_f(k, T) & \longrightarrow & H^1(k, T) & \longrightarrow & \mathrm{Hom}(\mathcal{I}, T)
\end{array}$$

If $S \to T$ is injective, then the right-hand vertical arrow is injective and it follows that the inverse image of $H^1_f(k, T)$ in $H^1(k, S)$ is equal to $H^1_f(k, S)$. □

Singular Quotient

(4.3.9) Suppose that k is non-archimedean and has characteristic different from l and T is an unramified representation of G_k. We define the *singular quotient* $H^1_s(k, T)$ by the exact sequence

$$0 \longrightarrow H^1_f(k, T) \to H^1(k, T) \longrightarrow H^1_s(k, T) \longrightarrow 0. \tag{4.3.10}$$

Remarks 4.3.11

(i) The above definitions of local conditions are straightforward extensions of those of [54, 62] and [30] to general global fields and suffice for this book.

However, these authors Rubin, Mazur, and Howard and also Bloch and Kato go further for number fields and define the finite part $H^1_f(k, T)$ in the remaining instances when T is ramified or when k is archimedean or when k is non-archimedean with residue characteristic equal to the residue characteristic of R. In the latter case the definition of $H^1_f(k, T)$ is not obvious and involves the ring B_{cris} defined by Fontaine.

The subgroup $H^1_f(k, T)$ when k is archimedean has a straightforward definition given in [62, Chap. I, Remark 3.7].

If T is ramified then $H^1_f(k, T)$ can again be defined as $H^1_{\mathrm{unr}}(k, T)$ but if k is a mixed characteristic local field and R is the ring of integers of a finite extension of $\mathbb{Q}_l$, this in general does not coincide with the definition when T is ramified of $H^1_f(k, T)$ given in [62, Chap. I, Definition 3.4].

(ii) The relaxed, strict, and L-transverse local conditions are not especially stable under change of field k or coefficient ring R. The finite condition is an exception as shown by Proposition 4.3.8.

4.4 A Normal Form for Endomorphisms

(4.4.1) In this section we give a normal form for a module endomorphism. This is an extension of the Jordan Normal Form, or the Jordan-Chevalley decomposition, of a vector space endomorphism to the case where the ground ring is no longer a field.

(4.4.2) Let

R be a coefficient ring;
$\mathfrak{m}$ be the maximal ideal of R;
κ be the finite residue field of R of characteristic l;
T be a finite free R-module;
$\sigma : T \to T$ be an R-endomorphism of T.

A polynomial $P(x) \in R[x]$, where x is an indeterminate over R, is said to be *monic* if the leading coefficient of $P(x)$ is equal to 1. Furthermore, two polynomials $P(x), Q(x) \in R[x]$ are said to be *strictly coprime* if the ideal $<P, Q>$ of $R[x]$ generated by P and Q is equal to $R[x]$.

As T is free of finite rank, there is the characteristic polynomial of σ

$$P(x) = \det(x.\mathrm{Id} - \sigma | T).$$

We have $P(\sigma) = 0$, by the Cayley-Hamilton theorem, and $P(x) \in R[x]$ is a monic polynomial.

Preliminaries

(4.4.3) Let $P(x) = \det(x.\mathrm{Id} - \sigma | T)$ and $\overline{P}(x)$ be the reduction mod $\mathfrak{m}$ of $P(x)$. Then $\overline{P}(x)$ is a product of coprime polynomials

$$\overline{P}(x) \equiv \prod_i p_i(x)^{e_i} \mod \mathfrak{m}$$

where the $p_i(x) \in \kappa[x]$ are irreducible monic coprime polynomials and e_i are positive integers. By Hensel's lemma applied to the ring $R[x]/(P(x))$, there are monic polynomials $q_i(x) \in R[x]$ such that

$$q_i(x) \equiv p_i(x)^{e_i} \mod \mathfrak{m} \text{ for all } i$$

$$P(x) = \prod_i q_i(x)$$

and the $q_i(x)$ are strictly coprime in that the ideal $< q_i(x), q_j(x) >$ of $R[x]$ is equal to $R[x]$ for all $i \neq j$ [51, Chap. 1, §4]; furthermore the polynomials $q_i(x)$ satisfying these conditions are uniquely determined [51, Chap 1, §4].

Lemma 4.4.4 *Suppose that $Q(x) \in R[x]$ is any polynomial and $S(x) \in R[x]$ is a polynomial such that $S(x) \equiv x^r \mod \mathfrak{m} R[x]$ where $r \geq 1$. We have that $S(x)$ and $Q(x)$ are strictly coprime if and only if the constant term of $Q(x) \in R[x]$ is a unit of R.*

Proof By definition, that x^r and $Q(x)$ are strictly coprime means that we have $< S(x), Q(x) >= R[x]$.

Suppose then that $< S(x), Q(x) >= R[x]$ then there are $a(x), b(x) \in R[x]$ such that

$$a(x)S(x) + b(x)Q(x) = 1$$

and so $a(0)S(0) + Q(0)b(0) = 1$. As $S(0) \in \mathfrak{m}$, we have that $Q(0)b(0)$ is a unit of R.

Conversely if $Q(0)$ is a unit of R, then the reductions mod $\mathfrak{m}$ of $S(x)$ and $Q(x)$ are strictly coprime in $\kappa[x]$ so we have $< S(x), Q(x) > + \mathfrak{m} R[x] = R[x]$. It follows by Nakayama's lemma that we have $< S(x), Q(x) >= R[x]$. □

(4.4.5) Let $P(x) = \det(x.\text{Id} - \sigma)$, a monic polynomial in $R[x]$. The reduction $\overline{P}(x)$ mod $\mathfrak{m}$ of $P(x)$ is then of the form

$$\overline{P}(x) = x^s q(x)$$

where $s \in \mathbb{N}$ and $q(x) \in \kappa[x]$ is monic and not divisible by x; in particular $x^s, q(x)$ are coprime in $\kappa[x]$.

Assume that $\overline{P}(0) = 0$. Then we have $s \geq 1$. By Hensel's lemma as in (4.4.3) this factorization lifts to $R[x]$

$$P(x) = S(x)Q(x)$$

where $S(x), Q(x) \in R[x]$ are monic polynomials, $S(x) \equiv x^s$, $Q(x) \equiv q(x) \mod \mathfrak{m}$, and $S(x), Q(x)$ are strictly coprime polynomials in $R[x]$ (this latter also follows from Lemma 4.4.4). Furthermore the monic polynomials $S(x), Q(x)$ are uniquely determined by these conditions.

Lemma 4.4.6 *Suppose that $\overline{P}(0) = 0$. Then we have:*

(i) *The image $Q(\sigma)T$ is the submodule of T on which σ is topologically nilpotent for the $\mathfrak{m}$-adic topology.*
(ii) *There is an R-module decomposition stable under the action of σ*

4.4 A Normal Form for Endomorphisms

$$T \cong Q(\sigma)T \oplus T^{Q(\sigma)=0}$$

where $Q(\sigma)T$, $T^{Q(\sigma)=0}$ are finite free R-modules and where $T^{Q(\sigma)=0}$ is the kernel of $Q(\sigma)$.

(iii) There is an R-module isomorphism

$$T/\sigma T \cong Q(\sigma)T/\sigma Q(\sigma)T.$$

Proof

(i) By the Cayley-Hamilton theorem $S(\sigma)Q(\sigma)$ annihilates T and hence $S(\sigma)$ annihilates $Q(\sigma)T$. As $S(\sigma) = \sigma^r + \sum_{i=1}^{r} ia_{r-i}\sigma^{r-i}$, where $a_i \in \mathfrak{m}$ for all i and $r \geq 1$, it follows that the a_i are topologically nilpotent for the $\mathfrak{m}$-adic topology and hence the restriction of σ to $Q(\sigma)T$ is topologically nilpotent.

Conversely let $x \in T$ be such that $\sigma^t x \to 0$, $\mathfrak{m}$-adically as $t \to +\infty$. Let M be the R-submodule of T which is the $\mathfrak{m}$-adic closure of the R-submodule generated by $\sigma^i x$ for all $i \in \mathbb{N}$; then σ restricts to a topologically nilpotent endomorphism of the submodule M. We have that $Q(x)$ is a polynomial of the form

$$Q(x) = \sum_{i=0}^{u} b_i \sigma^i$$

where $b_i \in R$, b_0 and b_u are units of R and for some $u \in \mathbb{N}$, by Lemma 4.4.4. It follows from the power series expansion in $R[[x]]$ of $Q(x)^{-1}$ that there is a power series $U(x) \in R[[x]]$ such that

$$Q(\sigma)U(\sigma) = 1$$

as an endomorphism of M and where $U(\sigma)$ makes sense as an $\mathfrak{m}$-adic limit because σ is topologically nilpotent on M. Therefore the endomorphism $Q(\sigma): M \to M$ of multiplication by $Q(\sigma)$ is an isomorphism and hence $x \in Q(\sigma)M \subseteq Q(\sigma)T$. Therefore $Q(\sigma)T$ contains the submodule of T on which σ is topologically nilpotent, whence the result.

(ii) We have the exact sequence

$$0 \to T^{Q(\sigma)=0} \to T \to Q(\sigma)T \to 0$$

given by multiplication by $Q(\sigma)$ on T. Here $T^{Q(\sigma)=0}$ and $Q(\sigma)T$ are both submodules of T and we have $T^{Q(\sigma)=0} \cap Q(\sigma)T = 0$ because by part (i), $Q(\sigma)T$ is the submodule of T on which σ is topologically nilpotent, and hence, as shown in the proof of part (i), $Q(\sigma)$ is an invertible endomorphism of $Q(\sigma)T$.

Therefore this exact sequence provides a direct sum free R-module decomposition $T \cong Q(\sigma)T \oplus T^{Q(\sigma)=0}$ of T.

(iii) As shown in part (i), $Q(x)$ is a polynomial of the form $Q(x) = \sum_{i=0}^{u} b_i \sigma^i$ where $b_i \in R$, b_0 and b_u are units of R and for some $u \in \mathbb{N}$. We then have

$$\sigma\left(\sum_{i=0}^{u-1} b_{i+1}\sigma^i\right) = -b_0$$

on the submodule $T^{Q(\sigma)=0}$. As b_0 is a unit of R it follows that σ restricted to $T^{Q(\sigma)=0}$ is an isomorphism and hence $T^{Q(\sigma)=0}/\sigma T^{Q(\sigma)=0} = 0$. The result now follows from the decomposition $T \cong Q(\sigma)T \oplus T^{Q(\sigma)=0}$ of part (ii). □

The Normal Norm for a Nilpotent Endomorphism

Definition 4.4.7 Suppose that $\sigma : T \to T$ is a nilpotent R-endomorphism. A *block* for the nilpotent endomorphism σ is a finite-free R-submodule M of T which is stable under the action of the endomorphism σ and such that there is a free R-basis $v_1, \ldots, v_m$ of M where

$$\sigma v_i = \begin{cases} v_{i+1} & \text{if } i < m \\ \\ 0 & \text{if } i = m. \end{cases}$$

(4.4.8) Suppose that $\sigma : T \to T$ is a topologically nilpotent R-endomorphism for the m-adic topology; that is to say $\sigma^n \to 0$ for the m-adic topology as $n \to \infty$. The map $\sigma \otimes 1$ on $T \otimes_R \kappa$ is then nilpotent over the finite field $\kappa = R/(\pi)$. Therefore $\sigma \otimes 1$ has a Jordan Normal Form and there is a decomposition of $T \otimes_R \kappa$ into blocks for the endomorphism $\sigma \otimes 1$

$$T \otimes_R \kappa \cong \bigoplus_{i=1}^{t} B_i.$$

Equivalently, there are are vectors $v_1, \ldots, v_s$ in $T \otimes \kappa$ and positive integers $l_1, \ldots, l_s$ such that

$$\sigma^i v_j, \quad j = 1, \ldots, s, \quad i = 0, \ldots l_j - 1$$

forms a basis of $T \otimes_R \kappa$ over κ and

$$\sigma^{l_j} v_j = 0, \quad j = 1, \ldots, s.$$

4.4 A Normal Form for Endomorphisms

Lemma 4.4.9 *Suppose that $\sigma : T \to T$ is a topologically nilpotent R-endomorphism for the $\mathfrak{m}$-topology. Select $w_j \in T$ lifting $v_j \in T \otimes_R \kappa$ for all j where the v_j are as in (4.4.8). Then the set of elements of T*

$$\sigma^i w_j, \quad j = 1, \ldots, s, \; i = 0, \ldots l_j - 1.$$

forms a basis of a free submodule over R of T.

Proof For if not, then there are $\lambda_{ij} \in R$, not all zero, with

$$\sum_{ij} \lambda_{ij} \sigma^i w_j = 0.$$

Then there is a greatest integer t such that $\lambda_{ij} \in \mathfrak{m}^t$ for all i, j and where $0 \le t <$ length(R). Let λ'_{ij} be the image of λ_{ij} in $\mathfrak{m}^t/\mathfrak{m}^{t+1}$; then by construction not all the λ'_{ij} are zero. Let $m_1, \ldots, m_n$ be a basis of the vector space $\mathfrak{m}^t/\mathfrak{m}^{t+1}$ over κ. Then we have

$$\lambda'_{ij} = \sum_k \mu_{ijk} m_k$$

for some $\mu_{ijk} \in \kappa$ and we then have the relation in $R/\mathfrak{m}^{t+1} \otimes T$

$$\sum_{ijk} \mu_{ijk} m_k \sigma^i w_j = 0.$$

Not all the μ_{ijk} are zero so that selecting k for which μ_{ijk} is non-zero for some i, j we then have the relation

$$(\sum_{ij} \mu_{ijk} \sigma^i w_j) m_k = 0.$$

As not all the μ_{ijk} are zero and as m_k is a basis element of the vector space $\mathfrak{m}^t/\mathfrak{m}^{t+1}$ over κ we obtain the relation in $T \otimes_R \kappa$

$$\sum_{ij} \mu_{ijk} \sigma^i w_j = 0.$$

This gives the non-trivial relation $\sum_{ij} \mu_{ijk} \sigma^i v_j = 0$ in $T \otimes_R \kappa$ between the $\sigma^i v_j$, $j = 1, \ldots, s$, $i = 0, \ldots l_j - 1$, which is a contradiction. $\square$

Theorem 4.4.10 *Suppose that $\sigma : T \to T$ is a topologically nilpotent endomorphism for the $\mathfrak{m}$-adic topology and that $T/\sigma T$ is a free finitely generated R-module. Then σ is* *nilpotent and there is a decomposition*

$$T \cong \bigoplus_i B_i$$

into a finite number of blocks B_i for the endomorphism σ on T.

Proof **Step 1.** *Construction of a free basis of T.*

As in (4.4.8) on $T/\mathfrak{m}T$ over the field κ, the endomorphism $\sigma \otimes 1$ has a Jordan Normal Form and there are vectors $v_1, \ldots, v_s$ in $T \otimes_R \kappa$ and positive integers $l_1, \ldots, l_s$ such that

$$\sigma^i v_j, \quad j = 1, \ldots, s, \ i = 0, \ldots l_j - 1$$

form a basis of $T \otimes_R \kappa$ over κ and where

$$\sigma^{l_j} v_j = 0, \quad j = 1, \ldots, s.$$

Select elements $w_j \in T$ lifting v_j for all j. The set of elements in T

$$\sigma^i w_j, \quad j = 1, \ldots, s, \ i = 0, \ldots l_j - 1$$

forms a free basis of a free submodule over R of T, by Lemma 4.4.9.

If $w \in T$ then $\sigma w \mod \mathfrak{m}$ is a linear combination of

$$\sigma^i v_j, , \quad j = 1, \ldots, s, \ i = 1, \ldots l_j - 1.$$

So there are $\lambda_{ij} \in R$ such that

$$\sigma w = \sum_{j=1,\ldots,s, \ i=1,\ldots l_j-1} \lambda_{ij} \sigma^i w_j + x$$

where $x \in \mathfrak{m}T$.

Let M be the free R-submodule of T generated by

$$\sigma^i w_j, \quad j = 1, \ldots, s, \ i = 1, \ldots l_j - 1.$$

Let N be the free R-submodule of T generated by

$$w_j, \quad j = 1, \ldots, s.$$

Then $N \oplus M$ is a free R-submodule of T. We have just shown that

$$\sigma T \subseteq M + \mathfrak{m}T.$$

4.4 A Normal Form for Endomorphisms

and
$$T = (N \oplus M) + \mathfrak{m}T.$$

The elements of $T \otimes_R \kappa$
$$\sigma^{l_j-1} v_j, , \quad j = 1, \ldots, s$$

are in the kernel of $\sigma \otimes 1$ and are linearly independent over κ; we can therefore extend them by adjoining suitable elements $y_1, \ldots, y_m \in \ker(\sigma \otimes 1)$ to form a basis of $\ker(\sigma \otimes 1)$. Lift $y_1, \ldots, y_m$ to elements $z_1, \ldots, z_m \in T$.

The elements
$$\sigma^i w_j, \quad j = 1, \ldots, s, \; i = 0, \ldots l_j - 1,$$

$$z_1, \ldots, z_m \in T \tag{4.4.11}$$

then form a free basis of T. This holds because their reductions mod $\mathfrak{m}T$ form a basis of the vector space $T \otimes_R \kappa$ so by Nakayama's lemma $\sigma^i w_j, j = 1, \ldots, s, i = 0, \ldots l_j - 1$, $z_1, \ldots, z_m \in T$ generate T over R and it is evident by a similar argument to that above that they are linearly independent over R. Therefore they form a free basis of T.

It follows that $\sigma(T)$ is the submodule generated by the elements
$$\sigma^i w_j, j = 1, \ldots, s, i = 1, \ldots l_j, \; \sigma(z_1), \ldots \sigma(z_m).$$

Step 2. *The R-submodule E of T generated by*
$$z_1, \ldots, z_m, \; \sigma^{l_j-1} w_j, j = 1, \ldots, s,$$

satisfies $\sigma(E) = 0$.

We have that $\sigma \otimes 1$ is zero on the submodule of $T \otimes_R \kappa$ generated by
$$z_1 \otimes 1, \ldots, z_m \otimes 1, \; \sigma^{l_j-1} w_j \otimes 1, j = 1, \ldots, s.$$

The submodule E of T generated by
$$z_1, \ldots, z_m, \; \sigma^{l_j-1} w_j, j = 1, \ldots, s,$$

is free over R by construction in Step 1 and we have
$$\sigma(E) \subseteq \mathfrak{m}T.$$

The image $\sigma(T)$ is generated by $\sigma(E)$ and

$$\sigma^i w_j, j = 1, \ldots, s, i = 1, \ldots l_j - 1.$$

We then have the decomposition

$$\sigma(T) = M \oplus \sigma E$$

where

$$M = \left(\bigoplus_{j=1,\ldots,s, i=1,\ldots l_j - 1} \sigma^i w_j R \right).$$

As $T/\sigma(T)$ is a free finitely generated R-module, it follows that $\sigma(T)$ is a free R-module of finite rank. Evidently $M \cong \bigoplus_{j=1,\ldots,s, i=1,\ldots l_j - 1} \sigma^i w_j R$ is a free R-module of finite rank; hence we have that $\sigma(E)$ is a free R-module of finite rank. We then have the decomposition of T into free R-modules

$$T \cong \sigma(T) \oplus T/\sigma(T) \cong \sigma(E) \oplus M \oplus T/\sigma(T).$$

But we have already shown that $\sigma(E) \subseteq \mathfrak{m}T$; applying $- \otimes_R R/\mathfrak{m}$ to this decomposition, it follows that $\sigma(E) \otimes_R R/\mathfrak{m}$ is a zero-dimensional $R/\mathfrak{m}$-vector space and this entails that $\sigma(E) = 0$.

Step 3. *The block decomposition.*

We have shown in (4.4.11) of Step 1 that the elements

$$\sigma^i w_j, \quad j = 1, \ldots, s, \ i = 0, \ldots l_j - 1,$$

$$z_1, \ldots, z_m \in T$$

form a free R-basis of T. Furthermore in Step 2 we have shown that

$$\sigma^{l_j} w_j = \sigma(z_i) = 0 \text{ for all } i, j.$$

Therefore T decomposes as a direct sum of finite free R-modules

$$T \cong \bigoplus_{k=0}^{s} B_k$$

where each B_i is a block, B_0 is generated by $z_1, \ldots, z_m$, and

4.4 A Normal Form for Endomorphisms

$$\sigma(B_0) = 0$$

$$B_k = \bigoplus_{i=1,\ldots,l_k-1} \sigma^i w_k R \quad k = 1, \ldots s$$

where

$$\sigma(\sigma^{l_k-1} w_k) = 0, \quad k = 1, \ldots, s.$$

This completes the proof of Theorem 4.4.10. □

The Normal Form for a General Endomorphism

Definition 4.4.12 Suppose that $\sigma : T \to T$ is an R-endomorphism. Let $\lambda \in R$ be such that the kernel of $\sigma - \lambda.\mathrm{Id} : T \to T$ is non-zero. A λ-*block* for the endomorphism σ is a finite free R-submodule B of T which is stable under the action of the endomorphism σ and such that there is a free R-basis $v_1, \ldots, v_m$ of B such that

$$\sigma v_i = \begin{cases} v_{i+1} + \lambda v_i & \text{if } i < m \\ \lambda v_i & \text{if } i = m. \end{cases} \quad (4.4.13)$$

If σ is nilpotent and $\lambda = 0$ then a 0-block of σ coincides with a block of a nilpotent endomorphism in Definition 4.4.7.

Theorem 4.4.14 *Put $P(x) = \det(x.\mathrm{Id} - \sigma | T)$. Let $\lambda_1, \ldots, \lambda_s \in R$ be elements whose reductions mod $\mathfrak{m}$ are distinct. Suppose that $P(\lambda_i) = 0$ for all i and that $T/(\sigma - \lambda_i)T$ is a finitely generated free R-module for all i. Then there is an R-module decomposition stable under the action of σ*

$$T \cong \bigoplus_{j=1}^{s} \left(\bigoplus_k B_k(\lambda_j) \right) \oplus M$$

where the $B_k(\lambda_j)$ are λ_j-blocks for σ for all j, k and M is a free R-submodule of T of finite rank and stable under σ for which the restriction to M of $\sigma - \lambda_i$ is an automorphism for all i.

This decomposition into blocks of T is unique up to isomorphism.

Proof The reduction $\overline{P}(x)$ of $P(x)$ mod $\mathfrak{m}$ takes the form

$$\overline{P}(x) = q(x) \prod_{i=1}^{s} (x - \overline{\lambda}_i)^{e_i}$$

where $\bar{\lambda}_i \in \kappa$ are the reductions mod $\mathfrak{m}$ of the λ_i, distinct for all i, and $q(x) \in \kappa[x]$ does not vanish at $\bar{\lambda}_i$ for all i.

Let

$$P(x) = Q(x) \prod_i S_i(x)$$

be a lifting, via Hensel's lemma (see (4.4.3)) of this factorization of $\overline{P}(x)$ so that we have

$$Q(x) \equiv q(x) \mod \mathfrak{m}$$

$$S_i(x) \equiv (x - \bar{\lambda}_i)^{e_i} \mod \mathfrak{m} \quad \text{for all } i.$$

Put for all i

$$Q_i(x) = Q(x) \prod_{j \neq i} S_j(x).$$

By Lemma 4.4.6 we have

(i) The image $Q_i(\sigma)T$ is the submodule of T on which $\sigma - \lambda_i$ is topologically nilpotent for the $\mathfrak{m}$-adic topology;
(ii) There is an R-module decomposition into submodules stable under the action of σ

$$T \cong Q_i(\sigma)T \oplus T^{Q_i(\sigma)=0}$$

where $Q_i(\sigma)T$, $T^{Q_i(\sigma)=0}$ are finite free R-modules;
(iii) There is an R-module isomorphism

$$T/(\sigma - \lambda_i)T \cong Q_i(\sigma)T/(\sigma - \lambda_i)Q_i(\sigma)T.$$

From (iii) and the hypothesis that $T/(\sigma - \lambda_i)T$ is a finitely generated free R-module, we obtain that $Q_i(\sigma)T/(\sigma - \lambda_i)Q_i(\sigma)T$ is a finite free R-module. It follows from Theorem 4.4.10 applied to $\sigma - \lambda_i$ and $Q_i(\sigma)T$ that there is a decomposition

$$Q_i(\sigma)T \cong \bigoplus_k B_k(\lambda_i)$$

into a finite number of λ_i-blocks $B_k(\lambda_i)$ for the endomorphism σ on T.

Hence we have the decomposition stable under the action of σ

4.4 A Normal Form for Endomorphisms

$$T \cong \left(\bigoplus_k B_k(\lambda_i)\right) \oplus T^{Q_i(\sigma)=0} \qquad (4.4.15)$$

where $B_k(\lambda_i)$ are λ_i-blocks and $T^{Q_i(\sigma)=0}$ is a finitely generated free R-module.

We may now repeat this decomposition for $T^{Q_i(\sigma)=0}$ as follows. For $j \neq i$, we have that $T^{Q_i(\sigma)=0}/(\sigma - \lambda_j)T^{Q_i(\sigma)=0}$ is a finitely generated free R-module because from (4.4.15) we have the isomorphism

$$\frac{T}{(\sigma - \lambda_j)T} \cong \left(\bigoplus_k \frac{B_k(\lambda_i)}{(\sigma - \lambda_j)B_k(\lambda_i)}\right) \oplus \frac{T^{Q_i(\sigma)=0}}{(\sigma - \lambda_j)T^{Q_i(\sigma)=0}}$$

$$\cong \frac{T^{Q_i(\sigma)=0}}{(\sigma - \lambda_j)T^{Q_i(\sigma)=0}}.$$

Then by Lemma 4.4.6 applied to $T^{Q_i(\sigma)=0}$ we have for $j \neq i$

(i) The image $Q_j(\sigma)\left(T^{Q_i(\sigma)=0}\right)$ is the submodule of $T^{Q_i(\sigma)=0}$ on which $\sigma - \lambda_j$ is topologically nilpotent for the m-adic topology;

(ii) There is an R-module decomposition into submodules stable under the action of σ for $i \neq j$

$$T^{Q_i(\sigma)=0} \cong Q_j(\sigma)\left(T^{Q_i(\sigma)=0}\right) \oplus T^{Q_i(\sigma)=0, Q_j(\sigma)=0}$$

where $Q_j(\sigma)\left(T^{Q_i(\sigma)=0}\right)$, $T^{Q_i(\sigma)=0,.Q_j(\sigma)=0}$ are finite free R-modules;

(iii) There is an R-module isomorphism

$$T/(\sigma - \lambda_i)T \cong Q_j(\sigma)\left(T^{Q_i(\sigma)=0}\right)/(\sigma - \lambda_j)Q_j(\sigma)\left(T^{Q_i(\sigma)=0}\right).$$

It follows again from Theorem 4.4.10 that there is a decomposition

$$Q_j(\sigma)\left(T^{Q_i(\sigma)=0}\right) \cong \bigoplus_m B_m(\lambda_j)$$

into a finite number of λ_j-blocks $B_m(\lambda_j)$ for the endomorphism σ on T. Combining this with the decomposition (4.4.15), we obtain

$$T \cong \left(\bigoplus_k B_k(\lambda_i)\right) \oplus \left(\bigoplus_m B_m(\lambda_j)\right) \oplus T^{Q_i(\sigma)=Q_j(\sigma)=0}.$$

In this way it now follows by induction on s that we have a decomposition into finitely generated free R-modules stable under the action of σ

$$T \cong \bigoplus_{j=1}^{s} \left(\bigoplus_{k} B_k(\lambda_j) \right) \oplus T^{Q_1(\sigma)=0,\ldots,Q_s(\sigma)=0}$$

where the $B_k(\lambda_j)$ are λ_j-blocks for σ. The decomposition now follows where we put

$$M = T^{Q_1(\sigma)=\ldots=Q_s(\sigma)=0}. \qquad (4.4.16)$$

From Lemma 4.4.4, we have that $Q_i(x)$ may be written as a polynomial of the form

$$Q_i(x) = \sum_{k=0}^{u} b_i(\sigma - \lambda_i)^k$$

where $b_k \in R$, b_0 and b_u are units of R and for some $u \in \mathbb{N}$. Hence we have on $T^{Q_i(\sigma)=0}$ that

$$(\sigma_i - \lambda_i)(\sum_{k=0}^{u-1} b_{k+1}(\sigma - \lambda_i)^k) = -b_0$$

and hence the restriction of $\sigma_i - \lambda_i$ to $T^{Q_i(\sigma)=0}$ is an isomorphism; it follows that the restriction of $\sigma_i - \lambda_i$ to M is an isomorphism for all i.

The last part of the theorem, on the uniqueness up to isomorphism, follows from the Krull-Remak-Schmidt theorem for $R[\sigma]$-modules. □

A Comparison Map

(4.4.17) Suppose $\sigma : T \to T$ is an R-endomorphism. Put

$$P(x) = \det(x.\text{Id} - \sigma | T) \in R[x].$$

Assume that $P(0) = 0$ and that $T/\sigma T$ is a finite generated free F-module.

We have, by (4.4.3), that $P(x)$ factorizes as a product of monic polynomials of $R[x]$

$$P(x) = S(x)Q(x)$$

where

$$S(x) \equiv x^r \bmod \mathfrak{m}$$

4.4 A Normal Form for Endomorphisms

for some $r \geq 1$ and $S(x)$ and $Q(x)$ are strictly coprime. Then from Theorem 4.4.14 and Eq. (4.4.16), there is an R-module decomposition stable under the action of σ

$$T \cong \left(\bigoplus_i B_i\right) \oplus M \tag{4.4.18}$$

where the B_i are 0-blocks for σ for all i and $M = T^{Q(\sigma)=0}$ is a free R-submodule of T of finite rank, stable under σ and for which the restriction of σ to M is an isomorphism.

(4.4.19) From (4.4.18), we then obtain the isomorphisms compatible with the action of σ

$$T/\sigma T \cong \bigoplus_i \frac{B_i}{\sigma B_i}, \quad T^{\sigma=0} \cong \bigoplus_i B_i^{\sigma=0}. \tag{4.4.20}$$

Put

$$d_i = \mathrm{rank}_R B_i \quad \text{for all } i.$$

On each block B_i we have that

$$B_i/\sigma B_i$$

is a free module of rank 1 and the map $\sigma^{d_i-1} Q(\sigma) : B_i \longrightarrow B_i$ has kernel σB_i and has image $B_i^{\sigma=0}$ and this map induces an isomorphism $\sigma^{d_i-1} Q(\sigma) : B_i/\sigma B_i \longrightarrow B_i^{\sigma=0}$. Therefore taking the sum over i we have that

$$\bigoplus_i \sigma^{d_i-1} Q(\sigma) : \bigoplus_i B_i \longrightarrow \bigoplus_i B_i$$

provides an isomorphism of R-modules

$$\bigoplus_i \sigma^{d_i-1} Q(\sigma) : \bigoplus_i \frac{B_i}{\sigma B_i} \longrightarrow \left(\bigoplus_i B_i\right)^{\sigma=0}. \tag{4.4.21}$$

Definition 4.4.22 Suppose that $\sigma : T \to T$ is an R-endomorphism, $P(0) = 0$ and that $T/\sigma T$ is a finitely generated free R-module. The *comparison isomorphism*

$$\psi : T/\sigma T \longrightarrow T^{\sigma=0}$$

is defined as, using (4.4.20) and (4.4.21),

$$T/\sigma T \cong \bigoplus_{i=1}^s \frac{B_i}{\sigma B_i} \xrightarrow{\psi} T^{\sigma=0} \cong \left(\bigoplus_{i=1}^s B_i\right)^{\sigma=0} \tag{4.4.23}$$

where

$$\psi = \bigoplus_{i=1}^{s} \sigma^{d_i-1} Q(\sigma).$$

Remarks 4.4.24

(i) The comparison isomorphism ψ depends on the decomposition of T into blocks; nevertheless if the blocks of T in a decomposition are all of the same length as $R[\sigma]$-modules then the comparison map ψ is evidently canonical and independent of the decomposition.

(ii) The special case of the above comparison map when $T/\sigma T$ is free of rank 1 and $P(0) = 0$ is given in [54, Definition 1.2.2]. In that special case the characteristic polynomial $P(x)$ factorizes as

$$P(x) = \det(x.\mathrm{Id} - \sigma | T) = x f(x)$$

where $f(x) \in R$. The comparison isomorphism is then

$$f(\sigma) : T/\sigma T \longrightarrow T^{\sigma=0}.$$

In this case when $T/\sigma T$ is free of rank 1 and $P(0) = 0$, from Theorem 4.4.14 there is a single 0-block in the decomposition of T.

The comparison map of Mazur and Rubin in [54, Definition 1.2.2] is stated there as follows. Assume the representation T is equipped with an action by the arithmetic Frobenius Fr. Put $P_{MR} = \det(1 - x.\mathrm{Fr}|T)$. Suppose that T is free of finite rank over R and assume that $P_{MR}(1) = 0$. Then there is a unique polynomial $Q_{MR}(x) \in R[x]$ such that $(x-1)Q_{MR}(x) = P_{MR}(x)$. Therefore $P_{MR}(\mathrm{Fr}^{-1})$ annihilates T so that $Q_{MR}(\mathrm{Fr}^{-1})T \subseteq T^{\mathrm{Fr}=1}$. Then the finite-singular comparison map is the composite map

$$H^1_f(k, T) \cong T/(\mathrm{Fr}-1)T \xrightarrow{Q_{MR}(\mathrm{Fr}^{-1})} T^{\mathrm{Fr}=1} \cong H^1_s(k, T) \otimes \kappa^*.$$

In [54, Definition 1.2.2] it is not assumed that $T/(\mathrm{Fr}-1)T$ is free of rank 1 so that in general this composite map is not an isomorphism. If $T/(\mathrm{Fr}-1)T$ is free of rank 1 over R then as we have seen this comparison map is indeed an isomorphism.

The connection between the notation of [54, Definition 1.2.2] with the notation above when $T/(\mathrm{Fr}-1)T$ is free of rank 1 is that $\sigma = \mathrm{Fr}^{-1} - 1$, $y = x - 1$. We then have $P_{MR}(x) = \det(\mathrm{Fr}^{-1})(-1)^{\mathrm{rank}(T)} P(y)$ and $Q_{MR}(\mathrm{Fr}^{-1}) = \det(\mathrm{Fr})(-1)^{\mathrm{rank}(T)} f(\mathrm{Fr}^{-1} - 1)$.

(iii) As $Q(\sigma)$ is an automorphism on all the components $\bigoplus_{i=1}^{s} B_i/\sigma B_i$ in (4.4.23), it could be suppressed from the map ψ which could then be written simply as $\bigoplus_{i=1}^{s} \sigma^{d_i-1}$.

(iv) The comparison map ψ is built from components which are polynomials in σ over R. In this way ψ is a *replica* of σ, analogously to the usual concept of a replica for Lie algebras over a field.

4.5 The Finite-Singular Comparison Map

(4.5.1) Suppose that in this section

R is a coefficient ring with residue field of characteristic l;
k is a non-archimedean local field of characteristic different from l;
κ is the residue field of the valuation ring of k;
T is an R-representation of G_k with dual T^*;
I is the inertia subgroup of G_k;
Fr is an arithmetic Frobenius element of G_k.

Then we have the exact sequence of profinite groups

$$0 \longrightarrow I \longrightarrow G_k \longrightarrow G_\kappa \longrightarrow 0.$$

We recall that as the residue field κ of k has characteristic different from l, then the R-representation T of G_k is *unramified* if I acts trivially on T.

The Finite Local Condition and Singular Quotient of an Unramified Representation

Proposition 4.5.2 *Suppose V is a G_k-representation which is either a finitely generated R-module or a finite dimensional $\mathbb{Q}_l$-vector space or a discrete torsion module. Then there are canonical isomorphisms:*

(i) $H^1_{\mathrm{unr}}(k, V) \cong V^I/(\mathrm{Fr}-1)V^I$ *where the isomorphism is obtained by evaluation of a 1-cocycle at* Fr;
(ii) $H^1(k, V)/H^1_{\mathrm{unr}}(k, V) \cong H^1(I, V)^{\mathrm{Fr}=1}$ *where the isomorphism is obtained from restriction from G_k to I.*

Proof

(i) From the inflation-restriction sequences of continuous cohomology (Proposition 1.3.13 and Corollary 1.3.14), we have the isomorphism $H^1_{\mathrm{unr}}(k, V) \cong H^1(k^{\mathrm{unr}}/k, V^I)$. The isomorphism of the proposition now follows from Corollary 1.3.18.

(ii) The inflation-restriction sequences of Proposition 1.3.13, Corollary 1.3.14 together with the result of Exercise 1.3.23(iii) applied to G_k and I provide the exact sequence, where k^{unr} is the maximal unramified separable extension of k,

$$0 \to H^1(k^{\mathrm{unr}}/k, V^I) \to H^1(k^{\mathrm{sep}}/k, V) \to H^1(I, V)^{\mathrm{Gal}(k^{\mathrm{unr}}/k)} \to H^2(k^{\mathrm{unr}}/k, V^I).$$

As V^I is a finitely generated R-module or a finite dimensional $\mathbb{Q}_l$-vector space or a discrete torsion module, we have by Corollary 1.3.18 $H^2(k^{\mathrm{unr}}/k, V^I) = 0$ whence the result. □

Corollary 4.5.3 *Suppose that T is an unramified G_k-representation and that $|\kappa^*|.T = 0$. Then there are canonical functorial isomorphisms:*

(a) $$H^1_f(k, T) \cong T/(\mathrm{Fr} - 1)T$$

given by the evaluation of a 1-cocycle $c \mapsto c(\mathrm{Fr})$ at Fr;

(b) $$H^1_s(k, T) \cong \mathrm{Hom}(I, T)^{\mathrm{Fr}=1}$$

obtained by restriction from G_k to I;

(c) $$H^1_s(k, T) \otimes_{\mathbb{Z}} \kappa^* \cong T^{\mathrm{Fr}=1}$$

given by $c \otimes \alpha \mapsto c(\sigma_\alpha)$, for $\alpha \in \kappa^*$, which is the evaluation of a cocycle c at σ_α where $\sigma_\alpha \in \mathrm{Gal}(k^{\mathrm{ab}}/k^{\mathrm{unr}})$ is the image via the reciprocity isomorphism of any lift of α to k^* and where k^{ab} is the maximal separable abelian extension of k and k^{unr} is the maximal unramified abelian extension of k.

Proof The hypothesis $|\kappa^*|.T = 0$ implies that either l is different from the residue characteristic of k or that T is zero. If T is zero then the corollary is trivial so we may assume that l is different from the residue characteristic of k.

The first two isomorphisms (a) and (b) follow from Proposition 4.5.2. The group $I/|\kappa^*|I$ is canonically isomorphic to κ^* via the reciprocity isomorphism of local class field theory. As $|\kappa^*|.T = 0$, we then have the canonical functorial isomorphisms

$$H^1_s(k, T) \cong \mathrm{Hom}(I, T)^{\mathrm{Fr}=1} \cong \mathrm{Hom}(\kappa^*, T)^{\mathrm{Fr}=1} \cong \mathrm{Hom}(\kappa^*, T^{\mathrm{Fr}=1})$$

and the third isomorphism follows. □

Proposition 4.5.4 *Suppose that $|\kappa^*|.T = 0$ and T is an unramified G_k-representation over R. Let L/k be a totally and tamely ramified abelian extension of the local field k of*

4.5 The Finite-Singular Comparison Map

degree $|\kappa^*|$ and $H^1_{tr}(k, T)$ be the corresponding L-transverse subgroup. Then there is a splitting depending on L and functorial in T

$$H^1(k, T) = H^1_f(k, T) \oplus H^1_{tr}(k, T)$$

so that passing to the singular quotient, $H^1_{tr}(k, T)$ maps isomorphically to $H^1_s(k, T)$. Under the local Poitou-Tate pairing

$$H^1(k, T) \times H^1(k, T^*) \longrightarrow H^2(k, \mu_{l^\infty}) \stackrel{\text{inv}}{\cong} \mathbb{Q}_l/\mathbb{Z}_l$$

we have that

(a) *$H^1_f(k, T)$ and $H^1_f(k, T^*)$ are orthogonal complements;*
(b) *$H^1_{tr}(k, T)$ and $H^1_{tr}(k, T^*)$ are orthogonal complements;*
(c) *$H^1_{rel}(k, T)$ and $H^1_{str}(k, T^*)$ are orthogonal complements and similarly for $H^1_{str}(k, T)$ and $H^1_{rel}(k, T^*)$.*

Proof As in the proof of corollary 4.3, the hypothesis $|\kappa^*|.T = 0$ implies that either l is different from the residue characteristic of k or that T is zero so that we may assume that l is different from the residue characteristic of k.

As the field L/k is totally ramified and T is unramified, we have $T^{G_k} = T^{G_L} = T^{\text{Fr}=1}$. From Corollary 4.5.3 there is the canonical isomorphism $H^1_s(k, T) \cong \text{Hom}(I, T)^{\text{Fr}=1}$ and as $|\kappa^*|.T = 0$ we obtain the isomorphism $H^1_s(k, T) \cong \text{Hom}(I/|\kappa^*|I, T^{\text{Fr}=1})$. This gives the commutative diagram where the extension L/k has Galois group isomorphic to κ^* by local class field theory

$$\begin{array}{ccccc}
H^1_{tr}(k, T) & \hookrightarrow & H^1(k, T) & \rightarrow & H^1_s(k, T) \\
\cong \downarrow & & & & \uparrow \cong \\
H^1(L/k, T^{G_L}) & \cong & \text{Hom}(\text{Gal}(L/k), T^{\text{Fr}=1}) & \cong & \text{Hom}(I/|\kappa^*|I, T^{\text{Fr}=1})
\end{array}$$

so that $H^1_{tr}(k, T)$ maps isomorphically to $H^1_s(k, T)$ and the splitting of the first part follows.

For part (a), this follows from [48, theorem I.2.6]. Part (c) is obvious from the non-degeneracy of the pairing.

Part (b) follows from part (a) and the functorial splitting of the first part provided we show that $H^1_{tr}(k, T)$ and $H^1_{tr}(k, T^*)$ are orthogonal.

Since T is unramified and L/k is totally ramified we have $T^{G_L} = T^{G_k}$, $(T^*)^{G_L} = (T^*)^{G_k}$ and it follows that

$$H^1_{tr}(k, T) = H^1(L/k, T^{G_L}) = H^1(L/k, T^{G_k}) = H^1_{tr}(k, T^{G_k})$$

$$H^1_{\text{tr}}(k, T^*) = H^1(L/k, (T^*)^{G_L}) = H^1(L/k, (T^*)^{G_k}) = H^1_{\text{tr}}(k, (T^*)^{G_k}).$$

So we are reduced to the case where G_k acts trivially on T.

The residue characteristic of R is l; let the highest power of l dividing $|\kappa^*|$ be l^n. Then we have $\mu_{l^n} \subseteq k^*$. We also have by hypothesis that $|\kappa^*|$ annihilates T; therefore we have that l^n annihilates T and T is an $R/l^n R$-module of finite type.

It follows from the definition of continuous cohomology (see Sect. 1.3) that if T is a topological R-module with trivial G_k-action then

$$H^1_{\text{tr}}(L/k, T) = \text{Hom}(\text{Gal}(L/k), T)$$

where this Hom is the group of continuous homomorphisms from $\text{Gal}(L/k)$ to T. As $\text{Gal}(L/k)$ is a finite discrete group every group homomorphism from $\text{Gal}(L/k)$ to T is continuous. Therefore by writing T as a direct limit of finite $\mathbb{Z}/l^n\mathbb{Z}$-modules and as Hom commutes with direct limits, we may reduce to the case where T is a finite $\mathbb{Z}/l^n\mathbb{Z}$-module. Again by writing this T as a finite direct sum of cyclic abelian groups we may reduce to the case where $T = \mathbb{Z}/l^m\mathbb{Z}$ where $m \leq n$ and T has a trivial G_k-action.

We then have where the first line follows from class field theory and the second from Kummer theory

$$H^1_{\text{tr}}(L/k, T) = \text{Hom}(\text{Gal}(L/k), \mathbb{Z}/l^m\mathbb{Z}) \cong \text{Hom}(k^*/N_{L/k}L^*, \mathbb{Z}/l^m\mathbb{Z})$$

$$H^1_{\text{tr}}(L/k, T^*) = \text{Hom}(\text{Gal}(L/k), \mu_{l^m}) \cong \ker(k^*/(k^*)^{l^m} \to L^*/(L^*)^{l^m}).$$

The cup product pairing between $H^1_{\text{tr}}(L/k, T)$ and $H^1_{\text{tr}}((L/k, T^*)$ is induced from the natural map

$$\text{Hom}(k^*, \mathbb{Z}/l^m\mathbb{Z}) \times k^* \longrightarrow \mathbb{Z}/l^m\mathbb{Z}.$$

Suppose that $\alpha \in k^*$ where $\alpha \mod (k^*)^{l^m} \in \ker(k^*/(k^*)^{l^m} \to L^*/(L^*)^{l^m})$ so that we have $\alpha = \beta^{l^m}$ for some $\beta \in L^*$. It follows from Galois theory that $N_{L/k}(\beta) = \alpha^{|\kappa^*|/l^m}$. The group $k^*/N_{L/k}L^*$ is cyclic of order $|\kappa^*|$ and therefore α is divisible by l^m in $k^*/N_{L/k}L^*$ and hence α is mapped to zero by every element of $\text{Hom}(k^*/N_{L/k}L^*, \mathbb{Z}/l^m\mathbb{Z})$ which proves (b).

In this latter case where $T = \mathbb{Z}/l^m\mathbb{Z}$, the group $H^1_{\text{tr}}(L/k, T)$ maybe identified with a subgroup of $k^*/(k^*)^{l^m}$ and the group $H^1_{\text{tr}}(L/k, T^*)$ is evidently a subgroup of $k^*/(k^*)^{l^m}$. Then the orthogonality of these groups under cup product, which has been demonstrated, is equivalent to these groups being orthogonal under the Hilbert symbol $k^*/(k^*)^{l^m} \times k^*/(k^*)^{l^m} \to \mu_{l^m}$. $\square$

4.5 The Finite-Singular Comparison Map

The Comparison Map

Definition 4.5.5 Suppose that, along with the assumptions of (4.5.1):

- The residue field κ of k has characteristic different from l (the residue characteristic of R);
- $|\kappa^*|.T = 0$ and T is a finitely generated free R-module;
- T is unramified;
- $\det(1 - \mathrm{Fr}|T) = 0$;
- $T/(\mathrm{Fr} - 1)T$ is a finitely generated free R-module.

Write $P(x)$ for the characteristic polynomial

$$P(x) = \det(1 - \mathrm{Fr}.x|T).$$

By Definition 4.4.22, taking $\sigma = \mathrm{Fr}^{-1} - 1$, there is a comparison isomorphism

$$\psi : T/(\mathrm{Fr} - 1)T \longrightarrow T^{\mathrm{Fr}=1} \qquad (4.5.6)$$

that is constructed from $P(x)$ and a block decomposition of T. This isomorphism ψ is not in general uniquely determined (see also Remark 4.4.24(ii)).

The *finite-singular comparison isomorphism* ϕ^{fs} obtained from ψ is then defined to be the composite isomorphism making the following diagram commutative

$$\begin{array}{ccc} H_f^1(k, T) & \cong & T/(\mathrm{Fr} - 1)T \\ {\scriptstyle \phi^{fs}}\downarrow & & \downarrow{\scriptstyle \psi} \\ H_s^1(k, T) \otimes_{\mathbb{Z}} \kappa^* \cong & & T^{\mathrm{Fr}=1} \end{array}$$

obtained by combining the canonical isomorphisms of Corollary 4.5.3 with the comparison isomorphism ψ of (4.5.6) (from Definition 4.4.22).

The isomorphism ϕ^{fs} depends on the block decomposition of T, whereas the two horizontal isomorphisms of this diagram are canonical and functorial (see also Remark 4.5.9(iv)).

Definition 4.5.7 (Global Fields) Suppose now that F is a global field, T is an R-representation of G_F, v is a non-archimedean place of F, and T is unramified at v. Assume the conditions of the above Definition 4.5.5 are satisfied for the local field F_v, T, and an arithmetic Frobenius Fr_v at v. Then the finite-singular comparison isomorphism

$$\phi_v^{fs} : H_f^1(F_v, T) \longrightarrow H_s^1(F_v, T) \otimes_{\mathbb{Z}} \kappa(v)^*$$

is that obtained for the local field F_v.

Proposition 4.5.8 *Suppose that $|\kappa^*|.R = 0$, T is unramified and T is free of finite rank as an R-module. Suppose also that $T/(\mathrm{Fr}-1)T$ is free of rank $r \in \mathbb{N}$. Then we have $\det(1-\mathrm{Fr}.x|T) = (-1)^{\mathrm{rank}(T)}(x-1)^s Q(x)$ where $s \geq r$ and $Q(x) \in R[x]$ is monic. Furthermore we have the isomorphisms*

$$\psi : T/(\mathrm{Fr}-1)T \longrightarrow T^{\mathrm{Fr}=1}, \phi^{fs} : H^1_f(k,T) \longrightarrow H^1_s(k,T) \otimes \kappa^*$$

of Definition 4.5.5 and that $H^1_f(k,T)$ and $H^1_s(k,T)$ are free R-modules of rank r.

Proof As $T/(\mathrm{Fr}-1)$ is a free R-module, we have the decomposition into free R-modules $T \cong (\mathrm{Fr}-1)T \oplus T/(\mathrm{Fr}-1)T$ which is compatible with the action of Fr. Selecting an R-basis of T with respect to this decomposition, we then see that $\det(1-\mathrm{Fr}.x|T) = (-1)^{\mathrm{rank}(T)}(x-1)^s Q(x)$ where $s \geq r = \mathrm{rank}(T/(\mathrm{Fr}-1)T)$ and $Q(x) \in R[x]$ is a monic polynomial. The proposition now follows from Definition 4.5.5. □

Remarks 4.5.9

(i) The finite-singular comparison map is the basic linking mechanism of cohomology classes of Kolyvagin systems.
(ii) For the Kolyvagin systems of [54], only the case where $T/(\mathrm{Fr}-1)T$ is free of rank 1 is considered (see Remark 4.4.24(ii) and [54, definition 1.2.2]). This case suffices for the systems there considered, but does not apply to Heegner point systems.
(iii) For the case of the Heegner point Kolyvagin systems for elliptic curves over $\mathbb{Q}$ of [30], the Galois group G_k, for a suitable local field k, can be taken to act trivially on T which has rank 2 over R and then one puts $P(x) = \det(1-\mathrm{Fr}.x)|T) = (x-1)^2$; in this event one takes $Q(x)(x-1)^2 = P(x)$, $r=2$ so that $Q(x)=1$, $Q(\mathrm{Fr}^{-1})$ is the identity map, $T/(\mathrm{Fr}-1)T$ is free of rank 2, and and the finite-singular comparison map ϕ_v^{fs} is an isomorphism of rank 2 free R-modules [30, Definition 1.1.8].

This as we see later also applies to Heegner points on elliptic curves via Drinfeld modular curves over global function fields as well as CM points on Shimura curves over totally real number fields.
(iv) The isomorphisms of Corollary 4.5.3

$$H^1_f(k,T) \cong T/(\mathrm{Fr}-1)T, \quad H^1_s(k,T) \otimes_{\mathbb{Z}} \kappa^* \cong T^{\mathrm{Fr}=1}$$

are functorial and canonical. The comparison map ψ of (4.5.6) depends on the block decomposition of T so that $\phi^{fs} : H^1_f(k,T) \longrightarrow H^1_s(k,T) \otimes \kappa^*$ is not in general canonical.

If T has blocks all of the same size for $\mathrm{Fr}-1$, in particular if rank $T/(\mathrm{Fr}-1)T = 1$ when there is only one block, then ϕ^{fs} is functorial and canonical (Remarks 4.4.24(i),(ii)).

4.5 The Finite-Singular Comparison Map

(v) In the definition of the finite-singular comparison map, the isomorphism $H^1_s(k, T) \otimes \kappa^* \to T^{\mathrm{Fr}=1}$ is given by $c \otimes \alpha \mapsto c(\sigma_\alpha)$ where (see Corollary 4.5.3) $\sigma_\alpha \in \mathrm{Gal}(k^{\mathrm{ab}}/k^{\mathrm{unr}})$ is the element associated by local class field theory to a lift to k^* of $\alpha \in \kappa^*$.

The hypothesis $|\kappa^*|.T = 0$ implies that $H^1_s(k, T) \otimes \kappa^*$ is trivially isomorphic to $H^1_s(k, T)$ but the presence of the factor κ^* "rigidifies" $H^1_s(k, T)$ and at least in the case where rank $T/(\mathrm{Fr}-1) = 1$ the comparison map ϕ^{fs} can then be made canonical.

(vi) From local class field theory, the maximal tamely ramified abelian extensions of a non-archimedean local field k have degree $|\kappa^*|$ where κ is the residue field of k. This explains why L/k has order $|\kappa^*|$ in the L-transverse condition of (4.3.4)(d) and Proposition 4.5.4.

Global Fields and Selmer Structures

Contents

5.1	Selmer Structures, Selmer Modules	105
5.2	Modification of Selmer Structures	107
5.3	Global Duality	107
5.4	Local to Global Principles for Rational Points	110
5.5	The Classical Selmer Structure of an Abelian Variety	117
5.6	Finiteness of Classical Selmer Modules of Abelian Varieties	123

5.1 Selmer Structures, Selmer Modules

(5.1.1) Suppose for this section that:

k is a global field;
R is a coefficient ring with residue field of characteristic l where this is different from the characteristic of k;
T is an R-representation of G_k;
T^* is the dual of T (see (4.1.9)).

A *Selmer structure* $\mathcal{F}$ on T consists of the following data:

(a) A finite set $S(\mathcal{F})$ of places of k containing the archimedean places, if any, all places above the prime number l, if k does not have characteristic l, and all places where T is ramified;
(b) For every place v of $S(\mathcal{F})$, a local condition on T that is to say a choice of R-submodule for each $v \in S(\mathcal{F})$

$$H^1_{\mathcal{F}}(k_v, T) \subseteq H^1(k_v, T);$$

(c) for every place $v \notin S(\mathcal{F})$, the finite local condition is imposed on T at v, namely, the R-submodule

$$H^1_f(k_v, T) \subseteq H^1(k_v, T).$$

There is an evident partial order on Selmer structures on T where we write $\mathcal{F} \leq \mathcal{G}$ if $H^1_{\mathcal{F}}(k_v, T) \subseteq H^1_{\mathcal{G}}(k_v, T)$ for all places v of k.

Definition 5.1.2 If $\mathcal{F}$ is a Selmer structure on T then the *Selmer module* $H^1_{\mathcal{F}}(k, T)$ is the submodule of $H^1(k, T)$ which is the kernel of the natural map

$$H^1(k, T) \to \bigoplus_{v \in \Sigma_k} H^1(k_v, T)/H^1_{\mathcal{F}}(k_v, T)$$

where the sum runs over all places, archimedean and non-archimedean, of k.

In other words, the group $H^1_{\mathcal{F}}(k, T)$ is the subgroup of $H^1(k, T)$ of those cohomology classes which are unramified outside $S(\mathcal{F})$ and which at each place $v \in S(\mathcal{F})$ the classes localized at v belong to $H^1_{\mathcal{F}}(k_v, T)$.

(5.1.3) For the dual $T^* = \operatorname{Hom}(T, \mu_{l^\infty})$ of T ((4.2.11) and (4.2.18)), put $S(T^*) = S(T)$; then the dual local conditions $\mathcal{F}^*$ for all places $v \in S(T^*)$ (Definition 4.3.6) adjoined with the finite local condition on T^* for all $v \notin S(T^*)$ defines the *dual Selmer structure* $\mathcal{F}^*$ on T^*.

Definition 5.1.4 Suppose the coefficient ring R is the ring of integers of a finite extension of $\mathbb{Q}_l$ and that the R-representation T of G_k is a finite free R-module (so that l is necessarily distinct from the characteristic of k). The *canonical Selmer structure* $\mathcal{F}_{\mathrm{can}}$ on T of G_k is given by the following data.

- Let $S(\mathcal{F}_{\mathrm{can}})$ be the finite set of places

$$S(\mathcal{F}_{\mathrm{can}}) =$$

{ramified places of T} ∪ {places dividing l and archimedean places};

- If $\mathfrak{p} \in S(\mathcal{F}_{\mathrm{can}})$ is a non-archimedean place not dividing l and $k_{\mathfrak{p}}^{\mathrm{sh}}$ is the field of fractions of the strict henselization of the valuation ring of $k_{\mathfrak{p}}$ then

$$H^1_{\mathcal{F}_{\mathrm{can}}}(k_{\mathfrak{p}}, T) = \ker(H^1(k_{\mathfrak{p}}, T) \longrightarrow H^1(k_{\mathfrak{p}}^{\mathrm{sh}}, T \otimes_R \operatorname{fract}(R));$$

5.3 Global Duality

- if $\mathfrak{p}$ is a place of k where $\mathfrak{p}$ divides l or is archimedean then

$$H^1_{\mathcal{F}_{can}} = H^1(k_\mathfrak{p}, T);$$

- if $\mathfrak{p}$ is a place of k not in $S(\mathcal{F}_{can})$ then

$$H^1_{\mathcal{F}_{can}}(k_\mathfrak{p}, T) = H^1_f(k_\mathfrak{p}, T).$$

If I is an ideal of R then the canonical Selmer structure on T/IT is defined by propagation of $\mathcal{F}_{can}$ via the map $T \to T/IT$.

5.2 Modification of Selmer Structures

(5.2.1) With the notation of the last Sect. 5.1, suppose that $\mathcal{F}$ is a Selmer structure on T. Then $\mathcal{F}$ can be modified at places of k by replacing a local condition at a given place by a different local condition at that place. In particular we shall consider the modification $\mathcal{F}_a^b(c)$ of $\mathcal{F}$ defined as follows.

Suppose that a, b, c are cycles of k whose supports are relatively coprime and without archimedean components and that c is not divisible by any prime in $S(\mathcal{F})$. Write $\mathcal{F}_a^b(c)$ for the Selmer structure on T given by

$S(\mathcal{F}_a^b(c)) = S(\mathcal{F}) \cup \mathrm{Supp}(abc)$

$$H^1_{\mathcal{F}_a^b(c)}(k_v, T) = \begin{cases} H^1_{\mathcal{F}}(k_v, T) & \text{if } v \in S(\mathcal{F}) \text{ and } v \notin \mathrm{Supp}(abc) \\ 0 & \text{if } v|a \\ H^1(k_v, T) & \text{if } v|b \\ H^1_{tr}(k_v, T) & \text{if } v|c. \end{cases}$$

That is to say at places dividing a the strict condition is imposed; at places dividing b the relaxed condition is imposed and at places dividing c one has the transverse condition. If one of a, b, c is equal to 1 then it is omitted from $\mathcal{F}_a^b(c)$. For the case of places v dividing c a choice of field extension L/k_v is made for each $v|c$ where L is a maximal tamely ramified abelian extension of k_v depending on v; this is in order to define the L-transverse condition at v.

These modified Selmer structures are especially important for Kolyvagin systems.

5.3 Global Duality

(5.3.1) In this section

k is a global field;

R is a coefficient ring with residue field of characteristic l where this is different from the characteristic of k;
T is an R-representation of G_k;
T^* is the dual of T (see (4.1.9)).

Theorem 5.3.2 *Suppose that $\mathcal{F} \leq \mathcal{G}$ are Selmer structures on T.*

(i) There are exact sequences obtained by localization at places of k

$$0 \longrightarrow H^1_{\mathcal{F}}(k, T) \longrightarrow H^1_{\mathcal{G}}(k, T) \xrightarrow{f} \bigoplus_v H^1_{\mathcal{G}}(k_v, T)/H^1_{\mathcal{F}}(k_v, T)$$

$$0 \longrightarrow H^1_{\mathcal{G}^*}(k, T^*) \longrightarrow H^1_{\mathcal{F}^*}(k, T^*) \xrightarrow{g} \bigoplus_v H^1_{\mathcal{F}^*}(k_v, T^*)/H^1_{\mathcal{G}^*}(k_v, T^*).$$

(ii) The sum of the local Poitou-Tate pairings $\sum_{v \in \Sigma} <,>_v$ (see (4.3.5)), where Σ is the finite set of places of k where $\mathcal{F}, \mathcal{G}$ differ, defines a bilinear pairing on the product

$$B : (\bigoplus_v H^1_{\mathcal{G}}(k_v, T)/H^1_{\mathcal{F}}(k_v, T)) \times (\bigoplus_v H^1_{\mathcal{F}^*}(k_v, T^*)/H^1_{\mathcal{G}^*}(k_v, T^*)) \to \mathbb{Q}_l/\mathbb{Z}_l.$$

The images of the maps f, g are orthogonal complements under B.

Proof

(i) The two exact sequences follow immediately from the definition of the Selmer modules $H^1_{\mathcal{F}}(k, T)$.
(ii) That the images being orthogonal complements is a consequence of Poitou-Tate global duality and is derived in Exercise 5.3.3. □

Exercise 5.3.3 The aim of this exercise is to derive Theorem 5.3.2(ii) from Poitou-Tate global duality. In parts (i)–(vi) of this exercise, let T be an R-representation that is annihilated by some power of l. For a topological R-module (Hausdorff and locally compact) M we denote by $M^\wedge$ its Pontryagin dual $\mathrm{Hom}_{\mathrm{cts}}(M, \mathbb{T})$ where $\mathbb{T}$ is the circle group (see (1.2)).

(i) Let $\mathfrak{F}$ be the Selmer structure on T given by

$$H^1_{\mathfrak{F}}(k_v, T) = H^1_f(k_v, T) \text{ for all places } v \text{ of } k.$$

Replacing here T by T^* in this formula defines $\mathfrak{F}$, the corresponding Selmer structure on T^* written with the same symbol $\mathfrak{F}$. Let Σ be a finite set of places of k that contains all archimedean places, all primes above l (if any), and all places where T is ramified.

5.3 Global Duality

Writing k_Σ for the maximal separable extension of k that is unramified outside Σ, show that
$$H^1_{\mathfrak{F}^\Sigma}(k, T) \cong H^1(k_\Sigma/k, T), \quad H^1_{\mathfrak{F}^\Sigma}(k, T^*) \cong H^1(k_\Sigma/k, T^*).$$

As l is coprime to the characteristic of k and T has order a power of l, conclude from the Tate-Poitou global duality long exact sequence [58, Theorem 8.6.10, p. 489] that there is an exact sequence
$$H^1_{\mathfrak{F}^\Sigma}(k, T) \longrightarrow \bigoplus_{v \in \Sigma} H^1(k_v, T) \longrightarrow H^1_{\mathfrak{F}^\Sigma}(k, T^*)^\wedge.$$

(ii) Suppose that $\mathcal{F} \le \mathcal{G}$ are Selmer structures on T. Let Σ be any finite set of places containing all places where the local conditions $\mathcal{F}$ and $\mathcal{G}$ differ. Using Tate local duality show that there is a natural isomorphism
$$\bigoplus_{v \in \Sigma} \frac{H^1_\mathcal{G}(k_v, T)}{H^1_\mathcal{F}(k_v, T)} \cong \bigoplus_{v \in \Sigma} \frac{H^1_{\mathcal{F}^*}(k_v, T^*)^\wedge}{H^1_{\mathcal{G}^*}(k_v, T^*)^\wedge}.$$

Show that the exact sequences of Theorem 5.3.2(i), taking the $\wedge$-dual of the second, can be arranged into a sequence of homomorphisms of R-modules
$$0 \to H^1_\mathcal{F}(k, T) \to H^1_\mathcal{G}(k, T) \to \bigoplus_{v \in \Sigma} \frac{H^1_\mathcal{G}(k_v, T)}{H^1_\mathcal{F}(k_v, T)} \to H^1_{\mathcal{F}^*}(k, T^*)^\wedge \to H^1_{\mathcal{G}^*}(k_v, T^*)^\wedge \to 0 \quad (*)$$

which is exact everywhere except possibly at the middle term. Using part (i) of this exercise, show that the statement of Theorem 5.3.2(ii) is equivalent to the exactness of this sequence $(*)$ in the middle.

(iii) Let $\Sigma_1 \subseteq \Sigma_2$ be finite sets of places of k. Assume that Σ_2 contains all archimedean places, all primes above l (if any), and all places where T is ramified. Using part (i), show that $(*)$ is exact for the pair $\mathfrak{F}^{\Sigma_1}, \mathfrak{F}^{\Sigma_2}$, that is to say $(*)$ is exact when $\mathcal{F} = \mathfrak{F}^{\Sigma_1}$ and $\mathcal{G} = \mathfrak{F}^{\Sigma_2}$.

(iv) Let $\Sigma_1 \subseteq \Sigma_2$ be any pair of finite sets of places of k. Derive from part (iii) that $(*)$ is exact for the pair $\mathfrak{F}^{\Sigma_1}, \mathfrak{F}^{\Sigma_2}$.

[Hint: Select a finite set of places Σ of k such that $\Sigma_2 \subseteq \Sigma$ and that Σ contains all archimedean places, all primes above l (if any), and all places where T is ramified. Apply (iii) to the two pairs $\mathfrak{F}^{\Sigma_1}, \mathfrak{F}^\Sigma$ and $\mathfrak{F}^{\Sigma_2}, \mathfrak{F}^\Sigma$ to show that $(*)$ is exact for these two pairs. Hence deduce that $(*)$ is then exact for the pair $\mathfrak{F}^{\Sigma_1}, \mathfrak{F}^{\Sigma_1}$.]

(v) Let $\mathcal{F}$ be a Selmer structure on T. Select a finite site of places Σ such that $\mathcal{F} \le \mathfrak{F}^\Sigma$. Using part (iv), show that $(*)$ is exact for the pair $\mathcal{F}, \mathfrak{F}^\Sigma$.

(vi) Let $\mathcal{F}, \mathcal{G}$ be Selmer structures on T such that $\mathcal{F} \leq \mathcal{G}$. Show that (*) is exact for the pair $\mathcal{F}, \mathcal{G}$ and hence that Theorem 5.3.2(ii) holds for T which is an R-representation of G_k of finite l-power order.
[Hint: Select a finite set of places Σ of k such that $\mathcal{F} \leq \mathcal{G} \leq \mathfrak{F}^\Sigma$. Apply (v) to show that the two pairs $\mathcal{F}, \mathfrak{F}^\Sigma$ and $\mathcal{G}, \mathfrak{F}^\Sigma$ both satisfy (*). Imitating the argument of (iv) deduce that (*) is exact for the pair $\mathcal{F}, \mathcal{G}$.]

(vii) By passing to inverse limits using Proposition 1.3.10, show that Theorem 5.3.2 holds when T is an R-representation (not necessarily finite) of G_k.

5.3.4. Bibliographical Remarks. Part (i) of Theorem 5.3.2 follows immediately from the definitions and is stated for number fields in [54, Theorem 2.3.4] and [30, Theorem 1.1.11]. Theorem 5.3.2(ii), in the case when k is a number field, is stated without proof in [54, Theorem 2.3.4] and [30, Theorem 1.1.11]. Part (i) of this exercise is proved for the number field case in [62, lemma 1.5.3, p. 12]. Parts (ii), (iii), and (iv) of this Exercise 5.3.3 for the case of a number field are demonstrated in [62, Theorem 1.7.3, pp.17–19].]

5.4 Local to Global Principles for Rational Points

(5.4.1) For a scheme X of finite type over a global field k, a fundamental question is to determine the set of k-rational points $X(k)$ of k. This may be a difficult problem but one can replace the determination of $X(k)$ by simpler problems which may themselves be solvable. One such question is that of finding obstructions to the existence of rational points. For example, if k_v is the completion of k at a place v and if $X(k_v)$ is empty then evidently $X(k)$ will be empty so that $X(k_v)$ being empty is an obstruction to the existence of points in $X(k)$. It is usually easy to determine whether or not $X(k_v)$ is empty, for a given place v; for example, if v is a non-archimedean place and X/k is smooth then Hensel's lemma may possibly be used to determine $X(k_v)$.

(5.4.2) Let X/k be a scheme of finite type over k and let $\mathbb{A}_k$ be the adèle ring of k equipped with the natural ring homomorphism $k \to \mathbb{A}_k$ and the projection ring homomorphisms $\mathbb{A}_k \to k_v$ for all places v of k. We then have the injections of sets, where the product runs over all places v of k,

$$X(k) \to X(\mathbb{A}_k) \to \prod_v X(k_v).$$

A k-rational point $x \in X(k)$ then provides a point $x_\mathbb{A} \in X(\mathbb{A}_k)$ and also points $x_v \in X(k_v)$ for all places v of k and hence a point $(x_v)_v$ of $\prod_v X(k_v)$. In this way if $\prod_v X(k_v)$ is empty then X has no k-rational points.

5.4 Local to Global Principles for Rational Points

Definition 5.4.3 Let X be a k-scheme of finite type where k is a global field.

(i) The scheme X/k satisfies the *Hasse principle* if $X(k_v) \neq \emptyset$ for all places v of k implies that $X(k) \neq \emptyset$.

(ii) The scheme X satisfies the *weak approximation property* if $X(k)$ is dense in $\prod_v X(k_v)$, where the product runs over all places and is equipped with the product topology where each $X(k_v)$ is equipped with its v-topology either non-archimedean or archimedean.

(5.4.4) More generally, let $\mathscr{S}ch/k$ be the category of schemes of finite type over k and let
$$F : \mathscr{S}ch/k \longrightarrow \mathscr{S}ets$$
be a contravariant functor from $\mathscr{S}ch/k$ to the category of sets. Let $f \in F(X)$. Then a point $x : \operatorname{Spec} k \to X$ gives a map of sets $F(x) : F(X) \to F(k)$ and the image $F(x)(f)$ of f in $F(k)$ provides a map
$$\phi : X(k) \to F(k), \quad x \mapsto F(x)(f).$$

Similarly f provides a map $\Phi : X(\mathbb{A}_k) \to F(\mathbb{A}_k)$ where $y \mapsto F(y)(f)$. We then have the commutative diagram, which depends on f,

$$\begin{array}{ccc} X(k) & \xrightarrow{\phi} & F(k), \quad \phi : x \mapsto F(x)(f) \\ \psi_X \downarrow & & \downarrow \psi_F \\ X(\mathbb{A}_k) & \xrightarrow{\Phi} & F(\mathbb{A}_k), \quad \Phi : y \mapsto F(y)(f). \end{array}$$

In this way $X(k)$ fits into this diagram depending on f.

Taking F to be the functor $\operatorname{Morph}_k(-, X)$ from $\mathscr{S}ch/k$ to $\mathscr{S}ets$ and f to be the identity morphism from X to itself, we reobtain the arrow $X(k) \to X(\mathbb{A}_k)$ of paragraph (5.4.2).

(5.4.5) Let $f \in F(X)$. Define the subset of $X(\mathbb{A}_k)$ given by
$$(X(\mathbb{A}_k))^f = \Phi^{-1}\psi_F(F(k)).$$

Then by this diagram above $\psi_X(X(k))$ is a subset of $(X(\mathbb{A}_k))^f$. Put
$$(X(\mathbb{A}_k))^F = \bigcap_{f \in F(X)} (X(\mathbb{A}_k))^f.$$

Again $\psi_X(X(k))$ is a subset of $(X(\mathbb{A}_k))^F$.

Definition 5.4.6 Let X be a k-scheme of finite type where k is a global field.

(i) If $X(\mathbb{A}_k) \neq \emptyset$ and $(X(\mathbb{A}_k))^F = \emptyset$ then X has a F *obstruction to the Hasse principle*.
(ii) If $(X(\mathbb{A}_k))^F \neq X(\mathbb{A}_k)$ and $(X(\mathbb{A}_k))^F$ is closed in $X(\mathbb{A}_k)$ then X has a F *obstruction to weak approximation*.

(5.4.7) Apart from the functor $F = \mathrm{Morph}_k(-, X)$, there are at least two other choices of the functor F which may be applied for this rational point question. We explain these two in this section.

The first is the method of descent, where the functor F is the group of torsors of a group scheme over the ground scheme. This last has a tenuous link with the "method of infinite descent" of Fermat who by this method demonstrated, for example, that the affine elliptic curve $y^2 = x^4 + 1$ has only the rational solutions $x = 0$, $y = \pm 1$.

The second is the Brauer-Manin obstruction, using the Brauer group for the functor F.

Remarks 5.4.8

(i) There is an evident inclusion of sets $X(\mathbb{A}_k) \subseteq \prod_v X(k_v)$. This inclusion becomes a bijection if, for example, X/k is proper and of finite type over k.

Furthermore, if the functor F takes values in the category of abelian groups then the natural map $F(\mathbb{A}_k) \to \prod_v X(k_v)$ may reduce to an isomorphism of abelian groups $F(\mathbb{A}_k) \to \bigoplus_v F(k_v)$ where the sum run over all places of k. This last is the case where F is the Brauer group functor (see Proposition 5.4.16).
(ii) When X is a projective smooth quadric hypersurface over k, then the Hasse principle holds for X and this is known as the Hasse-Minkowski principle; an equivalent way of stating this is that a quadratic form over k which has everywhere locally a non-trivial zero must have a global non-trivial zero.

The Hasse principle also holds for the torsors over a number field k of a semi-simple simply connected algebraic group G/k.
(iii) The Hasse principle and weak approximation both fail in general for curves of genus 1, cubic surfaces, and torsors of abelian varieties. The Shafarevich-Tate group (see Sect. 5.5) of an abelian variety over a global field is a measure of the failure of the Hasse principle for the torsors of the abelian variety.

The Method of Descent

(5.4.9) Let G be a group scheme flat and locally of finite type over a locally noetherian scheme U.

Recall that a (right) action of G on a locally noetherian U-scheme V is a morphism $V \times_U G \to V$ which satisfies the usual commutative diagrams for an action on a scheme;

5.4 Local to Global Principles for Rational Points

this last is equivalent to the property that for any U-scheme T the group $G(T)$ acts on the set $V(T)$ and this action is functorial in T.

Recall that a (right) G-torsor over U is a scheme V/U, faithfully flat and locally of finite type over U and equipped with a (right) G-action, such that the natural morphism

$$V \times_U G \longrightarrow V \times_U V, (y, g) \mapsto (y, yg)$$

is an isomorphism [51, p. 120].

(5.4.10) The G-torsors over U are classified by a cohomology set $H^1(U, G)$ (see (5.5.3) for the case where G is an abelian variety). This pointed set $H^1(U, G)$ may be defined via Čech cohomology for the flat topology on U by taking it as a limit over all open affine coverings of U of sets of 1-cocycles with values in G (see [51, pp.120–124] for more details).

It is a non-trivial problem to determine which elements of $H^1(U, G)$ correspond to G-torsors but if G is affine over U or G is regular and is also smooth and proper over U with geometrically connected fibres then there is a bijection between G-torsors over U and elements of $H^1(U, G)$ [51, Chap III, Propn. 4.3(a) and (c)].

If G is non-commutative then $H^1(U, G)$ is only a pointed set with a distinguished zero element, whereas if G is commutative then $H^1(U, G)$ is a group.

(5.4.11) Suppose now that $U = \operatorname{Spec} k$ where k is a global field, and for simplicity we can even assume that G/k is of finite type and smooth. For any k-scheme X locally of finite type, then $G_X = G \times_k X$ is a group scheme flat and locally of finite type over X.

Suppose that $\phi : Z \to X$ is a G_X-torsor over X, where X is a k-scheme locally of finite type. Let $x \in X$. Then the fibre of ϕ over x is a G-torsor Z_x over $\operatorname{Spec} k$. Hence we obtain a map

$$X(k) \to H^1(k, G), x \mapsto [Z_x]$$

which depends on Z where $[Z_x]$ is the cohomology class of the torsor Z_x. Similarly for any place v of k and where k_v is the completion of k at v, we obtain a map

$$X(k_v) \to H^1(k_v, G), y_v \mapsto [Z_{y_v}]$$

(strictly speaking, $H^1(k_v, G)$ with our notation should be written $H^1(k_v, G_{\operatorname{Spec} k_v})$). Taking $F = H^1(-, G)$ in the diagram of paragraph (5.4.4), we hence obtain the commutative diagram depending on Z where adèle groups are replaced by products

$$\begin{array}{ccc} X(k) & \xrightarrow{\phi} & H^1(k, G) \\ \psi_X \downarrow & & \downarrow \psi_{\text{torsor}} \\ \prod_v X(k_v) & \xrightarrow{\Phi} & \prod_v H^1(k_v, G) \end{array} \quad \begin{array}{l} \phi : x \mapsto [Z_x] \\ \\ \Phi : y = (y_v)_v \mapsto \prod_v [Z_{y_v}]. \end{array}$$

Here $[Z_x], [Z_{y_v}]$ denote the cohomology classes of the torsors Z_x, Z_{y_v} and y is the collection of points $y_v \in X(k_v)$ for all v.

We may then define $(\prod_v X(k_v))^Z$ analogously to (5.4.5) as

$$(\prod_v X(k_v))^Z = \Phi^{-1}\psi_{\text{torsor}}(H^1(k, G)).$$

We then define the Selmer set contained in $H^1(k, G)$ and depending on Z, where c_v denotes the restriction of the class c at v,

$$\mathcal{S}_Z = \psi_{\text{torsor}}^{-1}\Phi((\prod_v X(k_v))^Z)$$

$$= \{c \in H^1(k, G) \mid c_v \in \text{Image}(X(k_v) \to H^1(k_v, G)) \text{ for all } v\}.$$

This Selmer set $\mathcal{S}_Z$ extends the notion of Selmer module of Definition 5.1.2.

Example 5.4.12 With the above notation, if $\phi : Z \to X$ is an isogeny of abelian varieties over k, G is the finite group scheme $\ker(\phi)/k$, and hence Z is a G_X-torsor over X, then $\mathcal{S}_Z$ is a group and is the usual Selmer group of the abelian variety X with respect to the isogeny ϕ (see Sect. 5.5 or [74, Chap X,§4]).

In this case of a torsor given by an isogeny of abelian varieties over a global field, the Selmer set $\mathcal{S}_Z$ can be shown to be finite (Proposition 5.6.3).

The Brauer-Manin Obstruction

(5.4.13) Let X be a separated scheme. The *(cohomological) Brauer group* $\text{Br}(X)$ is defined to be

$$\text{Br}(X) = H^2_{\text{ét}}(X, \mathbb{G}_m).$$

The classical Brauer group of a separated scheme X is defined to be the group of Morita equivalence classes of Azumaya algebras over X; it can be shown that this classical Brauer group is naturally isomorphic to a subgroup of the cohomological Brauer group $\text{Br}(X)$ [26] and for a scheme X equipped with an ample invertible sheaf the two groups coincide [15].

(5.4.14) Suppose now that k is a global field and let $\mathscr{S}\!\mathit{chsep}/k$ be the category of separated schemes of finite type over k. Let Br: $\mathscr{S}\!\mathit{chsep}/k \to \mathscr{A}\!\mathit{b}\,\mathscr{G}\!\mathit{ps}$, $X \rightsquigarrow \text{Br}(X)$ be the Brauer group functor from the category of separated k-schemes of finite type to the category of abelian groups.

Let X be a separated k-scheme of finite type and $f \in \text{Br}(X)$. Exactly as in paragraphs (5.4.4)–(5.4.5), taking $F = \text{Br}$ we obtain the commutative diagram, which depends on f,

5.4 Local to Global Principles for Rational Points

$$\begin{array}{ccc} X(k) & \xrightarrow{\phi} & \mathrm{Br}(k), \quad \phi : x \mapsto \mathrm{Br}(x)(f) \\ \psi_X \downarrow & & \downarrow \psi_{\mathrm{Br}} \\ X(\mathbb{A}_k) & \xrightarrow{\Phi} & \mathrm{Br}(\mathbb{A}_k), \quad \Phi : y \mapsto \mathrm{Br}(y)(f). \end{array} \tag{5.4.15}$$

Proposition 5.4.16 *We have*

$$\mathrm{Br}(\mathbb{A}_k) \cong \bigoplus_v \mathrm{Br}(k_v)$$

where in the sum v runs over all places of k.

Proof The adèle ring $\mathbb{A}_k$ of k is the direct limit $\varinjlim T_S$ where S runs over all finite subsets of places of k containing all infinite places and where T_S is the ring

$$T_S = \prod_{v \text{ infinite}} k_v \prod_{v \notin S} O_v \prod_{v \in S \text{ finite}} k_v$$

where O_v is the ring of valuation integers of k_v for v a finite place. As the Brauer group of a discrete valuation ring with finite residue field is zero [26], the Brauer group of T_S is equal to

$$\mathrm{Br}(T_S) \cong \prod_{v \text{ infinite}} \mathrm{Br}(k_v) \prod_{v \in S \text{ finite}} \mathrm{Br}(k_v).$$

As an affine scheme is quasi-compact and quasi-separated, the direct limit $\varinjlim T_S$ commutes with the Brauer functor; hence we have that the Brauer group $\mathrm{Br}(\mathbb{A}_k)$ is isomorphic to $\varinjlim \mathrm{Br}(T_S)$ and hence there is a natural group isomorphism

$$\mathrm{Br}(\mathbb{A}_k) \cong \mathrm{Br}(\varinjlim T_S) \cong \varinjlim \mathrm{Br}(T_S) \cong \bigoplus_v \mathrm{Br}(k_v)$$

where in the sum v runs over all places of k.

(5.4.17) The Brauer group of a local field k_v is canonically isomorphic to $\mathbb{Q}/\mathbb{Z}$ if k_v is non-archimedean, it is equal to the subgroup $\{0, \frac{1}{2}\}$ of order 2 of $\mathbb{Q}/\mathbb{Z}$ if k_v equals $\mathbb{R}$, and it equals the subgroup 0 of $\mathbb{Q}/\mathbb{Z}$ if k_v is equal to $\mathbb{C}$.

For the global field k, we then have canonical homomorphisms ("invariant maps") for all places v of k

$$\mathrm{inv}_v : \mathrm{Br}(k_v) \to \mathbb{Q}/\mathbb{Z}.$$

The relation between Br(k) and the groups Br(k_v), for all the local fields k_v associated to k, is given by the exact sequence obtained from class field theory

$$0 \longrightarrow \mathrm{Br}(k) \longrightarrow \bigoplus_v \mathrm{Br}(k_v) \xrightarrow{\sum \mathrm{inv}_v} \mathbb{Q}/\mathbb{Z} \longrightarrow 0$$

where $\sum \mathrm{inv}_v$ denotes the sum of all the homomorphisms inv_v [66, Chapitre X, §7, p.171].

(5.4.18) Referring to diagram (5.4.15) and exactly as in (5.4.5), for any $f \in \mathrm{Br}(X)$ define, where for $x \in X(\mathbb{A}_k)$ the element $x_v \in X(k_v)$ is obtained restricting to k_v,

$$X(\mathbb{A}_k)^f = \{x \in X(\mathbb{A}_k) | \sum_v \mathrm{inv}((x_v)(f)) = 0\}.$$

Define

$$X(\mathbb{A}_k)^{\mathrm{Br}} = \bigcap_{f \in \mathrm{Br}(X)} X(\mathbb{A}_k)^f.$$

Then we have the inclusion

$$X(k) \subseteq X(\mathbb{A}_k)^{\mathrm{Br}}.$$

Definition 5.4.19 Let X be a separated k-scheme of finite type where k is a global field.

(i) If $X(\mathbb{A}_k) \neq \emptyset$ and $X(\mathbb{A}_k)^{\mathrm{Br}} = \emptyset$ then X has a *Brauer-Manin obstruction to the Hasse principle*.
(ii) If X/k is projective, $X(k) \neq \emptyset$, and $X(\mathbb{A}_k)^{\mathrm{Br}} \neq X(\mathbb{A}_k)$ then X has a *Brauer-Manin obstruction to weak approximation*.

Remarks 5.4.20

(i) The Brauer-Manin obstruction provides a workable method of showing that the Hasse principle fails or that weak approximation fails for an algebraic variety (see [39] for the effectivity of this method).

A classical example of a failure of the Hasse principle is the curve of Reichardt-Lind X which is the smooth compactification of the curve over the rational numbers $\mathbb{Q}$ defined by

$$2y^2 = x^4 - 17.$$

It is easily checked that $X(\mathbb{A}_\mathbb{Q}) \neq \emptyset$, where $\mathbb{A}_\mathbb{Q}$ is the adèle ring of $\mathbb{Q}$. Furthermore one can show that $X(\mathbb{A}_\mathbb{Q})^{\mathrm{Br}} = \emptyset$ so that X has a Brauer-Manin obstruction to the

Hasse principle and hence $X(\mathbb{Q}) = \emptyset$. More precisely, if $(y, 17) \in \mathrm{Br}(X)$ is the quaternion algebra defined by $i^2 = y$, $j^2 = 17$, $ij = -ji$ then one can show that $X(\mathbb{A}_\mathbb{Q})^{(y,17)} = \emptyset$ (see [13, Chapter 12, p. 260] for more details).

This curve X is a torsor under the action of its Jacobian J, which is an elliptic curve over $\mathbb{Q}$. The curve X then represents a non-trivial element of the Shafarevich-Tate group $\text{III}(J/\mathbb{Q})$ of J.

(ii) The Brauer-Manin method is equivalent to the descent method applied to all PGL_n-torsors for all positive integers n [76, Propn. 5.3.4]. In this way, the descent method applied to torsors of all group schemes G is stronger than the Brauer-Manin method. See also [13].

Exercise 5.4.21 The Reichardt Lind curve $X/\mathbb{Q}$ of Remark 5.4.20(i) can be shown to contravene the Hasse principle via quadratic reciprocity as follows. The projective smooth curve $X/\mathbb{Q}$ is defined to be the smooth compactification of its affine open subscheme U cut out in the affine plane by the equation $2y^2 = x^4 - 17$.

(i) Show that $X/\mathbb{Q}$ is isomorphic to the smooth projective curve in $\mathbb{P}_3/\mathbb{Q}$ given by the homogeneous equations $2D^2 = A^2 - 17B^2$, $AB = C^2$, and the open immersion $U \to X$ is obtained by $(y, x) \mapsto [A : B : C : D] = [x^2 : 1 : x : y]$. Show that $X(\mathbb{Q}_p)$ is non-empty for all places p of $\mathbb{Q}$ including the real place $p = \infty$. Show that $X(\mathbb{Q}) = U(\mathbb{Q})$.

(ii) Assume that P is a point of $X(\mathbb{Q})$. Show that there are integers $y_0, z_0, z_1 \in \mathbb{Z}$ provided by the coordinates of P such that

$$2y_0^2 = z_0^4 - 17z_1^4$$

where $\gcd(z_0, z_1) = 1$.

(iii) Show that every odd prime number p dividing y_0 satisfies $\left(\frac{17}{p}\right) = 1$ where $\left(\frac{a}{b}\right)$ is the Jacobi symbol.

(iv) As $\left(\frac{-1}{17}\right) = \left(\frac{2}{17}\right) = 1$, obtain from (iii) using quadratic reciprocity that every prime dividing y_0 and y_0 itself are quadratic residues mod 17. Conclude from $2y_0^2 = z_0^4 - 17z_1^4$ that 2 is a 4th power mod 17 which is false and so $X(\mathbb{Q})$ is empty.

5.5 The Classical Selmer Structure of an Abelian Variety

(5.5.1) Let

k be a global field;
A/k be an abelian variety;
l be a prime number different from the characteristic of k;

$T = T_l(A)$ be l-adic Tate module of A/k with coefficient ring $R = \mathbb{Z}_l$;
$A[\phi]/k$ be the finite subgroup scheme of A which is the kernel of an
    isogeny $\phi : A \to B$ of abelian varieties over k;
$[n]$ be the isogeny $[n] : A \to A$ of multiplication by n for any integer $n \in \mathbb{Z}$.

The Shafarevich-Tate Group

(5.5.2) The *Shafarevich-Tate group* $\text{III}(A/k)$ of A/k is defined as

$$\text{III}(A/k) = \ker\{H^1(k, A) \to \prod_{v \in \Sigma_k} H^1(k_v, A)\}.$$

The groups $H^1(k, A)$, $H^1(k_v, A)$ can be taken to be either Galois cohomology, étale cohomology, or fppf cohomology as these groups are canonically isomorphic for this case where A is an abelian variety over a field k (for étale and fppf cohomologies this follows from [51, Theorem III.3.9]).

(5.5.3) The group $H^1(k, A)$ is that of the torsors, that is to say principal homogenous spaces, over k of the group scheme A/k (cf. (5.4.10)). The group $\text{III}(A/k)$ is then the group of torsors over k of A/k which are trivialized (i.e. have a rational point) over every completion of k. In this way, $\text{III}(A/k)$ is the group of torsors over k of A/k which all fail the Hasse principle except for the trivial torsor (Definition 5.4.3).

This group $\text{III}(A/k)$ is known to be torsion of cofinite type, that is to say $\text{III}(A/k)_m$ is a finite group for every non-zero integer m, and the union of these subgroups is the whole of $\text{III}(A/k)$ (see Corollary 5.6.4 and Proposition 5.6.5 or [60]).

Remarks 5.5.4

(i) The group $\text{III}(A/k)$ of an abelian variety A/k can be generalized in many ways. For example, if M is a G_k-module (not necessarily an R-representation of the type of Sect. 4.1) then for all non-negative integers we can define

$$\text{III}^i(M) = \ker\{H^i(k, M) \to \prod_{v \in \Sigma_k} H^i(k_v, M)\}.$$

For a generalization of $\text{III}(A/k)$ to motives, see [7].

(ii) An example of (i) is the following. Let k be an algebraic number field. Let R be the integral closure in k^{sep} of the ring of integers of k, that is to say R is the ring of all algebraic integers. For a non-archimedean place v of k let $R(v)$ be the (non-discrete) valuation ring of k_v^{sep}, the separable closure of k_v; for an archimedean place v of k let $R(v) = k_v^{\text{sep}}$.

5.5 The Classical Selmer Structure of an Abelian Variety

The group of units R^* of the ring R is a discrete G_k-module and the group of units $R(v)^*$ of $R(v)$ is a discrete G_{k_v}-module, where we forget the v-topology on $R(v)^*$. Define

$$\text{III} = \ker\{H^1(k, R^*) \to \prod_{v \in \Sigma_k} H^1(k_v, R(v)^*)\}$$

where the H^1s refer to the profinite group cohomology of discrete Galois modules. It is not difficult to show that there is a canonical isomorphism

$$\text{III} \cong \text{Cl}(k)$$

where $\text{Cl}(k)$ is the divisor class group of k. In this way the Shafarevich-Tate group of an abelian variety is an analogue of the divisor class group of a number field. [For more details, see http://math.uchicago.edu/~may/REU2016/REUPapers/Kailasa.pdf.]

(iii) If M is a commutative finite group scheme over k then

$$\text{III}^1(M) = \ker\{H^1(k, M) \to \prod_{v \in \Sigma_k} H^1(k_v, M)\}$$

is a finite group (for the proof, see [48, Chap. I, Theorem 4.10] when k is a number field and [24, Propn. 4.6] when k is a function field).

The Finite Condition for Abelian Varieties

Proposition 5.5.5 *Let $\phi : A \to B$ be an isogeny of abelian varieties over k where the degree of ϕ is coprime to the characteristic of k and $A[\phi]/k$ is the finite group scheme kernel of ϕ. Let $\mathfrak{p}$ be a non-archimedean place of k where A/k has good reduction and assume that $\mathfrak{p}$ does not divide the degree of ϕ. Then the image of the Kummer map*

$$B(k_\mathfrak{p})/\phi(A(k_\mathfrak{p})) \longrightarrow H^1(k_\mathfrak{p}, A[\phi])$$

is the finite condition subgroup $H^1_f(k_\mathfrak{p}, A[\phi])$.

Furthermore for the l-adic Tate module T of A where $\mathfrak{p}$ does not divide l then the image of the Kummer map

$$A(k_\mathfrak{p}) \otimes_\mathbb{Z} \mathbb{Z}_l \to H^1(k_\mathfrak{p}, T)$$

is the finite condition subgroup $H^1_f(k_\mathfrak{p}, T)$.

Proof Note that the excluded cases where $\mathfrak{p}$ divides the degree of ϕ or that $\mathfrak{p}$ divides l only occur when k is a number field. Let the place $\mathfrak{p}$ of k be as in the statement of the lemma. The Kummer map of the isogeny ϕ on A provides the exact sequence of cohomology

$$0 \longrightarrow B(k_\mathfrak{p})/\phi(A(k_\mathfrak{p})) \longrightarrow H^1(k_\mathfrak{p}, A[\phi]) \longrightarrow H^1(k_\mathfrak{p}, A)_\phi \longrightarrow 0 \quad (5.5.6)$$

where $H^1(k_\mathfrak{p}, A)_\phi$ is the subgroup of $H^1(k_\mathfrak{p}, A)$ annihilated by ϕ. Let $x \in H^1(k_\mathfrak{p}, A[\phi])$ be a cohomology class which is in the image of $B(k_\mathfrak{p})/\phi(A(k_\mathfrak{p}))$ under the Kummer map. Then the image of x in $H^1(k_\mathfrak{p}, A)_\phi$ is zero. Therefore there is a point $y \in A(k_\mathfrak{p}^{\text{sep}})$ such that the cohomology class x is represented by the 1-cocycle

$$\sigma \mapsto y^\sigma - y, \quad G_{k_\mathfrak{p}} \to A(k_\mathfrak{p}^{\text{sep}}).$$

Here we have $y^\sigma - y \in A[\phi]$ for $\sigma \in G_{k_\mathfrak{p}}$ because $x \in H^1(k, A[\phi])$.

Let $A^0/\kappa(\mathfrak{p})$ be the reduction of A/k at the place $\mathfrak{p}$. Denote the reduction map on points by

$$z \mapsto z^0, A(k^{\text{sep}}) \longrightarrow A^0(\kappa(\mathfrak{p})^{\text{sep}}).$$

Under the hypotheses of the proposition, the finite group scheme $A^0[\phi]/\kappa(\mathfrak{p})$ is étale. Then the reduction map on ϕ-torsion points

$$f : A[\phi](k^{\text{sep}}) \longrightarrow A^0[\phi](\kappa(\mathfrak{p})^{\text{sep}})$$

is an isomorphism. Furthermore, letting I be the inertia subgroup of $G_{k_\mathfrak{p}}$, we have that I acts trivially on A^0. Therefore we have

$$(y^\sigma - y)^0 = (y^0)^\sigma - y^0 = 0, \text{ for } \sigma \in I.$$

Therefore $y^\sigma - y$ for $\sigma \in I$ is in the kernel of the reduction at $\mathfrak{p}$. But $y^\sigma - y \in A[\phi]$ and by the isomorphism f we obtain that $y^\sigma - y = 0$ for all $\sigma \in I$. Therefore y is defined over some unramified field extension of $k_\mathfrak{p}$. That is to say x is an unramified cohomology class and therefore the cohomology classes of $H^1(k_\mathfrak{p}, A[\phi])$ in the image of $B(k_\mathfrak{p})/\phi(A(k_\mathfrak{p}))$ under the Kummer map are unramified.

Suppose now that $x \in H^1(k, A[\phi])$ is an unramified cohomology class, that is to say x is in the kernel of the homomorphism

$$H^1(k_\mathfrak{p}, A[\phi]) \longrightarrow H^1(k_\mathfrak{p}^{\text{unr}}, A[\phi]).$$

By the inflation-restriction sequence, the kernel of this homomorphism is isomorphic to

$$H^1(k_\mathfrak{p}^{\text{unr}}/k_\mathfrak{p}, A[\phi]).$$

Then x can be represented by a cocycle

$$\sigma \mapsto x_\sigma, \quad \text{Gal}(k_\mathfrak{p}^{\text{unr}}/k_\mathfrak{p}) \longrightarrow A[\phi](k_\mathfrak{p}^{\text{unr}}).$$

5.5 The Classical Selmer Structure of an Abelian Variety

Here $\text{Gal}(k_{\mathfrak{p}}^{\text{unr}}/k_{\mathfrak{p}})$ is isomorphic to $\widehat{\mathbb{Z}}$, the profinite completion of $\mathbb{Z}$, and the group $\text{Gal}(k_{\mathfrak{p}}^{\text{unr}}/k_{\mathfrak{p}})$ is topologically generated by the arithmetic Frobenius element Fr and this provides the element

$$x_{\text{Fr}} \in A[\phi](k_{\mathfrak{p}}^{\text{unr}}).$$

Let $A^0/\kappa(\mathfrak{p})$, $B^0/\kappa(\mathfrak{p})$ be the reductions of the Néron models of A, B at $\mathfrak{p}$ so that ϕ induces an isogeny $\phi: A^0 \to B^0$. The map of reduction at $\mathfrak{p}$

$$A[\phi](k^{\text{unr}}) \longrightarrow A^0[\phi](\kappa(\mathfrak{p})^{\text{sep}})$$

is then an isomorphism of groups.

The Frobenius relative to $\kappa(\mathfrak{p})$ is then the reduction at $\mathfrak{p}$ of Fr. We then have the Lang isogeny of the group scheme $A^0/\kappa(\mathfrak{p})$

$$A^0 \longrightarrow A^0 : z \mapsto (\text{Fr} - \text{id})z.$$

This isogeny is surjective so there is $z \in A^0(\kappa(\mathfrak{p})^{\text{sep}})$ so that

$$(\text{Fr} - \text{id})z = x_{\text{Fr}}^0$$

where $x_{\text{Fr}}^0 \in A[\phi](\kappa(\mathfrak{p})^{\text{sep}})$ is the reduction at $\mathfrak{p}$ of the point x_{Fr}. It follows that $(\text{Fr} - \text{id})\phi(z) = \phi(x_{\text{Fr}}^0) = 0$ and so $\phi(z)$ is a point of B^0 rational over $\kappa(\mathfrak{p})$ that is to say $\phi(z) \in B^0(\kappa(\mathfrak{p}))$.

We can then lift $z \in A^0(\kappa(\mathfrak{p})^{\text{sep}})$ to an element $y \in A(k_{\mathfrak{p}}^{\text{unr}})$. We then have

$$y^{\text{Fr}} - y - x_{\text{Fr}}$$

is in the kernel of the reduction at $\mathfrak{p}$ and that $\phi(y) \in B(k_{\mathfrak{p}})$. Furthermore as $\phi(y) \in B(k_{\mathfrak{p}})$ we have that $\phi(y^{\text{Fr}} - y) = (\phi(y))^{\text{Fr}} - \phi(y) = \phi(y) - \phi(y) = 0$ so that $y^{\text{Fr}} - y \in A[\phi](k_{\mathfrak{p}}^{\text{unr}})$. As $x_{\text{Fr}} \in A[\phi](k_{\mathfrak{p}}^{\text{unr}})$ and $y^{\text{Fr}} - y - x_{\text{Fr}}$ is in the kernel of the reduction at $\mathfrak{p}$ and the reduction map $A[\phi](k_{\mathfrak{p}}^{\text{unr}}) \to A^0(\kappa(\mathfrak{p})^{\text{sep}})$ is a group isomorphism, it follows that

$$y^{\text{Fr}} - y - x_{\text{Fr}} = 0.$$

Therefore the cocycle $\sigma \mapsto x_\sigma$ is equal to the cocycle $\sigma \mapsto y^\sigma - y$ which represents a cohomology class in $H^1(k, A)_\phi$. Therefore by the Kummer exact sequence (5.5.6), the cohomology class of the cocycle $\sigma \mapsto x_\sigma$ is in the image of $B(k_{\mathfrak{p}}) \to H^1(k_{\mathfrak{p}}, A[\phi])$, as required.

By Proposition 1.3.10, there are isomorphisms

$$H^1(k_{\mathfrak{p}}, T) = \varprojlim H^1(k_{\mathfrak{p}}, A[l^m]), \; H^1(k_{\mathfrak{p}}^{\text{unr}}, T) = \varprojlim H^1(k_{\mathfrak{p}}^{\text{unr}}, A[l^m])$$

where m runs over all positive integers. By definition, we have

$$H^1_f(k_\mathfrak{p}, T) = \ker(H^1(k_\mathfrak{p}, T) \longrightarrow H^1(k_\mathfrak{p}^{\mathrm{unr}}, T))$$

where $k_\mathfrak{p}^{\mathrm{unr}}$ is the maximal unramified extension of k in k^{sep}.

By the first part, the image of the Kummer map $A(k_\mathfrak{p})/l^m A(k_\mathfrak{p}) \longrightarrow H^1(k_\mathfrak{p}, A[l^m])$ is the finite condition subgroup $H^1_f(k_\mathfrak{p}, A[l^m])$. We then obtain the commutative diagram

$$\begin{array}{ccccccc}
0 & \longrightarrow & A(k_\mathfrak{p})/l^m A(k_\mathfrak{p}) & \longrightarrow & H^1(k_\mathfrak{p}, A[l^m]) & \longrightarrow & H^1(k_\mathfrak{p}^{\mathrm{unr}}, A[l^m]) \\
& & \uparrow & & \uparrow & & \uparrow \\
0 & \longrightarrow & A(k_\mathfrak{p}) \otimes_\mathbb{Z} \mathbb{Z}_l & \longrightarrow & H^1(k_\mathfrak{p}, T) & \longrightarrow & H^1(k_\mathfrak{p}^{\mathrm{unr}}, T)
\end{array}$$

where the top row is exact and the vertical and bottom arrows are the evident homomorphisms. Taking limits of directed inverse systems is left exact, and as the groups in the bottom row are the inverse limits of the groups in the top row, we obtain that the bottom row is exact which completes the proof. □

Selmer Structures for Abelian Varieties

(5.5.7) Let $\phi : A \to B$ be an isogeny of abelian varieties over the global field k. Define a Selmer structure $\mathcal{F}$ on the finite group scheme $A[\phi]/k$ by taking:

- $S(\mathcal{F})$ to be the finite set of places of k where $A[\phi]$ is ramified (i.e. the places of bad reduction of A) together with the places dividing $\deg(\phi)$, if any, and the archimedean places;
- For every place $\mathfrak{p} \notin S(\mathcal{F})$ of k then the local condition $H^1_\mathcal{F}(k_\mathfrak{p}, A[\phi])$ is defined to be the image of the Kummer map

$$A(k_\mathfrak{p}) \to H^1(k_\mathfrak{p}, A[\phi]);$$

note that if $\deg(\phi)$ is coprime to the characteristic of k then this local condition $H^1_\mathcal{F}(k_\mathfrak{p}, A[\phi])$ for $\mathfrak{p} \notin S(\mathcal{F})$ is just the unramified local condition $H^1_f((k_\mathfrak{p}, A[\phi])$ by Proposition 5.5.5;
- For places $\mathfrak{p} \in S(\mathcal{F})$ we again define the local condition $H^1_\mathcal{F}(k_\mathfrak{p}, A[\phi])$ to be the image of the respective Kummer map $A(k_\mathfrak{p}) \to H^1(k_\mathfrak{p}, A[\phi])$. (See Remark 4.3.11(i) for comments on local conditions at "bad" places $\mathfrak{p}$.)

(5.5.8) The Selmer module (Definition 5.1.2) $H^1_\mathcal{F}(k, A[\phi])$ for this Selmer structure is then the *classical ϕ-Selmer group* which fits into the exact sequence

$$0 \longrightarrow B(k)/\phi(A(k)) \longrightarrow H^1_\mathcal{F}(k, A[\phi]) \longrightarrow \Sha(A)_\phi \longrightarrow 0 \qquad (5.5.9)$$

where $Ш(A)_\phi$ is the kernel of the induced homomorphism $\phi : Ш(A) \to Ш(B)$ of Shafarevich-Tate groups.

(5.5.10) The Tate module T of A is $\mathbb{Z}_l$-representation of G_k and is a free $\mathbb{Z}_l$-module of rank equal to $2\dim(A)$.

Define a similar Selmer structure $\mathcal{F}$ on T by taking

- $S(\mathcal{F})$ to be the finite set of places of k where T is ramified (i.e. the places of bad reduction of A) together with the places above l, if k is a number field, and the archimedean places, if any;
- For every place $\mathfrak{p} \notin S(\mathcal{F})$ of k then the local condition $H^1_{\mathcal{F}}(k_\mathfrak{p}, T)$ is defined to be the image of the Kummer map

$$A(k_\mathfrak{p}) \otimes_\mathbb{Z} \mathbb{Z}_l \to H^1(k_\mathfrak{p}, T);$$

- For places $\mathfrak{p} \in S(\mathcal{F})$ we again define the local condition $H^1_{\mathcal{F}}(k_\mathfrak{p}, T)$ to be the image of the Kummer map $A(k_\mathfrak{p}) \otimes_\mathbb{Z} \mathbb{Z}_l \to H^1(k_\mathfrak{p}, T)$.

(5.5.11) Again the Selmer module $H^1_{\mathcal{F}}(k, T)$ for this Selmer structure fits into an exact sequence

$$0 \longrightarrow A(k) \otimes_\mathbb{Z} \mathbb{Z}_l \longrightarrow H^1_{\mathcal{F}}(k, T) \longrightarrow \varprojlim Ш_{l^n} \longrightarrow 0$$

and $H^1_{\mathcal{F}}(k, T)$ is the classical Selmer group. If $Ш$ is a finite group this exact sequence reduces to an isomorphism $A(k) \otimes_\mathbb{Z} \mathbb{Z}_l \cong H^1_{\mathcal{F}}(k, T)$.

We call this $\mathcal{F}$, in either of the two cases $A[\phi]$ or T, the *classical Selmer structure* of an abelian variety over a global field.

Remark 5.5.12 For the Selmer structure for the Tate module T of the abelian variety A/k defined above, if k is a number field and $\mathfrak{p}$ is a place of k lying over l so that $\mathfrak{p} \in S(\mathcal{F})$ then the local condition $H^1_{\mathcal{F}}(k_\mathfrak{p}, T)$ defined above as the image of the map $A(k_\mathfrak{p}) \otimes_\mathbb{Z} \mathbb{Z}_l \to H^1(k_\mathfrak{p}, T)$ coincides with the Bloch-Kato definition of $H^1_f(k_\mathfrak{p}, T)$ (see Remark 4.3.11(i)).

5.6 Finiteness of Classical Selmer Modules of Abelian Varieties

(5.6.1) Let $\phi : A \to B$ be an isogeny of abelian varieties over the global field k. Let $A[\phi]$ be the finite group scheme over k which is the kernel of ϕ. Let $\mathcal{F}$ be the classical Selmer structure on $A[\phi]$ (see (5.5.7)).

In this section, the groups $H^1(k, A)$, $H^1(k_v, A)$, $H^1(k, A[\phi])$, $H^1(k_v, A[\phi])$, etc. are Galois cohomology or étale cohomology, if k is a number field, and they are fppf cohomology if k is a global function field (sometimes this is explicitly stated, cf. (5.4.10)).

Lemma 5.6.2 *Let N be a finite commutative group scheme over a field F. There is a finite separable extension field F'/F over which $N \times_F F'$ has a composition series of F'-group schemes of the form $\mathbb{Z}/m\mathbb{Z}$, μ_m, and α_p where m is a positive integer and p is the characteristic of F.*

Proof The statement evidently holds for a finite étale F-group scheme. By passing to the Cartier dual it then also holds for finite commutative F-group schemes of multiplicative type.

By [53, proposition 11.4] there is an exact sequence of finite group schemes over F

$$0 \longrightarrow N^0 \longrightarrow N \longrightarrow \pi_0(N) \longrightarrow 0$$

where N^0 is a connected group scheme and $\pi_0(N)/F$ is an étale group scheme. Then by induction on the degree $\dim_F \Gamma(N, \mathcal{O}_N)$ we may reduce to the case where N is connected and has connected Cartier dual and in particular F is then a field of positive characteristic p. By [SGA3, Vol 2, Exposé XVII, Propn. 4.2.1] such a group scheme is a successive extension of group schemes isomorphic to α_p.

Proposition 5.6.3 *The Selmer module $H^1_\mathcal{F}(k, A[\phi])$ for the classical Selmer structure $\mathcal{F}$ on the finite group scheme $A[\phi]/k$ is finite.*

Proof The proof is in a number of steps. The longest part of this proof (Steps 2 to 7) is for the special case where k has positive characteristic which divides the degree of the isogeny ϕ.

Step 1. *If $A[\phi]$ is an étale k-group scheme then $H^1_\mathcal{F}(k, A[\phi])$ is finite.*

Let $\mathfrak{p}$ be a place of k not in the finite set $S(\mathcal{F})$; that is to say $\mathfrak{p}$ is a non-archimedean place of k not lying over the divisors of $\deg(\phi)$, when k is an algebraic number field, and is a place of good reduction of A/k.

The finite group scheme $A[\phi]/k$ is étale if k has characteristic zero or if it has characteristic $p > 0$ and p does not divide $\deg(\phi)$ [53, Proposition 11.7]. By Proposition 5.5.5, we then have that the cohomology classes of $H^1_\mathcal{F}(k, A[\phi])$ are unramified at all places of k outside $S(\mathcal{F})$. Furthermore, as $A[\phi]/k$ is étale, the group $H^1(K, A[\phi])$ can be taken to be Galois cohomology for any field K containing k.

The group scheme $A[\phi]/k$ is finite and étale; hence there is a finite Galois field extension k'/k for which $A[\phi](k^{\text{sep}})$ is a trivial $G_{k'}$-module. Take $\mathcal{F}'$ to be the classical Selmer structure on $A \times_k k'/k'$ so that $S(\mathcal{F}')$ is the finite set of places of k' lying over the elements of $S(\mathcal{F})$. The inflation-restriction sequence is in part

5.6 Finiteness of Classical Selmer Modules of Abelian Varieties

$$0 \longrightarrow H^1(\operatorname{Gal}(k'/k), A[\phi]^{G_{k'}}) \longrightarrow H^1(k, A[\phi]) \longrightarrow H^1(k', A[\phi])^{\operatorname{Gal}(k'/k)}.$$

Here $H^1(\operatorname{Gal}(k'/k), A[\phi]^{G_{k'}})$ is a finite group and we obtain that the homomorphism

$$H^1_{\mathcal{F}}(k, A[\phi]) \longrightarrow H^1_{\mathcal{F}'}(k', A[\phi])^{\operatorname{Gal}(k'/k)}$$

has finite kernel and to show that $H^1_{\mathcal{F}}(k, A[\phi])$ is finite it is sufficient to show that $H^1_{\mathcal{F}'}(k', A[\phi])$ is finite. Therefore we may replace k' by k. But then G_k acts trivially on $A[\phi]$ and therefore

$$H^1(k, A[\phi]) = \operatorname{Hom}(G_k, A[\phi]) = \operatorname{Hom}(G_k^{\operatorname{ab}}, A[\phi])$$

where G_k^{ab} is the abelianization of G_k and $H^1_{\mathcal{F}}(k, A[\phi])$ is a subgroup of $H^1(k, A[\phi])$ consisting of elements unramified outside $S(\mathcal{F})$. Let K be the maximal abelian field extension of k of exponent $\deg(\phi)$ which is unramified outside $S(\mathcal{F})$. Then we have an injection

$$H^1_{\mathcal{F}}(k, A[\phi]) \longrightarrow \operatorname{Hom}(\operatorname{Gal}(K/k), A[\phi]).$$

By the theory of global fields, the maximal abelian field extension K of k of exponent $\deg(\phi)$ which is unramified outside $S(\mathcal{F})$ is a finite extension of k. Therefore $\operatorname{Hom}(\operatorname{Gal}(K/k), A[\phi])$ is finite and hence $H^1_{\mathcal{F}}(k, A[\phi])$ is too.

Step 2. *Reduction to the case where k has characteristic $p > 0$ and $A[\phi]/k$ is isomorphic to one of the two group schemes μ_p, α_p.*

The finite group scheme $A[\phi]/k$ is étale if k has characteristic zero or if k has characteristic $p > 0$ and p does not divide $\deg(\phi)$ [53, Proposition 11.7]. Therefore by step 1, we may assume that k has characteristic $p > 0$ and p divides $\deg(\phi)$.

Let $\mathcal{F}$ be the classical Selmer structure on $A[\phi]$. Then, as k has positive characteristic, $S(\mathcal{F})$ consists exactly of the places of bad reduction of A and is therefore independent of the isogeny $\phi : A \to B$ and depends only on A.

By Lemma 5.6.2, there is a finite separable Galois field extension k'/k where $A[\phi] \times_k k'$ has a composition series consisting of components isomorphic to the k'-group schemes $\mathbb{Z}/m\mathbb{Z}, \mu_m$, and α_p, for positive integers m. Therefore ϕ over k' becomes an isogeny of abelian varieties $\phi \times \operatorname{id} : A \times_k k' \to B \times_k k'$ whose kernel is isomorphic to $A|\phi] \times k'$ which has the above composition series. Let $\mathcal{F}'$ be the classical Selmer structure on $A|\phi] \times k'$ induced by $\mathcal{F}$ so that $S(\mathcal{F}')$ consists of those places of k' lying over those of $S(\mathcal{F})$.

We then have the Selmer modules $H^1_{\mathcal{F}'}(k', A[\phi]), H^1_{\mathcal{F}}(k, A[\phi])$. Write $G = \operatorname{Gal}(k'/k)$; then there is a Hochschild-Serre spectral sequence for fppf cohomology (by Milne [51, Chapter 3, theorem 2.20] for the case of étale cohomology; the same demonstration holds for fppf cohomology [51, Chapter 3, Remark 2.21(a)])

$$H^i(G, H^j_{\text{fppf}}(k', A[\phi])) \Longrightarrow H^{i+j}_{\text{fppf}}(k, A[\phi]).$$

This gives an exact sequence of low degree terms

$$0 \longrightarrow H^1(G, H^0_{\text{fppf}}(k', A[\phi])) \longrightarrow H^1_{\text{fppf}}(k, A[\phi]) \longrightarrow H^1_{\text{fppf}}(k', A[\phi])^G \longrightarrow \ldots$$

The group $H^1(G, H^0_{\text{fppf}}(k', A[\phi])) = H^1(G, A[\phi](k'))$ is evidently finite and hence

$$f : H^1_{\text{fppf}}(k, A[\phi]) \longrightarrow H^1_{\text{fppf}}(k', A[\phi])^G$$

has finite kernel.

For any place v of k' we have the homomorphism

$$A(k'_v) \longrightarrow H^1(k', A[\phi])$$

whose image is the subgroup $H^1_{\mathcal{F}'}(k'_v, A[\phi])$ of $H^1(k'_v, A[\phi])$ for all places v of k'. As f has finite kernel, it follows that the homomorphism

$$H^1_{\mathcal{F}}(k, A[\phi]) \longrightarrow H^1_{\mathcal{F}'}(k', A[\phi])^G$$

has finite kernel. Therefore to show that $H^1_{\mathcal{F}}(k, A[\phi])$ is finite it is sufficient to show that $H^1_{\mathcal{F}'}(k', A[\phi])$ is finite. We may therefore reduce to the case where $k = k'$ and $A[\phi]/k$ has a composition series consisting of components isomorphic to the k-group schemes α_p, $\mathbb{Z}/m\mathbb{Z}$, μ_m, for positive integers m.

By [53, proposition 11.4] there is an exact sequence of finite group schemes over k

$$0 \longrightarrow A[\phi]^0 \longrightarrow A[\phi] \longrightarrow \pi_0(A[\phi]) \longrightarrow 0$$

where $A[\phi]^0$ is a connected group scheme and $\pi_0(A[\phi])/k$ is an étale group scheme. This provides the exact sequence of cohomology

$$H^1_{\mathcal{F}}(k, A[\phi]^0) \longrightarrow H^1_{\mathcal{F}}(k, A[\phi]) \longrightarrow H^1_{\mathcal{F}}(k, \pi_0(A[\phi]))$$

By Step 1, $H^1_{\mathcal{F}}(k, \pi_0(A[\phi]))$ is a finite group. Therefore to show $H^1_{\mathcal{F}}(k, A[\phi])$ is finite it suffices to show that $H^1_{\mathcal{F}}(k, A[\phi]^0)$ is a finite group and we may assume that $A[\phi]/k$ is connected and has a composition series over k consisting of components isomorphic to the k-group schemes μ_p and α_p.

The isogeny ϕ is then a composite of isogenies over k of degree p with the composition factors μ_p and α_p. Hence ϕ factors as $\phi = \chi \circ \psi$ where $\psi : A \to C$ and $\chi : C \to B$ are isogenies of abelian varieties where ψ has degree p and χ has degree p^{a-1}. Hence we have an exact sequence of finite group schemes over k

5.6 Finiteness of Classical Selmer Modules of Abelian Varieties

$$0 \to N \to A[\phi] \to M \to 0$$

where $N \to A[\phi]$ is a closed immersion, N has degree p, and M has degree p^{a-1}. We then obtain the exact sequence of cohomology

$$H^1_{\mathcal{F}}(k, N) \longrightarrow H^1_{\mathcal{F}}(k, A[\phi]) \longrightarrow H^1_{\mathcal{F}}(k, M).$$

If the two outside groups of this exact sequence are finite then $H^1_{\mathcal{F}}(k, A[\phi])$ is finite, and so by induction on a, we are reduced to showing the finiteness of $H^1_{\mathcal{F}}(k, A[\phi])$ where $\phi : A \to A'$ is an isogeny of abelian varieties over k of degree p and $A[\phi]/k$ is isomorphic to either μ_p or α_p.

Step 3. *Definition of the restricted topological product $\prod' H^1(k_v, N)$ for a finite commutative group scheme N/k.*

The field k is the function field of a smooth projective irreducible algebraic curve $X/\mathbb{F}$ for some finite field $\mathbb{F}$ of characteristic p. Let R_v be the ring of valuation integers of k_v at the place v.

Let $\mathcal{M}$ be a finite flat group scheme over R_v and let M/k_v be its generic fibre. Then the homomorphism

$$H^1(R_v, \mathcal{M}) \to H^1(k_v, M)$$

is injective. This holds because the elements of $H^1(k_v, M)$, $H^1(R_v, \mathcal{M})$ are isomorphic to the groups of principal homogeneous spaces for M, resp. $\mathcal{M}$, over the base schemes Spec k_v, resp. Spec R_v [51, Chapter III, Cor. 4.7, p.123]; it is clear that a principal homogeneous space for $\mathcal{M}$ over R_v which has a point in k_v already has a point in R_v; hence this homomorphism is injective.

The group $H^1(R_v, \mathcal{M})$ is then isomorphic to the Čech cohomology group $\varinjlim H^1(R'_v/R_v, \mathcal{M})$ where R'_v runs over the valuation rings of all finite extension fields of k_v by [51, Chap. III, Cor. 2.10]. The group $H^1(k_v, M)$ is equipped with its standard topology (Sect. 4.2, paragraph (4.2.2) *et seq.*) and it follows that $H^1(R_v, \mathcal{M})$ under the injection $H^1(R_v, \mathcal{M}) \to H^1(k_v, M)$ is an open subgroup of $H^1(k_v, M)$.

Suppose now that N/k is a finite commutative group scheme. For some open subscheme U of X, there is a finite commutative flat group scheme $\mathcal{N}/U$ such that the generic fibre of $\mathcal{N}/U$ is k-isomorphic to N. Define $\prod' H^1(k_v, N)$ to be the restricted product of the groups $H^1(k_v, N)$, for all places v of X, with respect to the family of open subgroups $H^1(R_v, \mathcal{N}_v)$ where $\mathcal{N}_v = \mathcal{N} \times_U \text{Spec } R_v$, where v runs through the closed points of U. It is clear that there are only finitely many places v of X which do not correspond to closed points of U.

The restricted product $\prod' H^1(k_v, N)$ is then well defined and independent of the choice of $U, \mathcal{N}$ because for another such pair $U', \mathcal{N}'$ then $\mathcal{N}, \mathcal{N}'$ become isomorphic on some non-empty open subscheme contained in $U \cap U'$.

Step 4. *Suppose that $A[\phi]/k$ is k-isomorphic to one of the group schemes μ_p, α_p. The restriction homomorphisms $f_v : H^1(k, A[\phi]) \to H^1(k_v, A[\phi])$ for all places v define an injective homomorphism*

$$f : H^1(k, A[\phi]) \to \prod_v' H^1(k_v, A[\phi]).$$

The field k is the function field of a smooth projective irreducible algebraic curve $X/\mathbb{F}$ for some finite field $\mathbb{F}$ of characteristic p. There is an open subscheme U of X and a finite flat commutative group scheme N/U whose generic fibre is isomorphic to $A[\phi]/k$.

The two group schemes α_p, μ_p are kernels of isogenies $F : G \to G$ of affine commutative smooth group schemes G/k; more precisely there are the following exact sequences of commutative group schemes over $\mathbb{F}_p$, where $\mathbb{G}_a = \operatorname{Spec} \mathbb{F}_p[x]$ and $\mathbb{G}_m = \operatorname{Spec} \mathbb{F}_p[x, x^{-1}]$,

$$0 \to \alpha_p \longrightarrow \mathbb{G}_a \xrightarrow{x \mapsto x^p} \mathbb{G}_a \longrightarrow 0$$

$$0 \longrightarrow \mu_p \longrightarrow \mathbb{G}_m \xrightarrow{x \mapsto x^p} \mathbb{G}_m \longrightarrow 0.$$

That is to say, G can be taken to be either $\mathbb{G}_m$ or $\mathbb{G}_a$ over $\mathbb{F}_p$ and $F : G \to G$ is one of the above isogenies given by taking pth powers. Taking fppf cohomology of these exact sequences and writing H^1 in place of H^1_{fppf}, we obtain topological and algebraic isomorphisms for any field K containing k, as $H^1(K, \mathbb{G}_a) = H^1(K, \mathbb{G}_m) = 0$

$$H^1(K, \alpha_p) \cong K/K^p, \quad H^1(K, \mu_p) \cong K^*/K^{*p}.$$

It follows that $H^1(K, A[\phi])$ is isomorphic to $G(K)/F(G(K))$ for any field K containing k. As $G(k) \cong k^*$ or k, it follows that that an element of $H^1(k, A[\phi])$, represented by an element of k^* or k, is contained in $H^1(R_v, N)$ for all but finitely many v. Hence the homomorphism f does exist.

To show that f is injective, for α_p the homomorphism f takes the form

$$f : k/k^p \to \prod_v' k_v/k_v^p$$

and for μ_p the homomorphism f takes the form

$$f : k^*/k^{*p} \to \prod_v' k_v^*/k_v^{*p}.$$

Both these maps are injective because if m is a positive integer then an element a of the global function field k belongs to k_v^m for every place v of k if and only if $a \in k^m$ [2, Chapt. IX, §1, Theorem 1.]

5.6 Finiteness of Classical Selmer Modules of Abelian Varieties

Step 5. *Under the hypotheses of Step 4, the injective homomorphism*

$$f : H^1(k, A[\phi]) \to \prod_v' H^1(k_v, A[\phi])$$

has discrete image.

We have topological and algebraic isomorphisms

$$\prod' H^1(k_v, \mathbb{A}[\phi]) \cong \prod' G(k_v)/FG(k_v) \cong G(\mathbb{A})/F(G(\mathbb{A}))$$

and

$$H^1(k, A[\phi]) \cong G(k)/F(G(k)).$$

The map f is then

$$f : G(k)/F(G(k)) \to G(\mathbb{A})/F(G(\mathbb{A})).$$

The image of f is isomorphic to $(G(k) + F(G(\mathbb{A})))/F(G(\mathbb{A}))$ where the $+$ here denotes the composition law on G.

Let R_v be the valuation ring of k_v and let $R = \prod_v R_v \subseteq \mathbb{A}$ be the subgroup of integral adèles which is an open subgroup of $\mathbb{A}$. Let

$$M = (G(R) + F(G(\mathbb{A}))) \cap G(k).$$

We have a map

$$M \to G(\mathbb{A})/(G(R) + G(k))$$

$$m = a + F(b) \mapsto b \mod G(R) + G(k), \quad a \in R, b \in \mathbb{A}, m \in M$$

This is well defined and has kernel $F(G(k))$; therefore $M/F(G(k))$ is isomorphic to a subgroup of $G(\mathbb{A})/(G(R) + G(k))$. Furthermore, the group $M/F(G(k))$ is finite which we check for the two group schemes α_p, μ_p:

- Suppose that $A[\phi] \cong \alpha_p$ (i.e. $G = \mathbb{G}_a$). The topological group $\mathbb{A}/k$ is compact in the quotient topology [10, Chapter 2, §14, Theorem, pp. 64-65] and R is an open subgroup of $\mathbb{A}$. Therefore the group $\mathbb{A}/(R + k) \cong G(\mathbb{A})/(G(R) + G(k))$ is finite. As this last group has a subgroup isomorphic to M/k^p then M/k^p is also finite;
- Suppose that $A[\phi] \cong \mu_p$ (i.e. $G = \mathbb{G}_m$). The group $\mathbb{A}^*/(R^*.k^*) \cong G(\mathbb{A})/(G(R) + G(k))$ lies in the exact sequence

$$0 \longrightarrow \mathrm{Jac}^0(k) \longrightarrow \mathbb{A}^*/(R^*.k^*) \xrightarrow{deg} \mathbb{Z} \longrightarrow 0$$

where deg is the degree homomorphism which associates to a divisor on $X/\mathbb{F}$ its degree and $\mathrm{Jac}^0(\mathbb{F})$ is the group of $\mathbb{F}$-points on the Jacobian of the curve $X/\mathbb{F}$ where $\mathrm{Jac}^0(\mathbb{F})$ is a finite group. As M/k^{*p}, as a subgroup of $\mathbb{A}^*/(R^*.k^*)$, is in the kernel of deg, it follows that M/k^{*p} is isomorphic to a subgroup of $\mathrm{Jac}^0(k)$ and is therefore finite.

Therefore we have shown in the two cases that $M/F(G(k))$ is a finite group. Therefore $F(G(\mathbb{A}))$ is open in

$$M + F(G(\mathbb{A}))$$

as it is closed and of finite index because

$$(M + F(G(\mathbb{A})))/F(G(\mathbb{A})) \cong M/M \cap F(G(\mathbb{A}))$$

is a homomorphic image of $M/F(G(k))$. Hence $M + F(G(\mathbb{A})) = (G(R) + F(G(\mathbb{A})) \cap (G(k) + F(G(\mathbb{A})))$ is open in $G(k) + F(G(\mathbb{A}))$. It follows that $F(G(\mathbb{A}))$ is an open subgroup of $G(k) + F(G(\mathbb{A}))$. Therefore the image of f, which is isomorphic to $(G(k) + F(G(\mathbb{A}))/F(G(\mathbb{A}))$, is discrete in $\prod' H^1(k_v, A[\phi])$.

Step 6 *Let $\phi : A \to B$ be an isogeny of abelian varieties over k. Then there is an induced homomorphism*

$$g : \prod_v B(k_v) \to \prod'_v H^1(k_v, A[\phi])$$

which is continuous and has compact image.

The field k is the function field of a smooth projective irreducible algebraic curve $X/\mathbb{F}$ for some finite field $\mathbb{F}$ of characteristic p. There is an open subscheme U of X and abelian schemes $\mathcal{A}/U, \mathcal{B}/U$ and an isogeny $\Phi : \mathcal{A} \to \mathcal{B}$ which induces the isogeny $\phi : A \to B$ on the generic fibres. For a place v of k which is a closed point of U and where R_v is the valuation ring of k at v, there is a commutative diagram where g_v is the natural homomorphism

$$\begin{array}{ccccc} \mathcal{B}(R_v) & \longrightarrow & H^1(R_v, \mathcal{A}[\Phi]) & \longrightarrow & H^1(R_v, \mathcal{A}) \\ \downarrow & & \downarrow & & \downarrow \\ B(k_v) & \xrightarrow{g_v} & H^1(k_v, A[\phi]) & \longrightarrow & H^1(k_v, A). \end{array}$$

The first vertical arrow is an isomorphism as $B, \mathcal{B}$ are abelian schemes over their base schemes. Hence the image of g_v is contained in the subgroup $H^1(R_v, \mathcal{A}[\Phi])$ of $H^1(k_v, A[\phi])$. This implies that the g_v, for all places v, together define a map $g : \prod_v B(k_v) \to \prod' H^1(k_v, A[\phi])$. The maps g_v are continuous with respect to the standard topologies on the cohomology groups as they arise via the exact sequence $0 \to A[\phi] \to A \to B \to 0$ of commutative group schemes (see (4.2.7)). Therefore the map g is

5.6 Finiteness of Classical Selmer Modules of Abelian Varieties

continuous. But the groups $B(k_v)$ are compact for all v and therefore $\prod_v B(k_v)$ is compact and hence so is its image in $\prod' H^1(k_v, A[\phi])$.

Step 7. End of Proof

By Step 2, we may assume that k has characteristic $p > 0$ and $A[\phi]/k$ is isomorphic to one of the two group schemes μ_p, α_p. We have the commutative diagram where $\prod'$ denotes the restricted topological product

$$\begin{array}{ccccc} A(k) & \longrightarrow & H^1(k, A[\phi]) & \longrightarrow & H^1(k, A) \\ \downarrow & & \downarrow f & & \downarrow \\ \prod A(k_v) & \xrightarrow{g} & \prod' H^1(k_v, A[\phi]) & \longrightarrow & \prod_v H^1(k_v, A) \end{array}$$

The group $H^1_{\mathcal{F}}(k, A[\phi])$ is a subgroup of $H^1(k, A[\phi])$ whose image under f is contained in the image of g that is to say $H^1_{\mathcal{F}}(k, A[\phi])$ is $f^{-1}g(\prod A(k_v))$. By Steps 4 and 5, the map f is injective and $f(H^1(k, A[\phi]))$ is a discrete subgroup of $\prod' H^1(k_v, A[\phi])$. By Step 6, the image of g is compact. Therefore $\text{Im}(g) \cap \text{Im}(f)$ is a compact subset of a discrete group and is therefore finite. Hence $H^1_{\mathcal{F}}(k, A[\phi])$ is finite. □

Corollary 5.6.4 (Tate, Lang, Milne) *If $n \in \mathbb{N}$ then the n-torsion subgroup $\text{III}(A/k)_n$ of the Shafarevich-Tate group is finite.*

Proof This follows from the exact sequence (5.5.9). □

Proposition 5.6.5 *The group $\text{III}(A/k)$ is torsion.*

Proof Let G be a profinite group and M be a G-module equipped with the discrete topology. Then G is a projective limit $\varprojlim G_\lambda$ of a projective system of finite groups $G_\lambda, \lambda \in \Lambda$ for some set Λ. Then for the cohomology of profinite groups with values in a discrete module [58, proposition 1.2.5, p. 21], we have

$$H^i(G, M) = \varinjlim H^i(G_\lambda, M^{G_i}).$$

The cohomology group $H^i(G_\lambda, M^{G_\lambda})$ of the finite group G_λ is annihilated by $|G_\lambda|$ for all $i \geq 1$. Hence $H^i(G, M)$ is a torsion group for all $i \geq 1$. The group $\text{III}(A/k)$ is a subgroup of $H^1(k, A)$, where the latter group can be taken to be Galois cohomology, étale cohomology, or fppf cohomology [51, Theorem III.3.9]. Therefore taking Galois cohomology of G_k with values in the discrete module $A(k^{\text{sep}})$, we have that $H^1(k, A)$ is a torsion group and hence so is $\text{III}(A/k)$. □

Conjecture 5.6.6 The group $\text{III}(A/k)$ is finite.

This is a standard conjecture, part of the Birch-Swinnerton Dyer conjecture, and which is known to hold in a small number of cases.

Remarks 5.6.7

(i) On the Tate module T, the classical Selmer structure $\mathcal{F}_{\text{class}}$ and the canonical Selmer structure $\mathcal{F}_{\text{can}}$ (see Definition 5.1.4) are related by $\mathcal{F}_{\text{class}} \leq \mathcal{F}_{\text{can}}$.

(ii) Conjecture 5.6.6 can be proved in various cases where the ground field k is a number field or a function field of positive characteristic. In Chap. 8 this finiteness of Shafarevich-Tate groups will be shown for certain elliptic curves over global fields via Heegner points and CM points.

(iii) The classical Selmer group $H^1_{\mathcal{F}}(k, A[\phi])$ of an abelian variety over a global field k, with an isogeny $\phi : A \to B$, is in principle effectively computable and we have demonstrated that it is finite. The exact sequence (5.5.9) relating this Selmer group to $B(k)/\phi(A(k))$ and $\text{III}(A)_\phi$ provides information on these latter two groups but there is no known algorithm for deciding which parts of the Selmer group arise from $B(k)/\phi(A(k))$ or contribute to $\text{III}(A)_\phi$.

There is a standard method of determining the groups $B(k)/\phi(A(k))$ and $\text{III}(A)_\phi$ called *the classical descent* and a detailed explanation of it for elliptic curves can be found in [74, Chap. X, §4] and it is related to the method of descent of Sect. 5.4. This descent method works well in practice, but, with present knowledge, it is not an effective algorithm that is guaranteed to provide a finite set of generators of the group $B(k)/\phi(A(k))$.

(iv) The Mordell-Weil theorem, namely, that the group $A(k)$ of k-rational points of A is finitely generated, is a consequence of the finiteness of the Selmer group (Proposition 5.6.3) combined with the theory of heights (see [69]).

(v) The method of descent (as in (iii)) can sometimes provide examples of non-zero elements of $\text{III}(A)$ for A an abelian variety over a global field k. A second question is then to how to represent explicitly such elements.

Suppose for the rest of this remark that A/k is an elliptic curve. A non-zero element τ of $\text{III}(A)$ corresponds to an A-torsor E/k with rational points over every completion of k but no rational point over k. If τ has order n in $\text{III}(A)$, where n is coprime to char. k, then the quotient of E under the action of the finite subgroup $A[n]$ has a k-rational point and is then isomorphic to A/k. In this way E can be viewed as an étale covering of degree n^2 of A.

Another method is to try to find explicit equations for a curve of genus 1 representing a class of $\text{III}(A)$. More precisely one can try to represent such a genus 1 curve as a curve of degree n in projective $n-1$-space [11, §2].

Suppose that the elliptic curve A is an abelian subvariety of an abelian variety J/k. There is then an exact sequence $0 \to A \to J \to B \to 0$ of abelian varieties over k. An element $\tau \in \text{III}(A)$ is said to be *visible in J* if it is in the kernel of the natural homomorphism $\text{III}(A) \to \text{III}(J)$. It τ is visible in J then τ is represented by a curve of genus 1 in J which is the inverse image of a point $\alpha \in B(k)$ under the homomorphism $J \to B$ and where α maps to τ under the connecting map $B(k) \to H^1(k, A)$. In this way visible elements of $\text{III}(A)$ can be represented explicitly as subvarieties of abelian varieties [1, 11].

(vi) Following on from (v), more generally, let V/k be a projective k-scheme. We may say that an element $\tau \in \text{III}(A)$ *visible in V* if the principal homogeneous space representing τ is k-isomorphic to a closed subscheme of V/k.

5.6.8. Bibliographical Remarks. The finiteness of the n-torsion subgroup of the Shafarevich-Tate group, Corollary 5.6.4, is due to Tate, Lang, and Milne. This was proved by Tate and Lang [44], for the case where the characteristic of k does not divide n, and the theorem was completed by Milne [52] for the remaining case where k has characteristic p which divides n.

The finiteness of Selmer groups, Proposition 5.6.3, is a classical result and it entails the finiteness of the n-torsion subgroup of the Shafarevich-Tate group. The proof here combines the case for k an algebraic number field with the case of k a function field due to Milne [52]. A proof of finiteness for a more general class of Selmer groups for finite commutative group schemes can be found in [8, theorem 3.2].

Euler Systems

Contents

6.1	General Euler Systems	135
6.2	Euler Systems	139
6.3	Euler Systems Are General Euler Systems	141
6.4	Morphisms	144
6.5	Heegner Systems	152
6.6	The Heegner Point General Euler System: Classical and Drinfeld Modular Curves	157
6.7	The CM Point General Euler System: Shimura Curves	160
6.8	The Heegner Point and CM Point Euler Systems	165
6.9	Appendix: Further Examples of Euler Systems	169

In this chapter, we present Euler systems, a concept due to Kolyvagin [34] and developed by Rubin [62]. The Euler systems arising from Heegner points that we consider do not entirely fit into the elegant framework of an Euler system defined by Rubin in [62, Chapter 2]. For this reason, we give a more general definition of an Euler system in Sect. 6.1.

6.1 General Euler Systems

(6.1.1) Let

k be a global field;
F be a finite Galois field extension of k;
l be a prime number different from the characteristic of k;
R be a coefficient ring with residue field of characteristic l;
T be an R-representation of $\text{Gal}(k^{\text{sep}}/k)$ unramified outside a finite set of primes of k;

$\mathcal{N}$ be a cycle of k divisible by all Archimedean places of k, all places of k where T is ramified, all places of k where F/k is ramified, and all places above the prime number l, if any.

(6.1.2) Let $\mathrm{Cyc}(k)$ be the set of cycles of k (see (1.1)).

Assume that for each cycle $a \in \mathrm{Cyc}(k)$ coprime to $\mathcal{N}$, we are given a finite abelian Galois field extension F_a of F such that

(i) $a \leq b$ implies that $F_a \subseteq F_b$;
(ii) $F_{\gcd(a,b)} = F_a \cap F_b$;
(iii) for $a, b \in \mathrm{Cyc}(k)$ coprime to $\mathcal{N}$ there is $c \in \mathrm{Cyc}(k)$ coprime to $\mathcal{N}$ such that $F_c \supseteq F_a.F_b$.

Put

$$\mathcal{F} = \bigcup_{a \text{ coprime to } \mathcal{N}} F_a$$

where the union is in some separable closure of F and $a \in \mathrm{Cyc}(k)$ runs over all cycles coprime to $\mathcal{N}$. Then $\mathcal{F}$ is an abelian Galois extension of F but not necessarily finite over F and not necessarily abelian over k.

The completed group algebra $R[[\mathrm{Gal}(\mathcal{F}/F)]]$ acts on $H^1(F_a, T)$ for all a coprime to $\mathcal{N}$ and this action on $H^1(F_a, T)$ factors through the quotient $R[\mathrm{Gal}(F_a/F)]$.

Definition 6.1.3 A *general Euler system* **c** for $(T, \{F_a\}_a, \mathcal{N})$ is a family of cohomology classes

$$\{c_a \in H^1(F_a, T) \mid a \in \mathrm{Cyc}(k) \text{ coprime to } \mathcal{N}\}$$

and a family of maps of sets for all $i \in \mathbb{N}$

$$h_{a,i} : \Sigma_k \setminus \mathrm{Supp}(\mathcal{N}) \to R[[\mathrm{Gal}(\mathcal{F}/F)]]$$

such that these cohomology classes satisfy the *distribution relation* for any place $\mathfrak{p}$ of k not dividing $\mathcal{N}$

$$N_{F_{a\mathfrak{p}}/F_a}(c_{a\mathfrak{p}}) = \sum_{i=0}^{r} (h_{a,i}(\mathfrak{p}) \circ \mathrm{res}_{F_a/F_{a\mathfrak{p}^{-i}}}) c_{a\mathfrak{p}^{-i}}$$

where $r \in \mathbb{N}$ is the order of the cycle a at $\mathfrak{p}$, that is to say $a = \mathfrak{p}^r$ (coprime to $\mathfrak{p}$).

6.1 General Euler Systems

If a is coprime to $\mathfrak{p}$ then $r = 0$ and the sum reduces to the formula

$$N_{F_{a\mathfrak{p}}/F_a}(c_{a\mathfrak{p}}) = h_{a,0}(\mathfrak{p})c_a.$$

Here $h_{a,i}(\mathfrak{p}) \in R[[\mathrm{Gal}(\mathcal{F}/F)]]$ is considered as an R-endomorphism of $H^1(F_a, T)$ and $\mathrm{res}_{F_a/F_{a\mathfrak{p}^{-i}}}$ are restriction homomorphisms so that $(h_{a,i}(\mathfrak{p}) \circ \mathrm{res}_{F_a/F_{a\mathfrak{p}^{-i}}})c_{a\mathfrak{p}^{-i}}$ is an element of $H^1(F_a, T)$; this formula for $N_{F_{a\mathfrak{p}}/F_a}(c_{a\mathfrak{p}})$ is then a sum of elements of $H^1(F_a, T)$.

Definition 6.1.4 The least integer n, if it exists, such that $h_{a,i}(\mathfrak{p}) = 0$ for all $i \geq n+1$, all a and all $\mathfrak{p}$, is the *dimension* of the Euler system **c**

Remarks 6.1.5

(i) This definition of a general Euler system is wide enough to cover all known Euler systems, including Heegner points on elliptic curves arising from classical modular curves, Shimura curves, and Drinfeld modular curves.

Rubin's Euler systems (see Sect. 6.2 below), although more restrictive and do not cover fully the Heegner point case, have corresponding finiteness theorems as Rubin demonstrates, whereas the above Definition 6.1.3, without supplementary hypotheses, is too wide to imply any finiteness result. See [46] for an introduction to Euler systems.

(ii) The dimension of an Euler system (see Sects. 6.2, 6.3 below) is 0 whereas the dimension of the Heegner point general Euler systems on classical modular curves, Shimura curves, and Drinfeld modular curves is equal to 1 (see Sect. 6.6 *et seq.*).

(iii) One can evidently define a general Euler system for any subset $\mathcal{C}$ of $\mathrm{Cyc}(k)$ instead of taking all cycles coprime to a fixed cycle $\mathcal{N}$. One would then have a general Euler system for $(T, \{F_a\}_{a \in \mathcal{C}}, \mathcal{C})$.

Zero Dimensional General Euler Systems When $k = F$

(6.1.6) Suppose that $k = F$. Let F_a be a family of abelian extensions of F satisfying (i), (ii), (iii) of (6.1.2) and where $\mathcal{F} = \bigcup_a F_a$ where a runs over all cycles of F coprime to $\mathcal{N}$ where the union is in some separable closure of F.

Taking $k = F$ in Definition 6.1.3, then to give a general Euler system of dimension zero

$$\{d_a \in H^1(F_a, T) \mid a \in \mathrm{Cyc}(F) \text{ coprime to } \mathcal{N}\}$$

for $(T, \{F_a\}_a, \mathcal{N})$ satisfying the distribution relation for any place $\mathfrak{p}$ of F not dividing $\mathcal{N}$

$$N_{F_{a\mathfrak{p}}/F_a}(d_{a\mathfrak{p}}) = h_a(\mathfrak{p})d_a.$$

is equivalent to giving a family of cohomology classes

$$\{c_K \in H^1(K, T) \mid F \subseteq_f K \subseteq \mathcal{F}\}$$

for all intermediate fields K with $F \subseteq_f K \subseteq \mathcal{F}$, satisfying the distribution relation, where $F \subseteq K \subseteq K' \subseteq \mathcal{F}$ and K'/F is finite,

$$N_{K'/K}(c_{K'}) = A_{K'/K} c_K.$$

where $A_{K'/K}$ is as follows. Let $\mathfrak{m}'$, $\mathfrak{m}$ be the least cycles ordered by multiplicativity, such that $K \subseteq F_\mathfrak{m}$, $K' \subseteq F_{\mathfrak{m}'}$ and if $\mathfrak{p}_1, \ldots, \mathfrak{p}_r$ are prime cycles such that

$$\mathfrak{m}' = \mathfrak{p}_1 \mathfrak{p}_2 \ldots \mathfrak{p}_r \mathfrak{m}$$

then

$$A_{K'/K} = \prod_{i=0}^{r-1} h_{\mathfrak{p}_1 \mathfrak{p}_2 \ldots \mathfrak{p}_i \mathfrak{m}}(\mathfrak{p}_{i+1}).$$

[Given the classes $d_\mathfrak{m}$, for all cycles $\mathfrak{m}$ coprime to $\mathcal{N}$, for any intermediate field $F \subseteq_f K \subseteq \mathcal{F}$, there is a unique least cycle $\mathfrak{m}$, ordered by multiplicativity, such that $K \subseteq F_\mathfrak{m}$, and then we put

$$c_K = N_{F_\mathfrak{m}/K} d_\mathfrak{m}$$

and this gives the correspondence.]

(6.1.7) Conversely, suppose that $k = F$ and let $\mathcal{F}$ be an abelian separable extension of F, not necessarily finite. For each cycle $\mathfrak{m} \in \mathrm{Cyc}(F)$ coprime to $\mathcal{N}$, write $F_\mathfrak{m} = F^\mathfrak{m} \cap \mathcal{F}$ where $F^\mathfrak{m}$ is the ray class field of F mod $\mathfrak{m}$. To give cohomology classes

$$\{c_K \in H^1(K, T) \mid F \subseteq_f K \subseteq \mathcal{F}\}$$

for all intermediate fields K where $F \subseteq_f K \subseteq \mathcal{F}$ and which satisfy the distribution relation, where $F \subseteq K \subseteq K' \subseteq \mathcal{F}$ and K'/F is finite,

$$N_{K'/K}(c_{K'}) = A_{K'/K} c_K.$$

where $A_{K'/K} \in R[[\mathrm{Gal}(\mathcal{F}/F)]]$ is equivalent to giving a general Euler system of dimension zero $(T, \{F_a\}_a, \mathcal{N})$ where F_a are the fields just defined.

[To see this, given the classes c_K, for all $F \subseteq_f K \subseteq \mathcal{F}$, one puts $d_\mathfrak{m} = c_{F_\mathfrak{m}}$.]

Remark 6.1.8 The paragraphs (6.1.6), (6.1.7) are a relatively simple variant of Proposition 6.3.2.

Note that the two filtrations of $\mathcal{F}$ in (6.1.6) and (6.1.7) are not necessarily the same where the filtration of $\mathcal{F}$ in (6.1.6) is given by the subfields F_a, and in (6.1.7), the filtration is given by $F^{\mathfrak{m}} \cap F$ in (6.1.7) where $F^{\mathfrak{m}}$ is the ray class field mod $\mathfrak{m}$.

6.2 Euler Systems

(6.2.1) In this section, Euler systems are defined essentially in the formulation of Rubin [62, Chapter 2] and where this is here extended to include global fields of any characteristic.

(6.2.2) The notation of the previous section Sect. 6.1 holds here, but k, F, R, T are changed slightly as follows: The field k is superfluous, and $\mathcal{N}$ is a cycle on F instead of k so that we have

F is a global field;

R is a coefficient ring with residue field of characteristic l, different from the characteristic of F;

T is an R-representation of $\mathrm{Gal}(F^{\mathrm{sep}}/F)$ that is a finite free R-module;

$\mathcal{N}$ is a cycle of F divisible by all Archimedean places of F, all places of F where T is ramified, and all places above the prime number l, if any.

Let $\mathcal{F}$ be a separable abelian extension field (not necessarily finite) of F. By the notation

$$F \subseteq_f K \subseteq \mathcal{F}$$

is meant that K is an intermediate field between F and $\mathcal{F}$ which is finite over F.

For each place $\mathfrak{p}$ of F coprime to $\mathcal{N}$, let $\mathrm{Fr}_{\mathfrak{p}}$ denote an arithmetic Frobenius at $\mathfrak{p}$ in $\mathrm{Gal}(F^{\mathrm{sep}}/F)$. Define the characteristic polynomial $P_{\mathfrak{p}}(x) \in R[x]$ of $\mathrm{Fr}_{\mathfrak{p}}$ acting on the dual T^*, relative to some R-basis of T^*, by

$$P_{\mathfrak{p}}(x) = \det(1 - \mathrm{Fr}_{\mathfrak{p}}^{-1}.x | T^*). \tag{6.2.3}$$

Definition 6.2.4 An *Euler system* **c** for $(T, \mathcal{F}, \mathcal{N})$ is a family of cohomology classes

$$\{c_K \in H^1(K, T) \mid F \subseteq_f K \subseteq \mathcal{F}\}$$

such that whenever $F \subseteq K \subseteq K' \subseteq \mathcal{F}$ and K'/F is finite then these classes satisfy the *distribution relation*

$$N_{K'/K}(c_{K'}) = \Big(\prod_{\mathfrak{p} \in \Sigma(K'/K)} P_\mathfrak{p}(\mathrm{Fr}_\mathfrak{p}^{-1}) \Big) c_K \qquad (6.2.5)$$

where the product here runs over the finite set of non-Archimedean places

$$\Sigma(K'/K) = \left\{ \mathfrak{q} \; \middle| \; \begin{array}{c} \mathfrak{q} \text{ a finite place of } F \text{ coprime to } \mathcal{N} \\ \text{and unramified in } K/F \text{ where some place of } K \\ \text{over } \mathfrak{q} \text{ is ramified in } K' \end{array} \right\}$$

Here $N_{K'/K}$ denotes the corestriction from K' to K and the factor $P_\mathfrak{p}(\mathrm{Fr}_\mathfrak{p}^{-1})$ is an element of $\mathrm{End}_R(H^1(K,T))$.

The Euler systems for $(T, \mathcal{F}, \mathcal{N})$, with the evident operations of addition and multiplication by elements of R, form an R-module denoted $\mathbf{ES}(T, \mathcal{F}, \mathcal{N})$.

Remarks 6.2.6

(i) The order of the factors in this product $\prod_{\mathfrak{p} \in \Sigma(K'/K)} P_\mathfrak{p}(\mathrm{Fr}_\mathfrak{p}^{-1})$ in the distribution relation (6.2.5) is immaterial, because the action of the elements $P_\mathfrak{p}(\mathrm{Fr}_\mathfrak{p}^{-1})$ on c_K factors through the commutative group ring $R[\mathrm{Gal}(K/F)]$.

We stress that the class c_K belongs to $H^1(K,T)$, whereas $P_\mathfrak{p}(X)$ in the distribution relation is the characteristic polynomial of the *dual* $T^* = \mathrm{Hom}(T, \mathbb{Z}_l(1))$ of T.

(ii) In the case where F is a number field, Rubin [62, pp.21–22] imposes the following further conditions on the field $\mathcal{F}$ in order for a system of cohomology classes satisfying Definition 6.2.4 to be an Euler system:

- R is the ring of integers of a finite field extension of $\mathbb{Q}_l$ where l is different from the characteristic of F;
- for every prime $\mathfrak{q}$ of F coprime to $\mathcal{N}$, the field $\mathcal{F}$ contains $F(\mathfrak{q})$, the maximal l-extension of F inside the ray class field of F modulo $\mathfrak{q}$;
- the field $\mathcal{F}$ contains an extension F_∞ of F such that $\mathrm{Gal}(F_\infty/F) \cong \mathbb{Z}_l^d$ for some $d \geq 1$ and no finite prime of F splits completely in F_∞/F.

The first hypothesis here is to ensure that $\mathcal{F}$ is "sufficiently large". The second hypothesis is to ensure the rigidity of the Euler system (for more details of this rigidity, see [62, Chapter IX, §1]).

The Definition 6.2.4 here with R any coefficient ring allows Euler systems in the important case, not covered by the above, when the coefficient rings R are artin local.

(iii) Rubin also defines "anticyclotomic Euler systems" in [62, Chapter IX, §4] in order to cover the case of Heegner points on elliptic curves over the rational numbers. For this, in the case of Heegner points on an elliptic curve over $\mathbb{Q}$ with Tate module T, an imaginary quadratic field K is taken, and the field extension $\mathcal{F}$ is taken to be the anticyclotomic extension of K; the cohomology classes arising from Heegner points are then elements

$$c_E \in H^1(E, T)$$

for all intermediate fields E with $K \subseteq_f E \subseteq \mathcal{F}$ where these classes c_E satisfy a distribution relation.

Nevertheless, the anticyclotomic Euler system of Rubin, which has dimension 0 as a general Euler system, does not fully cover the Heegner point system, because the Heegner point general Euler system has dimension 1 (see Sect. 6.6 *et seq.*).

Rubin [62, Chapter IX] also introduces a number of other variants of the definition of Euler system notably by changing the hypotheses in (ii) above.

(iv) Relatively few examples of Euler systems are known. Apart from the Heegner point systems with which this book is principally concerned, some other examples of Euler systems are explained briefly in Sect. 6.9.

(v) The factors

$$P_{\mathfrak{p}}(\mathrm{Fr}_{\mathfrak{p}}^{-1})$$

appearing in the distribution relation are formed from Euler factors of the dual T^* whence the name "Euler system" for this structure.

(vi) This Definition 6.2.4 of an Euler system may be modified by replacing the cycle $\mathcal{N}$ of (6.2.2) by a set, possibly infinite, of primes of F such that the set $\mathcal{N}$ contains all Archimedean places of F, all places of F where T is ramified and all places above the prime number l, if any. This modification will be used in Sect. 6.8 for the case of elliptic curves parameterized by Shimura curves where the set $\mathcal{N}$ will be infinite.

(vii) Rubin in [62] proves general finiteness theorems for Euler systems under hypotheses. These finiteness theorems apply to the Euler systems mentioned in Sect. 6.9. But this does not include the Heegner point systems with which we are concerned and for which a different proof of the finiteness is required and which is outlined in [62, Chap IX] for anticyclotomic Euler systems.

These finiteness theorems obtained from Euler systems have important consequences such as the main conjecture of Iwasawa theory and the finiteness of Shafarevich-Tate groups for certain abelian varieties (for more details, see [62, Chap. III]).

6.3 Euler Systems Are General Euler Systems

(6.3.1) The notation and hypotheses of (6.2.2) in the previous section hold in this section in particular we are given an R-representation T and a cycle $\mathcal{N}$ on F satisfying the hypotheses of (6.2.2).

For a cycle $\mathfrak{m} \in \mathrm{Cyc}(F)$, let $F^{\mathfrak{m}}$ be the ray class field of F mod $\mathfrak{m}$.

For $\mathfrak{p}$, a non-Archimedean place of F where T is unramified and not dividing l, let $P_{\mathfrak{p}}(X)$ be the characteristic polynomial of the Frobenius at $\mathfrak{p}$ on T^* as in (6.2.3).

Proposition 6.3.2 *Suppose given a field extension $F \subseteq \mathcal{F}$, where $\mathcal{F}/F$ is separable and abelian but not necessarily finite. For any cycle $\mathfrak{m}$ of F, write*

$$F_{\mathfrak{m}} = F^{\mathfrak{m}} \cap \mathcal{F}.$$

Then to give a family of cohomology classes

$$\{c_K \in H^1(K, T) \mid F \subseteq_f K \subseteq \mathcal{F}\}$$

for all intermediate fields K satisfying the distribution relation, as in (6.2.5)

$$N_{K'/K}(c_{K'}) = \Big(\prod_{\mathfrak{p} \in \Sigma(K'/K)} P_{\mathfrak{p}}(\mathrm{Fr}_{\mathfrak{p}}^{-1}) \Big) c_K$$

is equivalent to giving a system of cohomology classes

$$d_{\mathfrak{m}} \in H^1(F_{\mathfrak{m}}, T)$$

indexed by all cycles $\mathfrak{m}$ on F such that

$$N_{F_{\mathfrak{m}\mathfrak{p}}/F_{\mathfrak{m}}}(d_{\mathfrak{m}\mathfrak{p}}) = \begin{cases} P_{\mathfrak{p}}(\mathrm{Fr}_{\mathfrak{p}}^{-1}) d_{\mathfrak{m}} & \text{if } \mathfrak{p} \nmid \mathfrak{m}\mathcal{N} \\ d_{\mathfrak{m}} & \text{if } \mathfrak{p} \mid \mathfrak{m}\mathcal{N}. \end{cases} \qquad (6.3.3)$$

Proof Suppose given a set of cohomology classes $d_{\mathfrak{m}} \in H^1(F_{\mathfrak{m}}, T)$ indexed by cycles $\mathfrak{m}$ of F, and satisfying the distribution relation (6.3.3). Then for any intermediate field

$$F \subseteq_f K \subseteq \mathcal{F}$$

we define the cohomology class

$$c_K \in H^1(K, T)$$

by

$$c_K = N_{F_{\mathfrak{m}}/K} d_{\mathfrak{m}}$$

6.3 Euler Systems Are General Euler Systems

where $\mathfrak{m}$ is the conductor of the abelian extension K of F. Then c_K forms an Euler system for $(T, \mathcal{F}, \mathcal{N})$.

For the converse, suppose that $c_K \in H^1(K, T)$ is a family of cohomology classes indexed by intermediate fields K with

$$F \subseteq_f K \subseteq \mathcal{F}$$

and such that for any field extensions

$$F \subseteq K \subseteq K' \subseteq \mathcal{F}$$

where K'/F is finite then we have

$$N_{K'/K}(c_{K'}) = \Big(\prod_{\mathfrak{p} \in \Sigma(K'/K)} P_{\mathfrak{p}}(\mathrm{Fr}_{\mathfrak{p}}^{-1}) \Big) c_K$$

as in (6.2.5).

We define for any cycle $\mathfrak{m}$ of F the cohomology class $d_{\mathfrak{m}} \in H^1(F_{\mathfrak{m}}, T)$ by

$$d_{\mathfrak{m}} = \Big(\prod_{\mathfrak{p}} P_{\mathfrak{p}}(\mathrm{Fr}_{\mathfrak{p}}^{-1}) \Big) c_{F_{\mathfrak{m}}}$$

where the product runs over all places $\mathfrak{p}$ dividing $\mathfrak{m}$ not in $\mathcal{N}$ and which are unramified in $F_{\mathfrak{m}}/F$. It is immediately checked that these $d_{\mathfrak{m}}$ satisfy the relation (6.3.3) of the proposition. □

Corollary 6.3.4 *An Euler system is a general Euler system of dimension 0.*

Proof Let $\mathbf{c}$ be an Euler system for $(T, \mathcal{F}, \mathcal{N})$ (Definition 6.2.4). With the notation of Definition 6.1.3, it is only necessary to take the field k to be equal to F, and $\mathcal{F}_a = \mathcal{F} \cap F^a$ for any cycle a where F^a is the ray class field mod a and $h_{F_a,i}(\mathfrak{p})$ to be the endomorphism

$$h_{F_a,i}(\mathfrak{p}) = \begin{cases} P_{\mathfrak{p}}(\mathrm{Fr}_{\mathfrak{p}}^{-1}) & \text{if } i = 0 \text{ and } \mathfrak{p} \nmid a\mathcal{N} \\ 1 & \text{if } i = 0 \text{ and } \mathfrak{p} | a\mathcal{N} \\ 0 & \text{if } i \geq 1. \end{cases}$$

□

6.4 Morphisms

(6.4.1) By taking linear combinations of the elements of an Euler system, one can obtain a family of cohomology classes with a different distribution relation. We examine these transformations in this section for general Euler systems of dimension zero of which Euler systems are a special case (Corollary 6.3.4).

The notation and hypotheses of (6.2.2) of the section Sect. 6.2 hold in this section, and in particular, we assume that the field k of Sect. 6.1 is equal to F.

(6.4.2) Let **c** be a general Euler system of dimension zero (shortened to GES of dimension zero) for $(T, \mathcal{F}, \mathcal{N})$ where $k = F$ (Sect. 6.1, (6.1.6), (6.1.7)). That is to say **c** is a set of cohomology classes

$$\{c_K \in H^1(K, T) \mid F \subseteq K \subseteq \mathcal{F}, K/F \text{ finite}\}$$

satisfying a distribution relation of the form, for all intermediate fields $F \subseteq K \subseteq K' \subseteq \mathcal{F}$ where K'/F is finite,

$$N_{K'/K}(c_{K'}) = A_{K'/K} c_K$$

where

$$A_{K'/K} \in R[[\mathrm{Gal}(\mathcal{F}/F)]].$$

The elements $A_{K'/K}$ belong to $R[[\mathrm{Gal}(\mathcal{F}/F)]]$ and the action of this group ring $R[[\mathrm{Gal}(\mathcal{F}/F)]]$ on $H^1(K, T)$ for $F \subseteq_f K \subseteq \mathcal{F}$ factors through the quotient $R[\mathrm{Gal}(K/F)]$.

Definition 6.4.3 Let **c**, **d** be zero dimensional GESs given by $c_K, d_K \in H^1(K, T)$ for all $F \subseteq_f K \subseteq \mathcal{F}$ where their distribution relations are

$$N_{K'/K}(c_{K'}) = A_{K'/K} c_K, \quad N_{K'/K}(d_{K'}) = B_{K'/K} d_K$$

where $A_{K'/K}, B_{K'/K} \in R[[\mathrm{Gal}(\mathcal{F}/F)]]$ and for all intermediate fields $F \subseteq K \subseteq K' \subseteq \mathcal{F}$ where K'/F is finite.

A *morphism* $\mathbf{c} \to \mathbf{d}$ of the GESs **c**, **d** is a collection of elements

$$\alpha_{K',K} \in R[[\mathrm{Gal}(\mathcal{F}/F)]]$$

and where K, K' are any fields with $F \subseteq K \subseteq K' \subseteq \mathcal{F}$ where K' is finite over F such that

6.4 Morphisms

$$d_K = \sum_{F \subseteq H \subseteq K} \alpha_{K,H} \text{res}_{K/H}(c_H)$$

for any $F \subseteq_f K \subseteq \mathcal{F}$ and where the sum runs over intermediate fields $F \subseteq H \subseteq K$ and where this equation is an equality of elements of $H^1(K, T)$.

Proposition 6.4.4 *Let* **c**, **d** *be families of cohomology classes* $c_K, d_K \in H^1(K, T)$ *for* $F \subseteq_f K \subseteq \mathcal{F}$. *Suppose that* $A_{K'/K}, B_{K'/K}, \alpha_{K',K}$ *are elements*

$$A_{K'/K}, B_{K'/K}, \alpha_{K',K} \in R[[\text{Gal}(\mathcal{F}/F)]]$$

for all intermediate fields with $F \subseteq K \subseteq K' \subseteq \mathcal{F}$ *and where* K' *is finite over* F.
Suppose that **c** *is zero dimensional GES with the distribution relation*

$$N_{K'/K}(c_{K'}) = A_{K'/K} c_K$$

for all $F \subseteq K \subseteq K' \subseteq \mathcal{F}$ *where* K'/F *is finite.*
Suppose that **d** *satisfies for any* $F \subseteq_f K \subseteq \mathcal{F}$

$$d_K = \sum_{F \subseteq_f H \subseteq K} \alpha_{K,H} \text{res}_{K/H}(c_H).$$

where $d_K \in H^1(K, T)$ *for all* K *and* $\text{res}_{K/H}$ *is restriction.*
Assume that the homomorphisms $\alpha_{K,H'}$ *satisfy the relation in the group ring* $R[\text{Gal}(K/F)]$, *which is a quotient of* $R[[\text{Gal}(\mathcal{F}/F)]]$,

$$B_{K'/K} \alpha_{K,H'} = \sum_{F \subseteq H \subseteq K', H \cap K = H'} \frac{[K' : K]}{[H : H']} \alpha_{K',H} A_{H/H'} \tag{6.4.5}$$

for all intermediate fields $F \subseteq K \subseteq K' \subseteq \mathcal{F}$, *where* K' *is finite over* F, *and all* $F \subseteq H' \subseteq K$ *and where the coefficients* $[K' : K]/[H : H']$ *are positive integers. Then the classes* **d** $= \{d_K\}_K$ *satisfy the distribution relation*

$$N_{K'/K}(d_{K'}) = B_{K'/K} d_K$$

and in particular the elements $\alpha_{K',K}$ *define a morphism of GESs* **c** $\to$ **d**.

Proof In this proof and in the statement of the proposition, note that the elements $\alpha_{H',H}, A_{H'/H}, B_{H'/H}$ although defined to be elements of $R[[\text{Gal}(\mathcal{F}/F)]]$, they are applied to cohomology classes in $H^1(K, T)$ for some field K, and in that case, they are considered to be elements of the quotient ring $R[\text{Gal}(K/F)]$.

We have by the distribution relation satisfied by **c** that

$$N_{K'/K} c_{K'} = A_{K'/K} c_K$$

where

$$A_{K'/K} \in R[[\mathrm{Gal}(\mathcal{F}/F)]].$$

Furthermore, we are given the family **d** of elements $d_K \in H^1(K, T)$ satisfying

$$d_K = \sum_{F \subseteq_f H \subseteq K} \alpha_{K,H} \mathrm{res}_{K/H}(c_H).$$

Suppose that we have fields $F \subseteq H \subseteq K' \subseteq \mathcal{F}$ and $F \subseteq K \subseteq K'$ where K'/F is finite. Then there is the diagram of fields

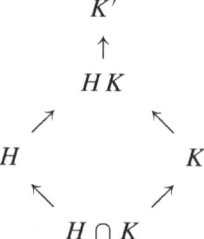

We have that, from the double coset formula [58, Propn. 1.5.6 and Cor. 1.5.8],

$$N_{KH/K} \circ \mathrm{res}_{KH/H}(c_H) = \mathrm{res}_{K/H \cap K} \circ N_{H/H \cap K}(c_H).$$

This gives the formula in $H^1(K, T)$, and where the elements $\alpha_{K',H}$ considered as elements of $R[\mathrm{Gal}(K/F)]$ commute with corestriction in the next formula,

$$N_{K'/K} d_{K'} = N_{K'/K} \Big(\sum_{F \subseteq_f H \subseteq K'} \alpha_{K',H} \mathrm{res}_{K'/H}(c_H) \Big)$$

$$= \sum_{F \subseteq_f H \subseteq K'} \alpha_{K',H} N_{K'/K}(\mathrm{res}_{K'/H}(c_H)).$$

$$= \sum_{F \subseteq H \subseteq K'} \frac{[K':K]}{[H:H \cap K]} \alpha_{K',H} \mathrm{res}_{K/K \cap H}(A_{H/K \cap H} c_{K \cap H}).$$

6.4 Morphisms

Collecting terms of the sum with the same intersection $H \cap K = H'$, we have

$$N_{K'/K} d_K = \sum_{F \subseteq H' \subseteq K} \left(\sum_{F \subseteq H \subseteq K', H \cap K = H'} \frac{[K' : K]}{[H : H']} \alpha_{K', H} \mathrm{res}_{K/H'} \circ A_{H/H'} \right) c_{H'}.$$

By the relation (6.4.5), considered as a relation in $\mathrm{End}_R(H^1(K, T))$, the coefficient

$$\sum_{F \subseteq H \subseteq K', H \cap K = H'} \frac{[K' : K]}{[H : H']} \alpha_{K', H} \mathrm{res}_{K/H'} \circ A_{H/H'}$$

of $c_{H'}$ may be identified with

$$B_{K'/K} \alpha_{K, H'}$$

again considered as an element of $\mathrm{End}_R(H^1(K, T))$. Therefore, we have the equality of cohomology classes in $H^1(K, T)$

$$N_{K'/K} d_{K'} = \sum_{F \subseteq H' \subseteq K} B_{K'/K} \alpha_{K, H'}(c_{H'})$$

$$= B_{K'/K} \sum_{F \subseteq H' \subseteq K} \alpha_{K, H'}(c_{H'}) = B_{K'/K} d_K.$$

That is to say $\mathbf{d}$ is a collection of cohomology classes satisfying the distribution relation $N_{K'/K} d_{K'} = B_{K'/K} d_K$. as required. □

The Category of Zero Dimensional GESs

(6.4.6) Suppose that $\mathbf{c}, \mathbf{d}, \mathbf{e}$ are zero dimensional GESs for $(T, \mathcal{F}, \mathcal{N})$ where their distribution relations are the following

$$N_{K'/K}(c_{K'}) = A_{K'/K} c_K, \; N_{K'/K}(d_{K'}) = B_{K'/K} d_K, \; N_{K'/K}(e_{K'}) = C_{K'/K} d_K$$

for all $F \subseteq K \subseteq K' \subseteq \mathcal{F}$ where K'/F is finite and where $A_{K'/K}, B_{K'/K}, C_{K'/K} \in R[[\mathrm{Gal}(\mathcal{F}/F)]]$. The composition of two morphisms $\mathbf{c} \to \mathbf{d}$ and $\mathbf{d} \to \mathbf{e}$ is well defined, and there is an evident identity morphism $\mathbf{c} \to \mathbf{c}$, and in this way, zero dimensional GESs for $(T, \mathcal{F}, \mathcal{N})$ form a category.

- A morphism $\alpha : \mathbf{c} \to \mathbf{d}$ is defined to be an *isomorphism* if there is a morphism $\beta : \mathbf{d} \to \mathbf{c}$ such that $\alpha \circ \beta$ and $\beta \circ \alpha$ are the identity morphisms $c_K \to c_K$ and $d_K \to d_K$ for all K;

- For a morphism $\mathbf{c} \to \mathbf{d}$, given by $d_K = \sum_{F \subseteq_f H \subseteq K} \alpha_{K,H} \mathrm{res}_{K/H}(c_H)$ where $\alpha_{K,H} \in R[[\mathrm{Gal}(\mathcal{F}/F)]]$, to be an isomorphism, it is sufficient that for all K with $F \subseteq_f K \subseteq \mathcal{F}$, then the endomorphism $\alpha_{K,K} \in R[[\mathrm{Gal}(\mathcal{F}/F)]]$ be invertible in $R[[\mathrm{Gal}(\mathcal{F}/F)]]$.

Proposition 6.4.7 *Let $(T, \mathcal{F}, \mathcal{N})$ be as in (6.2.2). Assume that the coefficient ring R is an integral domain. Suppose that there are maps of sets*

$$g, f : \Sigma_k \setminus \mathrm{Supp}(\mathcal{N}) \to R[[\mathrm{Gal}(\mathcal{F}/F)]], g, f : \mathfrak{q} \mapsto g_\mathfrak{q}, f_\mathfrak{q}.$$

Suppose also that for every place $\mathfrak{q}$ of k such that $\mathfrak{q}$ does not divide $\mathcal{N}$, we have the congruence in the ring $R[[\mathrm{Gal}(\mathcal{F}/F)]]$

$$f_\mathfrak{q} \equiv g_\mathfrak{q} \bmod N(\mathfrak{q}) - 1$$

where $N(\mathfrak{q})$ is the order of the residue field of k at $\mathfrak{q}$.

Let, where $\Sigma(K'/K)$ is as in Definition 6.2.4,

$$A_{K'/K} = \prod_{\mathfrak{q} \in \Sigma(K'/K)} f_\mathfrak{q}, \quad B_{K'/K} = \prod_{\mathfrak{q} \in \Sigma(K'/K)} g_\mathfrak{q}.$$

Let $\mathbf{c}, \mathbf{d}$ be zero dimensional GESs for $(T, \mathcal{F}, \mathcal{N})$, $c_K, d_K \in H^1(K, T)$ for $F \subseteq_f K \subseteq \mathcal{F}$, where their distribution relations are

$$N_{K'/K}(c_{K'}) = A_{K'/K} c_K, \quad N_{K'/K}(d_{K'}) = B_{K'/K} d_K$$

for all $F \subseteq K \subseteq K' \subseteq \mathcal{F}$ where K'/F is finite. Then there is an isomorphism $\mathbf{c} \longrightarrow \mathbf{d}$.

Proof For $F \subseteq_f K \subseteq \mathcal{F}$, write $\Sigma(K) = \Sigma(K/F)$, which is the set of places of F which are ramified in K/F and coprime to $\mathcal{N}$. For any finite set of places S of F, let K_S be the largest extension of F in K which is unramified outside S and $\mathcal{N}$.

We have

$$A_{K'/K} = \prod_{\mathfrak{q} \in \Sigma(K'/K)} f_\mathfrak{q}, \quad B_{K'/K} = \prod_{\mathfrak{q} \in \Sigma(K'/K)} g_\mathfrak{q}.$$

It is sufficient to check the equality (6.4.5) for some choice of elements $\alpha_{K,H}$. Put

$$d_\mathfrak{q} = g_\mathfrak{q} - f_\mathfrak{q}$$

If $F \subseteq H \subseteq K \subseteq \mathcal{F}$ and if $H = K_S$ for some $S \subseteq \Sigma(K)$, then we define $\alpha_{K,K_S} \in R[[\mathrm{Gal}(\mathcal{F}/F)]]$ by

6.4 Morphisms

$$\alpha_{K,K_S} = \frac{1}{[K:K_S]} \prod_{\mathfrak{q}\in\Sigma(K)\setminus S} d_\mathfrak{q} \prod_{\mathfrak{q}\in S\setminus\Sigma(K_S)} f_\mathfrak{q}.$$

We define

$$\alpha_{K,H} = 0$$

if H is not of the form $H = K_S$ for some $S \subseteq \Sigma(K)$.

The coefficient ring R is a local integral domain with residue characteristic l. We remark that for all primes $\mathfrak{q}$ of F where $\mathfrak{q}$ does not lie above l, then $N(\mathfrak{q})$ is a unit of R. Let $I_\mathfrak{q}$ be the inertia group of $\mathfrak{q}$ in $\mathrm{Gal}(K/F)$. The Galois group $\mathrm{Gal}(K/K_S)$ is generated, as a subgroup of $\mathrm{Gal}(K/F)$, by the subgroups $I_\mathfrak{q}$ for $\mathfrak{q} \in \Sigma(K) \setminus S$. Furthermore by the remark, the integer $|I_\mathfrak{q}|$ divides the integer $N(\mathfrak{q}) - 1$ in the ring R; therefore, $[K : K_S]$ divides $\prod_{\mathfrak{q}\in\Sigma(K)\setminus S}(N(\mathfrak{q})-1)$ in R. Since $\prod_{\mathfrak{q}\in\Sigma(K)\setminus S} d_\mathfrak{q}$ is a multiple of $\prod_{\mathfrak{q}\in\Sigma(K)\setminus S}(N(\mathfrak{q})-1)$ in $R[[\mathrm{Gal}(\mathcal{F}/F)]]$, it follows that $(\prod_{\mathfrak{q}\in\Sigma(K)\setminus S} d_\mathfrak{q})/[K : K_S]$ also belongs to $R[[\mathrm{Gal}(\mathcal{F}/F)]]$ and is a well-defined element.

The equality (6.4.5) to be verified is the following

$$B_{K'/K}\alpha_{K,H'} = \sum_{F\subseteq H\subseteq K', H\cap K=H'} \frac{[K':K]}{[H:H']} \alpha_{K',H} A_{H/H'} \tag{6.4.8}$$

The left-hand side here is zero unless H' is of the form K_U for some $U \subseteq \Sigma(K)$. In the right-hand side, $\alpha_{K',H}$ is zero unless H is of the form K'_U for some $U \subseteq \Sigma(K')$; but if $H = K'_U$ for some $U \subseteq \Sigma(K')$, then in the sum on the right-hand side, we have $H' = H \cap K = K'_U \cap K = K_{S'}$ where $S' = U \cap \Sigma(K)$. Therefore, if H' is not of the form K_U for some $U \subseteq \Sigma(K)$, then the right-hand side of (6.4.8) is zero; in conclusion, we have shown that the equality (6.4.8) holds and both sides are zero if H' is not of the form K_U for some $U \subseteq \Sigma(K)$.

Suppose then that

$$H' = K_U$$

for some $U \subseteq \Sigma(K)$. We then have as $\prod_{\mathfrak{q}\in\Sigma(K'/K)} g_\mathfrak{q} = B_{K'/K}$.

$$B_{K'/K}\alpha_{K,K_U} = \alpha_{K,K_U} \prod_{\mathfrak{q}\in\Sigma(K'/K)} g_\mathfrak{q} = \alpha_{K,K_U} \prod_{\mathfrak{q}\in\Sigma(K'/K)} (d_\mathfrak{q} + f_\mathfrak{q}) =$$

$$= \alpha_{K,K_U} \sum_{S'\subseteq\Sigma(K'/K)} \left(\prod_{\mathfrak{q}\in\Sigma(K'/K)\setminus S'} d_\mathfrak{q}\right)\left(\prod_{\mathfrak{q}\in S'} f_\mathfrak{q}\right).$$

As $\alpha_{K,K_U} = \frac{1}{[K:K_U]} \prod_{\mathfrak{q} \in \Sigma(K) \setminus U} d_\mathfrak{q} \prod_{\mathfrak{p} \in U \setminus \Sigma(K_U)} f_\mathfrak{p}$, this gives

$$B_{K'/K} \alpha_{K,K_U} = \frac{1}{[K:K_U]} \sum_{S' \subseteq \Sigma(K'/K)} \left(\prod_{\mathfrak{q} \in \Sigma(K') \setminus S' \setminus U} d_\mathfrak{q} \right) \left(\prod_{\mathfrak{q} \in S' \cup U \setminus \Sigma(K_U)} f_\mathfrak{q} \right)$$

$$= \frac{1}{[K:K_U]} \sum_{S' \subseteq \Sigma(K'/K)} \left(\prod_{\mathfrak{q} \in \Sigma(K') \setminus S' \setminus U} d_\mathfrak{q} \right) \left(\prod_{\mathfrak{q} \in S' \cup U \setminus \Sigma(K'_{S' \cup U})} f_\mathfrak{q} \right) \left(\prod_{\mathfrak{q} \in \Sigma(K'_{S' \cup U}/K_U)} f_\mathfrak{q} \right)$$

$$= \frac{1}{[K:K_U]} \sum_{S' \subseteq \Sigma(K'/K)} \left(\prod_{\mathfrak{q} \in \Sigma(K') \setminus S' \setminus U} d_\mathfrak{q} \right) \left(\prod_{\mathfrak{q} \in S' \cup U \setminus \Sigma(K'_{S' \cup U})} f_\mathfrak{q} \right) A_{K'_{S' \cup U}/K_U}$$

$$= \sum_{S' \subseteq \Sigma(K'/K)} \frac{[K':K]}{[K'_{S' \cup U} : K_U]} \frac{\prod_{\mathfrak{q} \in \Sigma(K') \setminus S' \setminus U} d_\mathfrak{q}}{[K' : K'_{S' \cup U}]} \left(\prod_{\mathfrak{q} \in S' \cup U \setminus \Sigma(K'_{S' \cup U})} f_\mathfrak{q} \right) A_{K'_{S' \cup U}/K_U}$$

$$= \sum_{S' \subseteq \Sigma(K'/K)} \frac{[K':K]}{[K'_{S' \cup U} : K_U]} \alpha_{K', K'_{S' \cup U}} A_{K'_{S' \cup U}/K_U}.$$

This is equal to the right-hand side of equation (6.4.8). This morphism $\mathbf{c} \to \mathbf{d}$ is then explicitly given by the equality

$$d_K = \sum_{S \subseteq \Sigma(K)} \left(\frac{\prod_{\mathfrak{q} \in \Sigma(K) \setminus S} d_\mathfrak{q}}{[K : K_S]} \right) \left(\prod_{\mathfrak{q} \in S \setminus \Sigma(K_S)} f_\mathfrak{q} \right) \mathrm{res}_{K/K_S}(c_{K_S}).$$

As $\alpha_{K,K} = 1$ for all K, this morphism $\mathbf{c} \longrightarrow \mathbf{d}$ is then an isomorphism by (6.4.6); alternatively that this is an isomorphism that follows from the symmetry between $\mathbf{c}$ and $\mathbf{d}$. □

Examples 6.4.9 Assume that the coefficient ring R is an integral domain. Suppose that $\mathbf{c}$ and $\mathbf{d}$ are zero dimensional GESs for $(T, \mathcal{F}, \mathcal{N})$ where their distribution relations are

$$N_{K'/K}(c_{K'}) = A_{K'/K} c_K, \quad N_{K'/K}(d_{K'}) = B_{K'/K} d_K$$

for all $F \subseteq K \subseteq K' \subseteq \mathcal{F}$ where K'/F is finite.

(i) Let $f_\mathfrak{q}, g_\mathfrak{q} \in R[x, x^{-1}]$ be Laurent polynomials for each place $\mathfrak{q}$ of F, and let $d \in \mathbb{Z}$. Let $u_\mathfrak{q} \in R^*$ be a collection of units indexed by all places $\mathfrak{q}$ of F where for all $\mathfrak{q}$

$$g_\mathfrak{q}(x) = u_\mathfrak{q} x^d f_\mathfrak{q}(x^{-1}).$$

6.4 Morphisms

Let $\mathcal{N}$ and $\Sigma(K'/K)$ be the finite set of places of F given in (6.2.2) and Definition 6.2.4.

Suppose that, where $\text{Fr}_\mathfrak{q} \in \text{Gal}(F^{\text{sep}}/F)$ are Frobenius elements,

$$A_{K'/K} = \prod_{\mathfrak{q} \in \Sigma(K'/K)} f_\mathfrak{q}(\text{Fr}_\mathfrak{q})$$

$$B_{K'/K} = \prod_{\mathfrak{q} \in \Sigma(K'/K)} g_\mathfrak{q}(\text{Fr}_\mathfrak{q}^{-1}).$$

Then there is an isomorphism $\mathbf{c} \longrightarrow \mathbf{d}$ given by

$$\alpha_{K,K} = \prod_{\mathfrak{q} \in \Sigma(K/F)} u_\mathfrak{q} \text{Fr}_\mathfrak{q}^{-d}, \quad \alpha_{K',K} = 0 \text{ if } K' \neq K.$$

[It is sufficient to check that $\alpha_{K,H}$ satisfies the relation (6.4.5). As $\alpha_{K,H}$ is zero if $K \neq H$, this relation (6.4.5) reduces to

$$B_{K',K} \alpha_{K,K} = \alpha_{K',K'} A_{K'/K}$$

and which is immediately checked. As $\alpha_{K,K} \in R[[\text{Gal}(\mathcal{F}/K)]]$ is a unit for all K, the morphism provided by the collection of $\alpha_{K',K}$s is an isomorphism.]

(ii) Define the following polynomials in $R[x]$

$$P_\mathfrak{q}(x) = \det(1 - \text{Fr}_\mathfrak{q}^{-1}.x|T^*), \ S_\mathfrak{q}(x) = \det(1 - \text{Fr}_\mathfrak{q}^{-1}.x|T)$$

Suppose that $A_{K'/K}$, $B_{K'/K}$ satisfy, whenever $F \subseteq K \subseteq K' \subseteq \mathcal{F}$ and where K'/F is finite,

$$A_{K'/K} = \prod_{\mathfrak{q} \in \Sigma(K'/K)} P_\mathfrak{q}(\text{Fr}_\mathfrak{q}^{-1})$$

$$B_{K'/K} = \prod_{\mathfrak{q} \in \Sigma(K'/K)} S_\mathfrak{q}(\text{Fr}_\mathfrak{q}).$$

In particular $\mathbf{c}$ is an Euler system for the G_F-representation T and not just a GES. Then there is an isomorphism $\mathbf{c} \to \mathbf{d}$.

[As $T^* = \text{Hom}(T, \mu_{l^\infty})$, we have the equalities of characteristic polynomials

$$\det(1 - \text{Fr}_\mathfrak{q}^{-1} x^{-1}|T) = \det(1 - N(\mathfrak{q})^{-1} \text{Fr}_\mathfrak{q} x^{-1}|T^*)$$

$$= (-xN(\mathfrak{q}))^{-\text{rank}(T)} \det(\text{Fr}_\mathfrak{q}|T^*) \det(1 - N(\mathfrak{q})\text{Fr}_\mathfrak{q}^{-1} x|T^*).$$

From (i) applied to

$$f_{\mathfrak{q}} = S_{\mathfrak{q}}(x^{-1}), u_{\mathfrak{q}} = (-N(\mathfrak{q}))^{-\text{rank}(T)} \det(\text{Fr}_{\mathfrak{q}}|T^*)$$

and then Proposition 6.4.7 applied to

$$f_{\mathfrak{q}} = P_{\mathfrak{q}}(N(\mathfrak{q})x), g_{\mathfrak{q}} = P_{\mathfrak{q}}(x),$$

we obtain the isomorphism **c** → **d**.]

6.5 Heegner Systems

(6.5.1) A Heegner system is a set of points on an elliptic curve which satisfies certain relations between the points. Examples of Heegner systems are the Heegner points or CM points on elliptic curves provided by a parameterization by a classical modular curve or a Drinfeld modular curve or a Shimura curve.

Let

k be a global field equipped with a special set of places ∞_k;

F be a finite Galois field extension of k which is imaginary quadratic with respect to the special set of places ∞_k; F is equipped with a special set of places ∞_F consisting of those places lying above ∞_k;

O_k, resp. O_F, be the ring of integers of k with respect to ∞_k, resp., the ring of integers of F with respect to ∞_F.

Let E/k be an elliptic curve with l-adic Tate module $T = T_l(E)$ where l is coprime to the characteristic of k, and let the coefficient ring be $R = \mathbb{Z}_l$.

A set of cycles $\mathcal{C}$ of k, a set of fields F_c for all $c \in C$, a set of points $x_c \in E(F_c)$ for all $c \in C$, and a map of sets $u : \mathcal{C} \to \mathbb{N}$, are a *Heegner system* if the following axioms hold, where each axiom Ax.n presupposes that all axioms Ax.m for $m < n$ hold.

Ax.1. The set of cycles $\mathcal{C}$ of k contains the trivial cycle 1 and is closed under the product of coprime cycles in $\mathcal{C}$ and such that if $c \in \mathcal{C}$ and d is a cycle on k which divides c then $d \in \mathcal{C}$. The map of sets $u : \mathcal{C} \to \mathbb{N}$ satisfies

$$u(c) = \begin{cases} 1 & \text{if } c \neq 1 \\ \geq 1 & \text{if } c = 1. \end{cases}$$

6.5 Heegner Systems

Ax.2. For each $c \in \mathcal{C}$, the field extension F_c/F is finite separable and abelian. If $c \leq c'$ are cycles in $\mathcal{C}$ then $F_c \subseteq F_{c'}$ and $F_{c'}/F_c$ are finite separable abelian field extension. Furthermore, for any $c, c' \in \mathcal{C}$, we have $F_c \cap F_{c'} = F_{\gcd(c,c')}$.

There is a cycle d_0 on k coprime to ∞_k and to all elements of $\mathcal{C}$ such that for all $c \in \mathcal{C}$, there is $d \in \mathcal{C}$ where the only primes dividing d are those dividing c such that $F_c \subseteq F[d_0 d]$ where $F[d_0 d]$ is the ring class field of F/k with conductor $d_0 d$.

Ax.3. For any square-free cycles $c, c' \in \mathcal{C}$, where $c < c'$, there is a canonical isomorphism $\mathrm{Gal}(F_{c'}/F_c) \cong (\prod_{\mathfrak{p}|c'/c} \frac{(O_F/\mathfrak{p}O_F)^*}{(O_k/\mathfrak{p}O_k)^*})/\Delta_c$ where the product runs over the distinct prime cycles $\mathfrak{p}$ in $\mathcal{C}$ dividing c'/c and where the finite abelian subgroup Δ_c of the product has order $|\Delta_c| = u(c)$ and Δ_c is determined up to isomorphism by c;

Furthermore, for any cycle $\mathfrak{n} \in \mathcal{C}$, we have:

(i) Let $\mathfrak{p}$ be a non-Archimedean prime of k. If $\mathfrak{n} \in \mathcal{C}$ and $\mathfrak{p} \nmid \mathfrak{n}$, then each prime of F over $\mathfrak{p}$ is unramified in $F_\mathfrak{n}/F$;
(ii) Let $\mathfrak{p}$ be a non-Archimedean prime of k. If k has characteristic zero, $\mathfrak{n} \in \mathcal{C}$, $\mathfrak{p} \nmid \mathfrak{n}$ and $\mathfrak{p}$ is inert in F/k then the prime place $\mathfrak{p}O_F$ of F splits completely in $F_\mathfrak{n}/F$;
(iii) If $\mathfrak{n}\mathfrak{p} \in \mathcal{C}$ where $\mathfrak{p}$ is a prime cycle, then each prime of $F_\mathfrak{n}$ above $\mathfrak{p}$ is totally ramified in $F_{\mathfrak{n}\mathfrak{p}}/F_\mathfrak{n}$;
(iv) Each prime in ∞_F splits completely in F_c/F for all $c \in \mathcal{C}$.

Ax.4. (Norm relation.) If $c, z \in \mathcal{C}$ where z is a prime cycle on k and c, z are coprime, then there is a norm relation

$$u(c)\mathrm{Tr}_{F_{cz}/F_c} x_{cz} = (a_z - \mathrm{Fr}_y - \left(\frac{z}{F}\right)\mathrm{Fr}_y^{-1}) x_c$$

where $\left(\frac{z}{F}\right)$ is the quadratic symbol equal to ± 1 or 0 depending on whether z is split, inert, or ramified in F/k, and Fr_y is the Frobenius element of $\mathrm{Gal}(F_c/F)$ corresponding to any place y of F over z, provided y is unramified in F_c/F (if Ax.3 holds then such a y is automatically unramified in F_c/F). Here a_z, for a non-Archimedean place z of k where T is unramified, is the trace of the Frobenius $\mathrm{Fr}_z \in \mathrm{Gal}(k^{\mathrm{sep}}/k)$ at z on the Tate module T where $a_z \in \mathbb{Z}$

$$a_z = \mathrm{Tr}(\mathrm{Fr}_z|T).$$

Ax.5. (Second norm relation.) If $c, z \in \mathcal{C}$ where z is a prime cycle on k and z divides c, then we have $\mathrm{Tr}_{F_{cz}/F_c} x_{cz} = a_z x_c - \mathrm{res}_{F_c/F_{c/z}} x_{c/z}$ where $\mathrm{res}_{F_c/F_{c/z}} : E(F_{c/z}) \to E(F_c)$ is the restriction homomorphism and Tr_{F_{cz}/F_c} is the corestriction, and $a_z = \mathrm{Tr}(\mathrm{Fr}_z|T)$, for a non-Archimedean place z of k where T is unramified, is the trace of the Frobenius $\mathrm{Fr}_z \in \mathrm{Gal}(k^{\mathrm{sep}}/k)$ at z on the Tate module T where $a_z \in \mathbb{Z}$.

Ax.6. (Congruence relation.) Let $cz \in \mathcal{C}$ where c, z are coprime and z is a prime divisor of k. For each prime $\mathfrak{q}|z$ where $\mathfrak{q}$ is a prime place of F_{cz}, then we have

$$x_{cz} \equiv \mathrm{Fr}_{z,\mathrm{arith}} x_c \equiv \mathrm{Fr}_{z,\mathrm{geom}} x_c \mod \mathfrak{q}$$

where this is an equality in the closed fibre of $E'/O_{k,z}$ where $E'/O_{k,z}$ is a proper smooth model of E/k.

Ax.7. For all prime numbers l, except finitely many, we have $E(F_c)_{l^\infty} = 0$ for all cycles $c \in \mathcal{C}$.

Basic Consequences of the Axioms

(6.5.2) Suppose for the rest of this section Sect. 6.5 that these axioms Ax.1–Ax.7 hold. We have the Kummer sequence for powers of l acting on the elliptic curve E/k, where l is the residue characteristic of $R = \mathbb{Z}_l$:

$$0 \longrightarrow E_{l^n}(F^{\mathrm{sep}}) \longrightarrow E(F^{\mathrm{sep}}) \xrightarrow{l^n} E(F^{\mathrm{sep}}) \longrightarrow 0.$$

Taking cohomology and passing to the limit as $n \to \infty$, we obtain the injective Kummer homomorphism

$$\delta : E(F_c) \to H^1(F_c, T)$$

for any $c \in \mathcal{C}$ where T is the l-adic Tate module of E/k and δ is the Kummer map.

Define e_c to be the image δx_c of the point $x_c \in E(F_c)$ in $H^1(F_c, T)$ under δ. These classes e_c, for all $c \in \mathcal{C}$, then evidently satisfy the two consequences (i) and (ii) in the next proposition which follow immediately from the axioms Ax.4 and Ax.5, respectively.

Proposition 6.5.3 *Suppose that axioms Ax.1–Ax.7 hold. Then the classes $e_c \in H^1(F_c, T)$ for each $c \in \mathcal{C}$ satisfy the following.*

(i) *If $c, z \in \mathcal{C}$ where z is a prime cycle on k and c, z are coprime, then there is the corestriction relation*

$$u(c)\mathrm{Tr}_{F_{cz}/F_c} e_{cz} = (a_z - \mathrm{Fr}_y - \left(\frac{z}{F}\right)\mathrm{Fr}_y^{-1}) e_c$$

where $\left(\frac{z}{F}\right)$ is the quadratic symbol equal to ± 1 or 0 depending on whether z is split, inert, or ramified in F/k and Fr_y is the Frobenius element of $\mathrm{Gal}(F_c/F)$ corresponding to any place y of F over z.

6.5 Heegner Systems

(ii) If $c, z \in \mathcal{C}$ where z is a prime cycle on k and z divides c then we have

$$\mathrm{Tr}_{F_{cz}/F_c} e_{cz} = a_z e_c - \mathrm{res}_{F_c/F_{c/z}} e_{c/z}$$

where $\mathrm{res}_{F_c/F_{c/z}} : H^1(F_{c/z}, T) \to H^1(F_c, T)$ is the restriction homomorphism and Tr_{F_{cz}/F_c} is the corestriction. □

Corollary 6.5.4 *Assume that $u(c)$ is a unit of the coefficient ring $\mathbb{Z}_l$ for all $c \in \mathcal{C}$. Let $z \in \mathcal{C}$ be a prime place of k. Then there is a collection of cohomology classes $d_c \in H^1(F_c, T)$ for all $c \in \mathcal{C}$, which coincide with the classes e_c of Proposition 6.5.3 for $c \ne 1$, and satisfying the distribution relation, where y is any prime place of F lying over z and Fr_y is a Frobenius element of $\mathrm{Gal}(F_c/F)$ at y if z is unramified in F_c/F*

$$\mathrm{Tr}_{F_{cz}/F_c} d_{cz} = \begin{cases} \left(a_z - \mathrm{Fr}_y - \left(\frac{z}{F}\right)\mathrm{Fr}_y^{-1}\right) d_c & \text{if } z \text{ is coprime to } \mathrm{Supp}(c) \\ \\ a_z d_c - u(c/z) \mathrm{res}_{F_c/F_{c/z}} d_{c/z} & \text{if } z \in \mathrm{Supp}(c). \end{cases}$$

The cohomology classes $d_c \in H^1(F_c, T)$, for all $c \in \mathcal{C}$, form a general Euler system for $(T, \{F_a\}_{a \in \mathcal{C}}, \mathcal{C})$ of dimension ≤ 1 (see Remark 6.1.5(iii)).

Proof Write e_c for the cohomology classes provided by Proposition 6.5.3. Define the cohomology classes $d_c \in H^1(F_c, T)$ by putting

$$d_c = u(c)^{-1} e_c \text{ for all } c.$$

Then the classes d_c evidently satisfy the stated relations, where we note that if z is coprime to c, then y is unramified in F_c/F by Ax.3(i). □

Torsion Points and Axiom Ax.7

(6.5.5) The next theorem is required to show that Ax.7 holds for the cases we consider because by Ax.2, every field F_c of the axioms is contained in a suitable ring class field.

Theorem 6.5.6 *Suppose that k, F are global fields such that k is equipped with a set of special places ∞_k and F/k is an imaginary quadratic extension field with respect to ∞_k. Write $F[c]$ for the ring class field of F/k corresponding to a cycle c on k. Let X/k be an elliptic curve.*

Suppose that if k has characteristic zero, then k is a totally real algebraic number field and that X/k does not have potential complex multiplication.

Then for all prime numbers l, except finitely many, we have $X(F[c])_{l^\infty} = 0$ for all cycles c of k coprime to the primes in ∞_k.

Proof of Theorem 6.5.6 The proof is divided into two cases depending on the characteristic of k.

Case 1. *Suppose that k is a global function field of positive characteristic.*

Let $n \geq 1$ be an integer coprime to the characteristic of k. Let $M(n, c) = F[c] \cap F(X[n])$, where $F(X[n])$ is the field over F generated by the n-torsion points of X. Then $M(n, c)/F$ is an abelian unramified extension. The unique place of ∞_F is split completely in $F[c]/F$; hence, the place of ∞_F is split completely in $M(n, c)/F$. It follows that the field $M(n, c)$ is contained in a maximal unramified abelian extension F' over F such that ∞_F is split completely in F'/F. Class field theory shows that there are only finitely many such extensions F'/F, and so there are only finitely many possibilities for the field extension $M(n, c)/F$ as n runs over all integers coprime to the characteristic of k and c runs over all cycles in $\mathcal{C}$.

It follows that for all prime numbers l, except finitely many, we have $X(F[c])_{l^\infty} = 0$ for all cycles c of k coprime to ∞_k. [See also [4, Propn. 7.3.8, p.347].]

Case 2. *Suppose that k is a totally real algebraic number field and X/k does not have potential complex multiplication (i.e. X/k does not acquire complex multiplication over any extension field of k).*

For all except finitely many prime numbers l, the Galois action on X_{l^∞} provides a surjective homomorphism (Theorem 2.7.3)

$$G_k \longrightarrow \mathrm{GL}_2(\mathbb{Z}_l).$$

Suppose that l is such a prime number. If $X(F[c])_l \neq 0$ then either $X(F[c])_l = \mathbb{F}_l$ or $X(F[c])_l = \mathbb{F}_l^2$. In the first case, the group scheme $X[l]/k$ would then have a cyclic subgroup scheme defined over k; hence, the image of $G_k \to \mathrm{GL}_2(\mathbb{F}_l)$ would be contained in a Borel subgroup of $\mathrm{GL}_2(\mathbb{F}_l)$, which is a contradiction. In the second case, $k(X[l])$ would be a subfield of $F[c]/k$; hence, there would be a surjection $\mathrm{Gal}(F[c]/k) \to \mathrm{GL}_2(\mathbb{F}_l)$ where $\mathrm{Gal}(F[c]/k)$ is a generalized dihedral group (Proposition 1.5.6). But this is impossible for $l > 2$ as $\mathrm{GL}_2(\mathbb{F}_l)$ is not a quotient of a group of generalised dihedral type (for $l \geq 5$ this holds because $\mathrm{PSL}_2(\mathbb{F}_l)$ is a non-abelian finite simple group; see, e.g. [4, proposition 7.3.6]; see also Exercise 6.5.8). Hence we must have $X(F[c])_l = 0$ for all but finitely many l and for all c. □

Question 6.5.7 With the notation of Theorem 6.5.6, suppose that k is a totally real field, F is a totally imaginary quadratic extension field of k, and that X/k has potential complex multiplication by an imaginary quadratic field extension $\mathfrak{K}/\mathbb{Q}$. Let $\mathcal{C}$ be the set of cycles on k coprime to the primes in ∞_k.

If $\mathfrak{K}$ is not contained in F, does it still hold that for all prime numbers l, except finitely many, we have $X(F[c])_{l^\infty} = 0$ for all cycles $c \in \mathcal{C}$? Suppose that $\mathfrak{K}$ is contained in F, again does it still hold that for all prime numbers l, except finitely many, we have $X(F[c])_{l^\infty} = 0$ for all cycles $c \in \mathcal{C}$?

Exercise 6.5.8 Show that $GL_2(\mathbb{F}_3)$ is not a quotient of a generalized dihedral group (cf. the proof of Theorem 6.5.6).

6.6 The Heegner Point General Euler System: Classical and Drinfeld Modular Curves

(6.6.1) We construct in this section the general Euler system for Heegner points arising from classical modular curves or for Drinfeld modular curves.

(6.6.2) Suppose that

k is a global field equipped with a set of special places ∞, and if k has characteristic zero, we assume that $k = \mathbb{Q}$;

A is the ring of integers of k with respect to ∞;

E/k is an elliptic curve where if k has positive characteristic E/k has split multiplicative reduction at the place in ∞;

$\mathfrak{J}$ is the conductor of E/k without the places contained in ∞;

$T = T_l(E)$ is the l-adic Tate module of E equipped with its action by $\mathrm{Gal}(k^{\mathrm{sep}}/k)$ where l is distinct from the characteristic of k;

F/k is an imaginary quadratic extension field with respect to ∞ where all places in the support of $\mathfrak{J}$ split completely (i.e. F/k satisfies the Heegner condition (3.2.2));

B is the integral closure of A in F; $\mathfrak{J}B = \mathfrak{J}_1\mathfrak{J}_2$ is a splitting of the ideal $\mathfrak{J}B$ (see (3.2.3));

$F[c]$ is the ring class field of F/k with conductor c, for any cycle $c \in \mathrm{Cyc}(k)$ supported only at primes outside ∞;

$\mathcal{N}$ be a cycle of k divisible by all places above the prime number l, if k is a number field, and by all places of k where T is ramified and all places of k where F/k is ramified and by all Archimedean places of k.

Note that $\mathcal{N}$ is divisible by all primes in ∞, because if k is a function field, then the single place in ∞ is a place of bad (more precisely, split multiplicative) reduction of E by assumption, and hence T is ramified at this place. Therefore, this above definition of the cycle $\mathcal{N}$ for this representation T attached to the curve E/k coincides precisely with the definition of the cycle $\mathcal{N}$ of (6.1.1) for more general representations T.

The Cohomology Classes $d_{F[c]}$

(6.6.3) We assume there is a finite surjective morphism of k-schemes

$$\pi : X_0^{\mathrm{Ell.Sp.}} \longrightarrow E.$$

As $X_0^{\text{Ell.Sp.}}$ has cusps in either the classical modular curve case or the Drinfeld modular curve case, we may then translate π so that π is rigidified by ξ (3.4.4).

(6.6.4) Let c be a cycle on k with support coprime to $\mathfrak{J}$ and ∞. Let O_c be the order of F with conductor c. Let $a \in \text{Pic}(O_c)$. Then we have the Heegner point

$$(a, \mathfrak{J}_1, c, \pi) \in E(F[c])$$

which is rational over the ring class field $F[c]$ and which is the image under π of the Heegner point $(a, \mathfrak{J}_1, c) \in X_0^{\text{Ell.Sp.}}(F[c])$ (see Sects. 3.2, 3.4 and Definition 3.4.5).

For z a non-Archimedean place of k where the Tate module T is unramified, we put

$$a_z = \text{Tr}(\text{Fr}_z | T_l(E))$$

which is the trace of a Frobenius element $\text{Fr}_z \in \text{Gal}(k^{\text{sep}}/k)$ at z acting on T, where a_z is an integer.

Proposition 6.6.5 *Assume that if k has characteristic zero, then E/k does not have potential complex multiplication. Take $\mathcal{C}$ to be the cycles on k coprime to $\mathcal{N}$, and for any cycle $c \in \mathcal{C}$, take $u(c) = |O_c^*|/|A^*|$. Take $F_c = F[c]$ and x_c to be the Heegner point $(1, \mathfrak{J}_1, c, \pi) = \pi(1, \mathfrak{J}_1, c) \in E(F[c])$ for all $c \in \mathcal{C}$ where $(1, \mathfrak{J}_1, c, \pi)$ is the Heegner point rational over the ring class field $F[c]$ and which is the image under π of the Heegner point $(a, \mathfrak{J}_1, c) \in X_0^{\text{Ell.Sp.}}(F[c])$, as in (6.6.4), where $a \in \text{Pic}(O_c)$ is taken to be the identity element 1 of this Picard group. Then the axioms Ax.1–Ax.7 all hold for E/k, $T = T_l(E)$, the Heegner points $(1, \mathfrak{J}_1, c, \pi)$, the ring class fields $F[c]$, the set $\mathcal{C}$ of cycles on k, and the integers $a_z, u(c)$. That is to say, these objects form a Heegner system.*

Proof For Ax.1, we take, as stated, $\mathcal{C}$ to be the cycles on k coprime to $\mathcal{N}$. For any cycle c on k, take $u(c) = |O_c^*|/|A^*|$.

For Ax.2, we take the field extension F_c/F to be the ring class field extension $F[c]/F$ for any $c \in \mathcal{C}$, and we put $d_0 = 1$. We have for any $c, c' \in \mathcal{C}$, $F_c \cap F_{c'} = F_{\gcd(c,c')}$ by Lemma 1.5.5.

To show that Ax.3 holds, we take $c, c' \in \mathcal{C}$ to be square-free cycles where $c < c'$. Then there is a canonical isomorphism $\text{Gal}(F_{c'}/F_c) \cong (\prod_{\mathfrak{p} | c'/c} \frac{(B/\mathfrak{p}B)^*}{(A/\mathfrak{p}A)^*})/\Delta_c$ where the product runs over the distinct prime cycles z in $\mathcal{C}$ dividing c'/c and where the subgroup Δ_c of the product is isomorphic to B^*/A^* if $c = 1$ and to 1 if $c \neq 1$ so that $|\Delta_c| = u(c)$; this isomorphism follows by the definition of ring class fields Sect. 1.5 (in the function field case it is given in [4, (2.3.8), (2.3.10)] and the proof in the number field case is identical).

For the four properties of Ax.3, we have that Ax.3(i) follows from (1.5)(a), Ax.3 (iii) follows from (1.5)(c), and Ax.3(iv) follows from (1.5)(b).

6.6 The Heegner Point General Euler System: Classical and Drinfeld...

For Ax.3(ii), we assume that k has characteristic zero and $\mathfrak{n} \in \mathcal{C}$. Then we have $k = \mathbb{Q}$; hence, any non-Archimedean prime $\mathfrak{p}$ of k is principal in A, and if it is inert in F/k, then the prime $\mathfrak{p}B$ of F is principal. Hence if $\mathfrak{p}$ is inert in F/k, the prime $\mathfrak{p}B$ of F splits completely in $F_\mathfrak{n}/F$.

For Ax.4, suppose that $c \in \mathcal{C}$ and $z \in \mathcal{C}$ is a place of k coprime to $\mathrm{Supp}(c)$.

Suppose first also that z is unramified and inert in F/k. Then z is unramified in the field extension $F[c]/k$, because $F[c]/F$ is ramified only at the places in $\mathrm{Supp}(c)$. We then have from Table 3.4.8, where we put throughout $a = 1$ the identity class of the Picard group,

$$\frac{|O_c^*|}{|A^*|}\mathrm{Tr}_{F[cz]/F[c]}(1, \mathfrak{J}_1, cz, \pi) = a_z(1, \mathfrak{J}_1, c, \pi) \tag{6.6.6}$$

and a_z is the trace of the Frobenius of T at z (see (6.6.4)).

Suppose now that z is unramified and split completely in F/k. Then for the order O_c, we have the factorization $\mathfrak{m}_z O_c = \mathfrak{p}_1 \mathfrak{p}_2$ where $\mathfrak{p}_1, \mathfrak{p}_2$ are two distinct prime ideals of O_c and $\mathfrak{m}_z$ is the ideal of A that cuts out z. From Table 3.4.8, we then have the following relation of Heegner points, putting $a = 1$,

$$\frac{|O_c^*|}{|A^*|}\mathrm{Tr}_{F[cz]/F[c]}(1, \mathfrak{J}_1, cz, \pi) = a_z(1, \mathfrak{J}_1, c, \pi) - ([[\mathfrak{p}_1]]^{-1}, \mathfrak{J}_1, c, \pi) - ([[\mathfrak{p}_2]]^{-1}, \mathfrak{J}_1, c, \pi). \tag{6.6.7}$$

From the main theorem of complex multiplication, the Heegner points $([[\mathfrak{p}_i]]^{-1}, \mathfrak{J}_1, c, \pi)$, for $i = 1, 2$, can be written as

$$([[\mathfrak{p}_i]]^{-1}, \mathfrak{J}_1, c, \pi) = \mathrm{Fr}_{y_i}(1, \mathfrak{J}_1, c, \pi)$$

where y_1, y_2 are the two points of F lying over z and Fr_{y_i} is the arithmetic Frobenius element of $\mathrm{Gal}(F[c]/F)$ corresponding to the prime y_i, $i = 1, 2$ where y_i are unramified in $F[c]/F$ as the only places ramified in $F[c]/F$ are those dividing $\mathrm{Supp}(c)$.

The relations (6.6.6) and (6.6.7) can then be combined into the following formula, where y is any point of F lying over z and Fr_y is the Frobenius element of $\mathrm{Gal}(F[c]/F)$ corresponding to the prime y and we put $x_{cz} = (1, \mathfrak{J}_1, cz, \pi)$,

$$\frac{|O_c^*|}{|A^*|}N_{F[cz]/F[c]}x_{cz} = \left(a_z - \mathrm{Fr}_y - \left(\frac{z}{F}\right)\mathrm{Fr}_y^{-1}\right)x_c. \tag{6.6.8}$$

Suppose finally that z is ramified in F/k. From Table 3.4.8, we obtain putting $a = 1$

$$\frac{|O_c^*|}{|A^*|}\mathrm{Tr}_{F[cz]/F[c]}(1, \mathfrak{J}_1, cz, \pi) = a_z(1, \mathfrak{J}_1, c, \pi) - ([[\mathfrak{m}_z']]^{-1}, \mathfrak{J}_1, c, \pi)$$

where $\mathfrak{m}_z'$ is the unique prime ideal of O_c lying above the ideal $\mathfrak{m}_z$ of A defining z and $[[\mathfrak{m}_z']]^{-1}$ is the inverse divisor class in $\mathrm{Pic}(O_c)$ corresponding to $\mathfrak{m}_z'$. From the main theorem of complex multiplication, we have

$$([[\mathfrak{m}'_z]]^{-1}, \mathfrak{I}_1, c, \pi) = \mathrm{Fr}_w(1, \mathfrak{I}_1, c, \pi)$$

where Fr_w is the Frobenius element of $\mathrm{Gal}(F[c]/F)$ corresponding to the unique prime w of F lying over the prime z of k. The place w is unramified in $F[c]/F$. Therefore, the formula (6.6.8) still holds when z is ramified in F/k, for then $\left(\frac{z}{F}\right) = 0$. This demonstrates Ax.4.

To show Ax.5 holds, suppose that $z \notin \mathrm{Supp}(\mathcal{N})$ is a place of k which belongs to $\mathrm{Supp}(c)$ and $c \in \mathcal{C}$. Then we have $|O_c^*|/|A^*| = 1$. From Table 3.4.8, we then obtain, putting $a = 1$,

$$\mathrm{Tr}_{F[zc]/F[c]}(1, \mathfrak{I}_1, cz, \pi) = a_z(1, \mathfrak{I}_1, c, \pi) - (1, \mathfrak{I}_1, c/z, \pi)$$

as required.

Ax.6 follows from Theorem 3.4.9 and Ax.7 follows from Theorem 6.5.6. □

Corollary 6.6.9 *Let* $\delta : E(F[c]) \to H^1(F[c], T)$ *be the Kummer homomorphism (see (6.5.2)). Then the classes* $e_c = \delta x_c$ *for all* $c \in \mathcal{C}$ *where* $x_c = (1, \mathfrak{I}_1, c, \pi) \in E(F[c])$ *form a general Euler system* $(T, \{F[c]\}_c, \mathcal{N})$ *of dimension* 1 *satisfying the distribution relations of Proposition 6.5.3.*

Proof This follows from Proposition 6.5.3. □

Corollary 6.6.10 *Assume that* $\frac{|B^*|}{|A^*|}$ *is a unit of the coefficient ring* $\mathbb{Z}_l$. *Then there is a collection of cohomology classes* $d_c \in H^1(F[c], T)$ *for all* $c \in \mathcal{C}$ *which satisfy the distribution relations of Corollary 6.5.4, and these* d_c *form a general Euler system for* $(T, \{F[c]\}_c, \mathcal{N})$ *of dimension* 1.

The classes d_c *coincide with the classes* e_c *of Corollary 6.6.9 for* $c \neq 1$.

Proof This follows from Proposition 6.6.5 and Corollary 6.5.4 where writing e_c for the cohomology classes of Corollary 6.6.9, we put

$$d_1 = \frac{|B^*|}{|A^*|} e_1, \quad d_c = e_c \text{ if } c \neq 1.$$

□

6.7 The CM Point General Euler System: Shimura Curves

(6.7.1) We construct in this section the general Euler system for CM points arising from Shimura curves.

6.7 The CM Point General Euler System: Shimura Curves

(6.7.2) The notation includes that of Sect. 2.3 on Shimura curves so that we have:

k is a totally real number field of degree d over $\mathbb{Q}$;
O_k is the ring of integers of k;
$\mathbb{Z}_l$ is the coefficient ring (before labelled R) where l is a prime number;
S_B is a finite set of non-Archimedean places of k such that $|S_B| + d$ is odd;
$\tau_1, \ldots, \tau_d : k \to \mathbb{R}$ are the real embeddings of k;
B is the quaternion algebra over k which ramifies precisely at the places $S_B \cup \{\tau_2, \ldots, \tau_d\}$ and $B \neq M_2(\mathbb{Q})$;
H is an open compact subgroup of $\widehat{B}^*$ where $\widehat{B}^* = (B \otimes_{\mathbb{Z}} \widehat{\mathbb{Z}})^*$;
M_H/k and its compactification M_H^*/k are the Shimura curves defined over the field k corresponding to H;
N_H/k, N_H^*/k are the quotient curves of M_H, M_H^* by the group $\widehat{k}^*/(\widehat{k}^* \cap H)$ where $\widehat{k}^* = k^* \otimes_{\mathbb{Z}} \widehat{\mathbb{Z}}$;
F is a totally imaginary quadratic extension field of k such that every prime in S_B is either ramified or inert in F/k (see (2.3));
O_F is the ring of integers of F;
$CM(N_H, F)$ is the set of CM points by F/k on N_H (Definition 2.3.16);
$I_0 \subseteq O_F$ is the nonzero ideal of O_F given by $\bigcap_{u \neq 1}(u-1)O_F$ where u runs over the roots of unity, different from 1, contained in F (see Proposition 3.3.9).

Note that the hypothesis above that $B \neq M_2(\mathbb{Q})$ excludes the case of classical modular curves.

(6.7.3) We recall some notation for CM points on Shimura curves as in (3.4.10)–(3.4.24).

- Suppose that E/k is an elliptic curve equipped with a parametrization by the projective Shimura curve N_H^*/k (see (2.3)), that is to say there is a finite surjective morphism of k-schemes

$$\pi : N_H^* \longrightarrow E \qquad (6.7.4)$$

and which is assumed rigidified by ξ (see (3.4.11)). Let T be the l-adic Tate module of E/k. For z a non-Archimedean place of k where the Tate module T is unramified, we put

$$a_z = \text{Tr}(\text{Fr}_z | T_l(E)) \in \mathbb{Z}$$

which is the trace of a Frobenius element $\text{Fr}_z \in \text{Gal}(k^{\text{sep}}/k)$ at z acting on T.
- Let

$$x = [w, b] \in CM(N_H, F)$$

be a CM point with modulator $c(x)$ (see Definitions 2.3.16 and 3.3.5).

- Let U be a finite set of non-Archimedean places of k containing the ramified places of the quaternion algebra B such that H can be written as $H = H_U H^U$ where

$$H_U \subseteq \prod_{v \in U} B_v^*, \quad H^U = \prod_{v \notin U} H_v \tag{6.7.5}$$

(see (3.4.14) for details).
- Let $\mathcal{S}_1$, $\mathcal{S}_r$ for $r \geq 2$, $\mathcal{S} = \bigcup_{r \geq 1} \mathcal{S}_r$ be the sets of cycles of k given in Definition 3.4.15 and which depend on the modulator $c(x)$ of x as well as U.
- We have the CM points $x(c)$ in $\mathrm{CM}(N_H, F)$ for all $c \in \mathcal{S}$ which are associated with the CM point x as in Definition 3.4.16, and also there are the CM points $\pi x(c)$ on the elliptic curve E.
- There are the positive integers $u(r, x) \in \mathbb{N}$ given in Definition 3.4.21.

Definition 6.7.6 Let $\mathcal{N}$ be the square-free cycle on k which is divisible by all places $\mathfrak{p}$ of k where one of the following holds

- $\mathfrak{p} \in U$;
- $\mathfrak{p}$ is a ramified place of F/k;
- $\mathfrak{p}$ is non-Archimedean and divides either l, or $c(x)$, or $I_0 \subseteq \mathfrak{p}O_F$ where l is the residue characteristic of the coefficient ring $\mathbb{Z}_l$.

Here $I_0 \subseteq O_F$ is the ideal as in (6.7.2) (see Proposition 3.3.9); $c(x)$ is the modulator of the CM point x (Definition 3.3.5), and l is the prime number which is the residue characteristic of the coefficient ring $\mathbb{Z}_l$.

Let $\mathcal{I}$ be the set of non-Archimedean places of k which are inert in F/k. Then $\mathcal{S}_1$ ((6.7.3) and Definition 3.4.15) are the following set

$$\mathcal{S}_1 = \mathcal{I} \setminus \mathrm{Supp}(\mathcal{N}).$$

The set $\mathcal{S}$ is that of all square-free cycles on k divisible only by primes in $\mathcal{S}_1$ (Definition 3.4.15).

[The corresponding definition of $\mathcal{N}$ for classical modular curves or Drinfeld modular curves is in (6.6.2).]

Proposition 6.7.7 *Assume that E/k does not have potential complex multiplication. Let $x = [w, b] \in \mathrm{CM}(N_H, F)$ be a CM point. Take*

- $\mathcal{C} = \{c \in \mathcal{S} | c \text{ coprime to the ramification of } T\}$.
- *For any cycle $c \in \mathcal{C}$, take $u(c) = u(r, x)$ where r is the number of prime cycles in the factorization of c and where $u(r, x)$ is defined in Definition 3.4.21.*

6.7 The CM Point General Euler System: Shimura Curves

- For any $c \in \mathcal{C}$, take $F_c = F(x(c))$ where $F(x(c))$ is the field of rationality of the CM point $x(c)$ (see Definitions 3.3.4, 3.4.12, 3.4.16 and 3.4.21);
- take x_c to be the CM point $\pi x(c) \in E(F(x(c)))$ for all $c \in \mathcal{C}$ where $\pi x(c)$ is the CM point which is the image under π of the CM point $x(c) \in N_H^*(F(x(c)))$ (see also Definition 3.4.12).

Then the axioms Ax.1–Ax.7 of Sect. 6.5 all hold for E/k, $T = T_l(E)$, the CM points $\pi x(c)$, the fields $F(x(c))$, and the set $\mathcal{C}$ of cycles on k and the integers $a_z, u(r, x)$. That is to say, these objects form a Heegner system.

Proof For Ax.1, we take $\mathcal{C}$ to be the square-free cycles on k coprime to $\mathcal{N}$ and coprime to the ramification of T. For any cycle c on k, take $u(c) = u(r, x)$ where is r be the number of prime cycles in the factorisation of c and where $u(r, x)$ is defined in Definition 3.4.21 where $u(r, x) = 1$ for $r \geq 1$. Here $u(0, x)$ equals $[F^* \cap \widehat{k}^* Z(x) : k^*]$ which equals either $[O_F^* \cap Z(x) : O_k^*]$ or $2[O_F^* \cap Z(x) : O_k^*]$.

For Ax.2, we have already taken the field extension F_c/F to be the field extension $F(x(c))/F$ for any $c \in \mathcal{C}$, and we put $d_0 = c(x)$ where $c(x)$ is the modulator of the CM point x.

Let c, c' be cycles on k, without Archimedean components. Let O_k be the ring of integers of k. We have for the O_k-orders O_c, $O_{c'}$ of F that $O_c.O_{c'} = O_{\gcd(c,c')}$. Hence we have, where $\widehat{O}_c = O_c \otimes_{\mathbb{Z}} \widehat{\mathbb{Z}}$ as in (2.3),

$$\widehat{O}_c^* . \widehat{O}_{c'}^* = \widehat{O}_{\gcd(c,c')}^*.$$

Therefore, we have

$$F^* \widehat{k}^* \widehat{O}_c^* . F^* \widehat{k}^* \widehat{O}_{c'}^* = F^* \widehat{k}^* \widehat{O}_{\gcd(c,c')}^*.$$

The reciprocity map then provides isomorphisms (see Definition 3.3.4)

$$\widehat{F}^*/(F^* \widehat{k}^* \widehat{O}_c^*) \cong \mathrm{Gal}(F(x(c))/F), \quad \widehat{F}^*/(F^* \widehat{k}^* \widehat{O}_{c'}^*) \cong \mathrm{Gal}(F(x(c'))/F)$$

$$\widehat{F}^*/(F^* \widehat{k}^* \widehat{O}_c^* . F^* \widehat{k}^* \widehat{O}_{c'}^*) \cong \widehat{F}^*/(F^* \widehat{k}^* \widehat{O}_{\gcd(c,c')}^*)$$

$$\cong \mathrm{Gal}(F(x(\gcd(c, c')))/F).$$

It follows that in a separable closure of F, we have

$$F(x(c)) \cap F(x(c')) = F(x(\gcd(c, c'))).$$

Furthermore, by Definition 3.3.5 and Proposition 3.4.19, $F_c = F(x(c))$ is contained in a ring class field $F[d_0c]$, which shows that Ax.2 holds.

To show that Ax.3 holds, we take cycles $c, c' \in \mathcal{C}$ where $c < c'$. By Proposition 3.3.7, where $Z((x(c))$, $Z(x(c'))$ are as in in Definition 3.3.4, there is the exact sequence of groups

$$0 \longrightarrow \frac{F^* \cap \widehat{k}^* Z(x(c))}{F^* \cap \widehat{k}^* Z(x(c'))} \longrightarrow \frac{Z(x(c))}{Z(x(c'))} \xrightarrow{f} \mathrm{Gal}(F(x(c'))/F(x(c))) \longrightarrow 0.$$

By Proposition 3.4.19, we have the decomposition

$$\frac{Z(x(c))}{Z(x(c'))} \cong \prod_{\mathfrak{p}|c'/c} \frac{Z(x(c))_\mathfrak{p}}{Z(x(c'))_\mathfrak{p}}, \quad \frac{Z(x(c))_\mathfrak{p}}{Z(x(c'))_\mathfrak{p}} \cong \frac{(O_F/\mathfrak{p}O_F)^*}{(O_k/\mathfrak{p}O_k)^*} \quad \text{(for } \mathfrak{p}|(c'/c)\text{)}.$$

By Definition 3.4.21, we have $u(r, x) = [F^* \cap \widehat{k}^* Z(x) : k^*]$ if $r = 0$, and $u(r, x) = 1$ if $r \geq 1$.

As $1 \leq c < c'$, we then have $F^* \cap \widehat{k}^* Z(x(c')) = k^*$ by Proposition 3.3.9 and as the prime cycles in $\mathcal{C}$ are coprime to the ideal I_0 (see Definitions 3.4.15 and 6.7.6).

Let r be the number of prime components in the support of c. If $r > 0$ then by Proposition 3.3.9, we again have $F^* \cap \widehat{k}^* Z(x(c)) = k^*$ and then f is an isomorphism. If $r = 0$, which only occurs if $c = 1$, then we have that $F^* \cap \widehat{k}^* Z(x(c))$ has order $u(0, x)$. We obtain that the kernel of f in the above exact sequence has order $u(r, x)$.

We have then a canonical isomorphism $\mathrm{Gal}(F_{c'}/F_c) \cong (\prod_{\mathfrak{p}|c'/c} \frac{(O_F/\mathfrak{p}O_F)^*}{(O_k/\mathfrak{p}O_k)^*})/\Delta_c$ where the product runs over the distinct prime cycles $\mathfrak{p}$ in $\mathcal{C}$ dividing c'/c where the subgroup Δ_c has order $u(c) = u(r, x)$ and $\Delta_c \cong F^* \cap \widehat{k}^* Z(x(c))$.

For the four properties of Ax.3, we have that Ax.3(i)(ii)(iii) follow, respectively, from Proposition 3.4.20(i)(ii)(iii), and Ax.3(iv) follows immediately from the definition of $F_c = F(x(c))$.

For Ax.4, suppose that $c \in \mathcal{C}$ and $z \in \mathcal{C}$ is a prime place of k coprime to $\mathrm{Supp}(c)$. First note that z is unramified and inert in F/k and hence z is unramified in the field extension $F(x(c))/k$ by Proposition 3.4.20(i) (or Ax.3(i)). Then this Ax.4 follows from Proposition 3.4.23(i) where we have $\left(\frac{z}{F}\right) = -1$.

Ax.5 is vacuous as the cycles in $\mathcal{C}$ are assumed here to be square-free.

Ax.6 follows from Theorem 3.4.23(ii).

As each field F_c, $c \in \mathcal{C}$ is contained in a ring class field $F[d]$ for some d coprime to ∞_k by Ax.2; it follows from Theorem 6.5.6 that Ax.7 holds. □

Corollary 6.7.8 *Let* $\delta : E(F[c]) \to H^1(F[c], T)$ *be the Kummer homomorphism (see (6.5.2)). Then the classes* $e_c = \delta \pi x(c) \in H^1(F(x(c)), T)$, *where* $x(c) \in N_H(F(x(c))$, *form a general Euler system for* $(T, \{F(x(c))\}_{c \in \mathcal{C}}, \mathcal{C})$ *of dimension 0 satisfying the distribution relation that if* $c, z \in \mathcal{C}$ *where* z *is a prime cycle on* k *and* c, z *are coprime then, where* r *is the number of distinct primes in the support of* c,

$$u(r, x) \mathrm{Tr}_{F_{cz}/F_c} e_{cz} = a_z e_c.$$

[This follows from Propositions 6.5.3, 6.7.7 and Remark 6.1.5(iii).] □

6.8 The Heegner Point and CM Point Euler Systems

Corollary 6.7.9 *Assume that $u(r, x)$ is a unit of the coefficient ring $\mathbb{Z}_l$ for all $r \in \mathbb{N}$. Then there is a collection of cohomology classes $d_c \in H^1(F_c, T)$ for all $c \in \mathcal{C}$, which coincide with the classes e_c of Corollary 6.7.8 for $c \neq 1$, and satisfying the distribution relation, for any prime place $z \in \mathcal{C}$ coprime to c,*

$$\mathrm{Tr}_{F_{cz}/F_c} d_{cz} = a_z d_c.$$

The cohomology classes $d_c \in H^1(F_c, T)$, for all $c \in \mathcal{C}$, form a general Euler system for $(T, \{F_a\}_{a \in \mathcal{C}}, \mathcal{C})$ of dimension 0.

[This follows from Propositions 6.5.3, Corollary 6.5.4, Proposition 6.7.7, and Remark 6.1.5(iii).] □

6.8 The Heegner Point and CM Point Euler Systems

(6.8.1) For a system of cohomology classes e_c, $c \in \mathcal{C}$, obtained from points satisfying the axioms of Sect. 6.5, we have seen in Sect. 6.5 that the e_c form a general Euler system provided that $u(c)$ is a unit for all c (Corollary 6.5.4). In this section, we show by restricting the range of the cycles c in this system e_c, we can extract an Euler system in the sense of Rubin of Sect. 6.2.

This is then applied to the Heegner point (resp., CM point) cohomology classes e_c of Corollary 6.6.10 for classical modular curves or Drinfeld modular curves (resp., Corollary 6.7.9 for Shimura curves) to show that the e_c, for a restricted range of c, provide an Euler system again assuming that $u(c)$ is a unit for all c.

(6.8.2) As in Sect. 6.5, let

k be a global field equipped with a special set of places ∞_k;
F be a finite Galois field extension of k which is imaginary quadratic with respect to the special set of places ∞_k;

We assume that the axioms Ax.1–Ax.7 of Sect. 6.5 hold in this section. We then have, where these objects are assumed to satisfy the axioms,

$\mathcal{C}$ a set of cycles on k;
F_c a finite abelian separable field extension of F for all $c \in \mathcal{C}$;
E/k an elliptic curve over k;
$T = T_l(E)$, the l-adic Tate module of E;
$a_z = \mathrm{Tr}(\mathrm{Fr}_z | T)$ is the trace of the Frobenius $\mathrm{Fr}_z \in \mathrm{Gal}(k^{\mathrm{sep}}/k)$ on T at a non-Archimedean place z of k where T is unramified;
$x_c \in E(F_c)$ a point of E defined over F_c for all $c \in \mathcal{C}$;
$u(c)$ a positive integer for all $c \in \mathcal{C}$.

We modify $\mathcal{C}$ and take $\widetilde{\mathcal{C}}$ to be the subset of all cycles of $\mathcal{C}$ without multiple components and coprime to the ramification of F/k:

$$\widetilde{\mathcal{C}} = \left\{ c \in \mathcal{C} \,\middle|\, \begin{array}{c} \text{all prime components of } c \text{ have multiplicity } 1 \\ \text{and are unramified in } F/k \end{array} \right\}$$

Let $\mathcal{F}$ be the separable abelian extension of F given by the union, where c runs over all cycles of $\widetilde{\mathcal{C}}$,

$$\mathcal{F} = \bigcup_{c \in \widetilde{\mathcal{C}}} F_c.$$

Definition 6.8.3 Let K be an intermediate field $F \subseteq_f K \subseteq \mathcal{F}$. The *director* of K is the unique smallest cycle $c \in \widetilde{\mathcal{C}}$ such that $K \subseteq F_c \subseteq \mathcal{F}$.

The existence of the director is evident. The uniqueness of the director c follows from Ax.2 with the equality $F_c \cap F_{c'} = F_{\gcd(c,c')}$.

Lemma 6.8.4 *Let K be an intermediate field $F \subseteq_f K \subseteq \mathcal{F}$, and let c be the director of K. Then the non-Archimedean primes of k which are unramified in F/k but are ramified in K/k are precisely those dividing c.*

Proof Suppose first that $\mathfrak{r}$ is a non-Archimedean prime of k which is unramified in F/k but which ramifies in K/k. Because F_c contains K and F_c/F is ramified only at the primes of F above those in $\mathrm{Supp}(c)$, by Ax.3, it follows that $\mathfrak{r}$ is an element of $\mathrm{Supp}(c)$.

For the converse, let $\mathfrak{r}$ be a non-Archimedean prime of k unramified in K/k. Taking $n_\mathfrak{r} \geq 0$ to be the multiplicity of $\mathfrak{r}$ in c, we have

$$c = n_\mathfrak{r} \mathfrak{r} + c'$$

where c' is a cycle on k which is $\leq c$ and coprime to $\mathfrak{r}$.

The abelian field extension of $F_{c'}/F$ is unramified at all places of F lying over $\mathfrak{r}$ by Ax.3. Furthermore, $F_c/F_{c'}$ is totally ramified at every prime of $F_{c'}$ over $\mathfrak{r}$ again by Ax.3. It follows that $F_{c'}$ is a maximal subfield with $F \subseteq F_{c'} \subseteq F_c$ such that F_c is totally ramified at every prime of $F_{c'}$ over $\mathfrak{r}$ and $F_{c'}/F$ is unramified at every prime of F lying over $\mathfrak{r}$. As K/F is unramified at every prime of F over $\mathfrak{r}$, it follows that the join $K.F_{c'}$ is unramified at every prime of F over $\mathfrak{r}$. By the maximality of $F_{c'}$, we must have $K.F_{c'} = F_{c'}$ and hence $K \subseteq F_{c'}$. Hence by the minimality of the director c, we must have $c = c'$ and hence $\mathfrak{r}$ does not divide c. □

(6.8.5) Assume that $u(c)$ is a unit of the coefficient ring $\mathbb{Z}_l$ for all $c \in \widetilde{\mathcal{C}}$.

Let $d_c \in H^1(F_c, T)$ for all $c \in \widetilde{\mathcal{C}}$ be the cohomology classes given by Corollary 6.5.4.

6.8 The Heegner Point and CM Point Euler Systems

Let K be any intermediate field $F \subseteq_f K \subseteq \mathcal{F}$, and let $c \in \widetilde{\mathcal{C}}$ be the director of K. Define the class $d_K \in H^1(K, T)$ by

$$d_K = N_{F_c/K} d_c \in H^1(K, T).$$

Let $z \in \widetilde{\mathcal{C}}$ be a prime place of k such that z is coprime to c. Let $y \in \Sigma_F$ be any prime place of F lying over z and Fr_y be the arithmetic Frobenius element at z of $\mathrm{Gal}(F_c/k)$, where F_c/k is unramified at z, by Lemma 6.8.4. Let $h_K(z)$ be the endomorphism of $H^1(K, T)$ given by

$$h_K(z) = a_z - \mathrm{Fr}_y - \left(\frac{z}{F}\right)\mathrm{Fr}_y^{-1}$$

where this endomorphism is independent of the choice of y over z.

Then the classes d_c, $c \in \widetilde{\mathcal{C}}$, satisfy the distribution relation (Corollary 6.5.4)

$$\mathrm{Tr}_{F_{cz}/F_c} d_{cz} = \left(a_z - \mathrm{Fr}_y - \left(\frac{z}{F}\right)\mathrm{Fr}_y^{-1}\right) d_c \text{ if } z \text{ is coprime to } \mathrm{Supp}(c).$$

By Corollary 6.5.4, the cohomology classes $d_c \in H^1(F_c, T)$, for all $c \in \mathcal{C}$, form a general Euler system for $(T, \{F_a\}_{a \in \mathcal{C}}, \widetilde{\mathcal{C}})$ of dimension ≤ 1 (see Remark 6.1.5(iii)).

Proposition 6.8.6 *Assume that $u(c)$ is a unit of the coefficient ring $\mathbb{Z}_l$ for all $c \in \widetilde{\mathcal{C}}$. Then the classes $d_K \in H^1(K, T)$, for all intermediate fields $F \subseteq_f K \subseteq \mathcal{F}$, form an Euler system (as in Sect. 6.2 and Remark 6.2.6(vi)) with the distribution relation, where $F \subseteq_f K \subseteq K' \subseteq \mathcal{F}$ and K'/F is finite,*

$$N_{K'/K} d_{K'} = \Big(\prod_{\mathfrak{p} \in \Sigma(K'/K)} h_K(\mathfrak{p}) \Big) d_K$$

where the product runs over the set $\Sigma(K'/K)$ of all non-Archimedean prime places $\mathfrak{p}$ of k such that some place of K above $\mathfrak{p}$ is ramified in the extension K'/K but $\mathfrak{p}$ is unramified in K/k.

Proof Let $k \subseteq F \subseteq \mathcal{F}$. The system of classes

$$\{d_K \in H^1(K, T) \mid \text{for all fields } F \subseteq_f K \subseteq \mathcal{F}\}$$

satisfies the following distribution relation. Suppose that

$$F \subseteq_f K \subseteq K' \subseteq \mathcal{F}$$

where K' is finite over F.

Let c be the director of K and c' that of K', so that we have $K \subseteq F_c$ and $K' \subseteq F_{c'}$.

As $K \subseteq F_c$ and $K \subseteq F_{c'}$, we have that $K \subseteq F_c \cap F_{c'} = F_{\gcd(c,c')}$, by Ax.2, and hence by the minimality of the director c, we have $c \leq \gcd(c, c') \leq c'$. Then we have shown that $c \leq c'$ and $F_c \subseteq F_{c'}$. We have by definition

$$d_{K'} = N_{F_{c'}/K'} d_{F_{c'}}, \quad d_K = N_{F_c/K} d_{F_c}.$$

Furthermore, we have the distribution relation from Corollary 6.5.4

$$N_{F_{c'}/F_c} d_{F_{c'}} = \left(\prod_{\mathfrak{p}} h_{F_c}(\mathfrak{p}) \right) d_{F_c}$$

where the product runs over all places $\mathfrak{p}$ of k which divide c' but do not divide c. By Lemma 6.8.4, the places $\mathfrak{p}$ of k which divide c' but do not divide c are exactly the places $\mathfrak{p}$ of k which ramify in $F_{c'}/k$ but are unramified in F_c/k, and this is the set $\Sigma(F_{c'}/F_c)$, in the notation of Definition 6.2.4.

Hence we have

$$N_{K'/K} d_{K'} = N_{K'/K} \circ N_{F_{c'}/K'} d_{F_{c'}} = N_{F_{c'}/K} d_{F_{c'}} = N_{F_c/K} \circ N_{F_{c'}/F_c} d_{F_{c'}}$$

$$= N_{F_c/K} \left(\prod_{\mathfrak{p} \in \Sigma(F_{c'}/F_c)} h_{F_c}(\mathfrak{p}) \right) d_{F_c}$$

where $\Sigma(F_{c'}/F_c)$ is the set of primes of k ramified in $F_{c'}/k$ but not in F_c/k. But then again by Lemma 6.8.4 $\Sigma(F_{c'}/F_c) = \Sigma(K'/K)$, the set of places $\mathfrak{p}$ of k ramified in K' but unramified in K.

Therefore, we have

$$N_{K'/K} d_{K'} = N_{F_c/K} \left(\prod_{\mathfrak{p} \in \Sigma(K'/K)} h_{F_c}(\mathfrak{p}) \right) d_{F_c} = \left(\prod_{\mathfrak{p} \in \Sigma(K'/K)} h_K(\mathfrak{p}) \right) N_{F_c/K} d_{F_c}$$

$$= \left(\prod_{\mathfrak{p} \in \Sigma(K'/K)} h_K(\mathfrak{p}) \right) d_K.$$

Therefore, we have shown that the classes d_K for all intermediate fields $F \subseteq_f K \subseteq \mathcal{F}$ form an Euler system. □

Definition 6.8.7 Assume that $u(c)$ is a unit of the coefficient ring $\mathbb{Z}_l$ for all $c \in \widetilde{\mathcal{C}}$.

The set of cohomology classes $d_K \in H^1(K, T)$ for all intermediate fields $F \subseteq_f K \subseteq \mathcal{F}$ is the *Heegner point (or CM point) Euler system of T*.

This in particular applies to the cases of elliptic curves parametrized by (a) classical modular curves and Drinfeld modular curves or (b) Shimura curves. The notation (6.6.1), (6.6.2) and (6.7.2), (6.7.3) of the previous two sections holds in the rest of this section Sect. 6.8.

In the case (a) of classical modular curves or Drinfeld modular curves as in Sect. 6.6, the axioms Ax.1–Ax.7 hold by Proposition 6.6.5. Assume that $\frac{|B^*|}{|A^*|}$ is a unit of the coefficient ring $\mathbb{Z}_l$ for the classical/Drinfeld case. Then the cohomology classes $d_K \in H^1(K, T)$ for all intermediate fields $F \subseteq_f K \subseteq \mathcal{F}$ form the *Heegner point Euler system of T*.

In the case (b) of Shimura curves as in Sect. 6.7, the axioms Ax.1–Ax.7 hold by Proposition 6.7.7. Assume that $u(r, x)$ is a unit of the coefficient ring $\mathbb{Z}_l$ for all $r \in \mathbb{N}$. Then the cohomology classes $d_K \in H^1(K, T)$ for all intermediate fields $F \subseteq_f K \subseteq \mathcal{F}$ form the *Heegner point Euler system of T* or also called the *CM point Euler system of T*.

6.9 Appendix: Further Examples of Euler Systems

(6.9.1) In this section, we give some notes on other examples of Euler systems, apart from the Heegner point systems. A complete exposition is beyond the scope of this book, and the reader is directed to the references for more details.

(6.9.2) *Cyclotomic units*. This is the basic example of an Euler system.

Let p be a prime number. For any finite extension field F of the rational field $\mathbb{Q}$, we have by Kummer theory

$$H^1(F, \mathbb{Z}_p(1)) = \varprojlim(F, \mu_n) = \varprojlim F^*/(F^*)^{p^n} = \widehat{F}^*$$

where $\widehat{F}^*$ is the p-adic completion of the abelian group F^*; denote by $\kappa_F : F^* \to H^1(F, \mathbb{Z}_p(1))$ the corresponding inclusion.

Let $\zeta_m = \exp(\frac{2\pi i}{m})$ be a primitive mth root of unity in $\mathbb{C}$, so that these satisfy the compatibility condition $\zeta_{mn}^n = \zeta_m$ for all m, n.

For every positive integer m and prime number l, there is the norm relation in the extension of cyclotomic fields $\mathbb{Q}(\zeta_{ml})/\mathbb{Q}(\zeta_m)$

$$N_{\mathbb{Q}(\zeta_{ml})/\mathbb{Q}(\zeta_m)}(\zeta_{ml} - 1) = \begin{cases} \zeta_m - 1 & \text{if } l \mid m \\ (\zeta_m - 1)^{1-\text{Fr}_l^{-1}} & \text{if } l \nmid m \text{ and } m > 1 \\ (-1)^{l-1} l & \text{if } m = 1. \end{cases}$$

It follows that $\zeta_m - 1$ is an algebraic integer which is a unit away from primes dividing m. The cyclotomic units of a cyclotomic field form a subgroup of the global units of the field constructed by multiplicative combinations of these algebraic integers $\zeta_m - 1$ which are units [43, Chap. 6,§3].

For every positive integer m, we define

$$\eta_{m\infty} = N_{\mathbb{Q}(\mu_{mp})/N(\mathbb{Q}_m)}(\zeta_{mp} - 1) \in \mathbb{Q}(\mu_m)^*$$

and

$$\eta_m = N_{\mathbb{Q}(\mu_m)/N(\mathbb{Q}_m^+)}(c_{m\infty}) \in (\mathbb{Q}(\mu_m)^+)^*$$

where $\mathbb{Q}(\mu_m)^+$ is the maximal real subfield of $\mathbb{Q}(\mu_m)$. Here $m\infty$ and m are Arakelov ideals of $\mathbb{Z}$, that is to say they are ideals that may have infinite places as well as finite places.

Put

$$\xi_m = \kappa_{\mathbb{Q}(\mu_m)}(\eta_m) \in H^1(\mathbb{Q}(\mu_m), \mathbb{Z}_p(1))$$

$$\xi_{m\infty} = \kappa_{\mathbb{Q}(\mu_m)^+}(\eta_{m\infty}) \in H^1(\mathbb{Q}(\mu_m)^+, \mathbb{Z}_p(1)).$$

Let $\mathbb{Q}^{ab}$ be the maximal abelian extension of $\mathbb{Q}$, so that $\mathbb{Q}^{ab} = \bigcup_m \mathbb{Q}(\mu_m)$ where m runs over all positive integers. Then for every non-Archimedean place $\mathfrak{q}$ of $\mathbb{Q}$ where $\mathfrak{q} \neq p$, we have

$$\det(1 - \mathrm{Fr}_\mathfrak{q}^{-1}.x | \mathbb{Z}_p(1)^*) = \det(1 - \mathrm{Fr}_\mathfrak{q}^{-1}.x | \mathbb{Z}_p) = 1 - x.$$

The norm relation above then shows that the cohomology classes ξ_m, $\xi_{m\infty}$ for all positive integers m form an Euler system for $(\mathbb{Z}_p(1), \mathbb{Q}^{ab}, \{p\})$ (Definition 6.2.4 and Proposition 6.3.2).

By means of this Euler system, the "Main Conjecture" of Iwasawa theory can be demonstrated, a result originally proved by Mazur and Wiles.

[For more details on this cyclotomic unit Euler system and its application to the "Main Conjecture", see [62, Chapter III, §2], [La2, Appendix by K. Rubin].]

(6.9.3) *Elliptic units.* Let p be a prime number and F be an imaginary quadratic extension of $\mathbb{Q}$ of class number 1 (this latter is not an essential hypothesis). Let $E/\mathbb{C}$ be an elliptic curve with complex multiplication by the ring of integers $\mathcal{O}$ of F. Fix a non-trivial ideal $\mathfrak{a}$ of $\mathcal{O}$ coprime to 6. Associated with $\mathfrak{a}$ is a rational function Θ on E whose divisor equals $(\Theta) = 12N(\mathfrak{a})[0] - 12\sum_{x \in E[\mathfrak{a}]}[x]$ and which has the following property [64, §7].

Let $\mathfrak{b}$ be an ideal of $\mathcal{O}$ coprime to $\mathfrak{a}$, Q be a point of E of exact order $\mathfrak{b}$, $\mathfrak{p}$ be a prime of $\mathcal{O}$ dividing $\mathfrak{b}$, π be a generator of $\mathfrak{p}$ and $\mathfrak{b}' = \mathfrak{b}/\mathfrak{p}$. If the reduction map $\mathcal{O}^* \to (\mathcal{O}/\mathfrak{b}')^*$ is injective, then we have (see [64, Theorem 7.4, Cor. 7.7])

6.9 Appendix: Further Examples of Euler Systems

- $\Theta(Q) \in F(\mathfrak{b})$ where $F(\mathfrak{b})$ is the ray class field of F mod $\mathfrak{b}$;
- $N_{K(\mathfrak{b})/K(\mathfrak{b}')}\Theta(Q) = \begin{cases} \Theta(\pi Q) & \text{if } \mathfrak{p}|\mathfrak{b}' \\ \Theta(\pi Q)^{1-\text{Fr}_\mathfrak{p}^{-1}} & \text{if } \mathfrak{p} \nmid \mathfrak{b}'; \end{cases}$
- If $\mathfrak{b}$ is not a prime power, then $\Theta(Q) \in F(\mathfrak{b})$ is a global unit. If $\mathfrak{b}$ is a power of a prime $\mathfrak{p}$, then $\Theta(Q) \in F(\mathfrak{b})$ is a unit at primes not dividing $\mathfrak{p}$.

These properties of $\Theta(Q)$ are the basis of the construction of an Euler system of elliptic units where the norm relation above is clearly analogous to that of cyclotomic units in (6.9.2).

Although the concept of an Euler system had not then been defined, the Euler system of elliptic units was, in effect, used by Rubin to give the first known examples of elliptic curves over the rational numbers with finite Shafarevich-Tate groups.

[For more details on elliptic units, see [64] and also [62, Chapter III, §3], [46].]

(6.9.4) *Stickelberger elements.* For every integer $m \geq 2$, put

$$\theta_m = \sum_{a \in (\mathbb{Z}/m\mathbb{Z})^*} \left(\frac{<a>}{m} - \frac{1}{2}\right) g_a^{-1} \in \mathbb{Q}[\text{Gal}(\mathbb{Q}(\mu_m)/\mathbb{Q})]$$

where $<a>$ is an integer depending on a such that $0 \leq\, <a>\, < m$ and $<a> \equiv a$ mod m and furthermore $g_a \in \text{Gal}(\mathbb{Q}(\mu_m)/\mathbb{Q})$ is the automorphism which sends every mth root of unity to its ath power. Define also $\theta_1 = 0$. The θ_m are *Stickelberger elements*.

If $b \in \mathbb{Z}$ is coprime to $2m$, then $(b - g_b)\theta_m \in \mathbb{Z}[\text{Gal}(\mathbb{Q}(\mu_m)/\mathbb{Q})]$, and if l is a prime number, then we have

$$\theta_{ml}|_{\mathbb{Q}(\mu_m)} = \begin{cases} (1 - \text{Fr}_l^{-1})\theta_m & \text{if } l \nmid m \\ \theta_m & \text{if } l|m. \end{cases} \tag{6.9.5}$$

Fix an integer b coprime to $2p$ and for every integer $m \geq 1$ coprime to b define the elements of $\mathbb{Z}[\text{Gal}(\mathbb{Q}(\mu_m)/\mathbb{Q})]$

$$\widehat{\theta}_m = \begin{cases} (b - g_b)\theta_m & \text{if } p|m \\ (b - g_b)(1 - \text{Fr}_p^{-1})\theta_m & \text{if } p \nmid m. \end{cases}$$

where this $\widehat{\theta}_m$ is defined so that we have

$$\widehat{\theta}_{mp}|_{\mathbb{Q}(\mu_m)} = \widehat{\theta}_m \tag{6.9.6}$$

for every m coprime to b. Note that $\widehat{\theta}_m$ depends on b unlike θ_m.

Let F be a finite extension field of $\mathbb{Q}$. Then class field theory gives the isomorphisms

$$H^1(F, \mathbb{Z}_p) \cong \text{Hom}(G_F, \mathbb{Z}_p) \cong \text{Hom}(\mathbb{A}_F^*/F^*, \mathbb{Z}_p) \tag{6.9.7}$$

where $\mathbb{A}_F^*/F^*$ is the idèle class group of F and Hom denotes the group of continuous homomorphisms. An element of $\text{Hom}(\mathbb{A}_F^*/F^*, \mathbb{Z}_p)$ necessarily vanishes on the subgroup $F^* \prod_{w|\infty} F_w^* \prod_{w|p} \{1\} \prod_{w \nmid p\infty} \mathcal{O}_{F,w}^*$ of the idèle group $\mathbb{A}_F^*$, where O_F is the ring of integers of F. Hence all homomorphisms in $\text{Hom}(\mathbb{A}_F^*/F^*, \mathbb{Z}_p)$ factor through the quotient B_F of $\mathbb{A}_F^*/F^*$ by this subgroup. Then B_F lies in the exact sequence

$$0 \longrightarrow U_F/\mathcal{E}_F \longrightarrow B_F \longrightarrow C_F \longrightarrow 0$$

where C_F is the divisor class group of F and U_F are the local units of $F \otimes_{\mathbb{Z}} \mathbb{Q}_p$ and $\mathcal{E}_F$ is the p-adic closure in U_F of the group of global units O_F^* of F.

For every positive integer n, let

$$\lambda_m : \mathbb{Z}_p[\mu_m]^* \longrightarrow \mathbb{Z}_p$$

be the homomorphism defined in [62, Appendix D, §1]. The maps λ_m have the following property [62, Appendix D, lemma 1.4]: for every integer $m \geq 1$ coprime to l, there is a commutative diagram

$$\begin{array}{ccc}
\mathbb{Z}_p[\mu_{ml}]^* & \xrightarrow{\lambda_{ml}} & \mathbb{Z}_p \\
{\scriptstyle 1 \text{ or } -\text{Fr}_l}\uparrow & \nearrow \lambda_m & \\
\mathbb{Z}_p[\mu_m]^* & &
\end{array} \tag{6.9.8}$$

where the vertical left arrow is one of the following:

- the inclusion $\mathbb{Z}_p[\mu_m]^* \hookrightarrow \mathbb{Z}_p[\mu_{ml}]^*$ if $l | m$ or $l = p$;
- the map $-\text{Fr}_p$ followed by the inclusion $\mathbb{Z}_p[\mu_m]^* \hookrightarrow \mathbb{Z}_p[\mu_{ml}]^*$ if $l \nmid mp$.

By Stickelberger's theorem, $\widehat{\theta}_m$ annihilates the divisor class group $C_{\mathbb{Q}(\mu_m)}$ so that we have $\widehat{\theta}_m C_{\mathbb{Q}(\mu_m)} = 0$. As $U_{\mathbb{Q}(\mu_m)} = \mathbb{Z}_p[\mu_m]^*$, the map of multiplication by $\widehat{\theta}_m$ on B_F can be considered by the above exact sequence to be a homomorphism $B_F \to \mathbb{Z}_p[\mu_m]^*/\mathcal{E}_F$.

Let ϕ_m be the composite homomorphism where c is complex conjugation, so that $(1-c)\mathcal{E}_{\mathbb{Q}_p(\mu_m)}$ is a finite group

$$\phi_m : B_{\mathbb{Q}(\mu_m)} \xrightarrow{\widehat{\theta}_m} \mathbb{Z}_p[\mu_m]^*/\mathcal{E}_{\mathbb{Q}_p[\mu_m]} \xrightarrow{1-c} \mathbb{Z}_p[\mu_m]^*/(\mathbb{Z}_p[\mu_m]^*)_{\text{tors}} \xrightarrow{\lambda_m} \mathbb{Z}_p.$$

The first map here is multiplication by $\widehat{\theta}_m$ on $B_{\mathbb{Q}(\mu_m)}$, and the last homomorphism $\lambda_m : \mathbb{Z}_p[\mu_m]^* \longrightarrow \mathbb{Z}_p$ evidently has $(\mathbb{Z}_p[\mu_m]^*)_{\text{tors}}$ in its kernel as $\mathbb{Z}_p$ is torsion free.

6.9 Appendix: Further Examples of Euler Systems

Let $c'_m \in H^1(\mathbb{Q}(\mu_m), \mathbb{Z}_p)$ be the element corresponding to $\phi_m : B_{\mathbb{Q}(\mu_m)} \to \mathbb{Z}_p$ under the canonical isomorphism $H^1(\mathbb{Q}(\mu_m), \mathbb{Z}_p) \cong \mathrm{Hom}(B_{\mathbb{Q}(\mu_m)}, \mathbb{Z}_p)$ obtained from (6.9.7).

Then these properties over $\widehat{\theta}_m$ show immediately that the elements c'_m satisfy the following distribution relation.

Proposition 6.9.9 *Suppose that m is prime to b and that l is a prime not dividing b. Then we have*

$$N_{\mathbb{Q}_p(\mu_{ml})/\mathbb{Q}_p(\mu_m)}(c'_{ml}) = \begin{cases} (1 - \mathrm{Fr}_l^{-1})c'_m & \text{if } l \nmid mp \\ \\ c'_m & \text{if } l \mid mp. \end{cases}$$

Proof Suppose first that $l \nmid mp$. By (6.9.8), we have

$$\lambda_{ml}|_{\mathbb{Z}_p[\mu_m]^*} = \lambda_m \circ (-\mathrm{Fr}_l).$$

By (6.9.5), we have $\widehat{\theta}_{ml}|_{\mathbb{Q}(\mu_m)} = (1 - \mathrm{Fr}_l^{-1})\widehat{\theta}_m$. It follows that

$$\phi_{ml}|_{B_{\mathbb{Q}_p(\mu_m)}} = \phi_m \circ (-\mathrm{Fr}_l)(1 - \mathrm{Fr}_l^{-1}) = \phi_m \circ (1 - \mathrm{Fr}_l) = (1 - \mathrm{Fr}_l^{-1})\phi_m$$

from which it follows immediately that $N_{\mathbb{Q}_p(\mu_{ml})/\mathbb{Q}_p(\mu_m)}(c'_{ml}) = (1 - \mathrm{Fr}_l^{-1})c'_m$. The case where $l \mid mp$, using (6.9.6) and (6.9.8), follows similarly. □

(6.9.10) The last proposition shows that $\mathbf{c}'$ is a zero dimensional generalized Euler system but is not quite an Euler system for the trivial representation $\mathbb{Z}_p$.

We have for every prime number $l \neq p$

$$\det(1 - \mathrm{Fr}_l^{-1}.x|\mathbb{Z}_p^*) = \det(1 - \mathrm{Fr}_l^{-1}.x|\mathbb{Z}_p(1)) = 1 - l^{-1}x.$$

As we have

$$1 - l^{-1}x \equiv 1 - x \mod (l-1)\mathbb{Z}_p[x]$$

from Proposition 6.4.7 and Example 6.4.9(ii), there is an isomorphism of zero dimensional generalized Euler systems $\mathbf{c}' \to \mathbf{c}$ where the collection of classes $c_m \in H^1(\mathbb{Q}(\mu_m), \mathbb{Z}_p)$, for all $m > 1$ coprime to b, satisfies the distribution relation

$$N_{\mathbb{Q}_p(\mu_{ml})/\mathbb{Q}_p(\mu_m)}(c_{ml}) = \begin{cases} (1 - l^{-1}\mathrm{Fr}_l^{-1})c_m & \text{if } l \nmid mp \\ \\ c_m & \text{if } l \mid mp \end{cases}$$

where $\det(1 - \mathrm{Fr}_l^{-1}.x|\mathbb{Z}_p^*) = 1 - l^{-1}x$. Therefore, **c** is an Euler system (Definition 6.2.4) for $(\mathbb{Z}_p, \mathcal{F}, bp)$ where $\mathcal{F}$ is the maximal abelian extension of $\mathbb{Q}$ unramified above every prime dividing b.

[Sse also [62, Chapter III, §4].]

(6.9.11) *Kato's Euler system.* Kato's Euler system is formed for modular elliptic curves.

Let $E/\mathbb{Q}$ be a modular elliptic curve equipped with a parameterization

$$\phi : X_0(N) \to E$$

where the positive integer N is the conductor of E. Let p be a prime number not dividing $2N$ and $T = T_p(E)$ be the p-adic Tate module of E equipped with its Galois action by $\mathrm{Gal}(\mathbb{Q}^{\mathrm{sep}}/\mathbb{Q})$.

We choose two auxiliary integers $D, D' > 1$ which are coprime to $6Np$. Define the sets of positive integers

$$\mathcal{R}'_p = \{\text{square-free positive integers coprime to } 6NpDD'\}$$

$$\mathcal{R}_p = \{r = r_0 p^m \mid r_0 \in \mathcal{R}'_p, m \geq 1\}.$$

Then Kato, by means of Beilinson elements in the K-theory of modular curves, constructs a system of cohomology classes

$$\xi_r \in H^1(\mathbb{Q}(\mu_r), T)$$

satisfying the following norm relations:

(a) For every $r \in \mathcal{R}_p$,

$$N_{\mathbb{Q}(\mu_{rp})/\mathbb{Q}(\mu_r)} \xi_{rp} = \xi_r;$$

(b) If l is a prime number and $\gcd(l, NDD'r) = 1$ then

$$N_{\mathbb{Q}(\mu_{lp})/\mathbb{Q}(\mu_r)} \xi_{lr} = Q_l(\mathrm{Fr}_l) \xi_r$$

where $Q_l(x) = \det(1 - \mathrm{Fr}_l x \mid T^*(1)) = (1 - l^{-1} a_l \mathrm{Fr}_l + l^{-1} \mathrm{Fr}_l^2)$, $\mathrm{Fr}_l \in \mathrm{Gal}(\mathbb{Q}(\mu_r)/\mathbb{Q})$ is a geometric Frobenius and a_l is the trace of the Frobenius on the p-adic Tate module T of E.

To obtain an Euler system from this in the exact form given in Sect. 6.2, one takes the norm (corestriction) of ξ_r to $\mathbb{Q}_{m-1}(\mu_{r_0})$ where $r = r_0 p^m$ and $\mathbb{Q}_{m-1}/\mathbb{Q}$ is the unique extension of degree p^{m-1} contained in the cyclotomic $\mathbb{Z}_p$-extension of $\mathbb{Q}$.

6.9 Appendix: Further Examples of Euler Systems

Using this Ruler system, Kato is able to demonstrate, under hypotheses, the finiteness of the Shafarevich-Tate groups of the elliptic curve E over abelian extensions of $\mathbb{Q}$. In particular, this reproved a theorem of Kolyvagin without requiring auxiliary analytic results due to Gross, Zagier, Bump, Friedberg, and Hoffstein.

[For more details on Kato's Euler system, see [32,65,70], [62, Chap. III, §§5.2–5.5].

(6.9.12) Some other examples of Euler systems are: (a) Siegel unit Euler systems; (b) Flach elements for the symmetric square of an elliptic curve [20]; (c) Gauss sums over algebraic number fields [34,63] and [41]; and (d) various Euler systems for higher K-groups.

Kolyvagin Systems

Contents

7.1	Kolyvagin Systems	177
7.2	Euler Systems Induce Kolyvagin Systems	188
7.3	Preliminaries on Heegner Point and CM Point Kolyvagin Systems	190
7.4	Derived Cohomology Classes	192
7.5	Properties of the Derived Classes	198
7.6	Heegner Point and CM Point Kolyvagin Systems	206

The objective of this chapter is to construct a Kolyvagin system starting from the Heegner systems defined in Sect. 6.5.

Special cases of these Heegner systems are the Heegner points and CM points on elliptic curves parameterized by classical modular curves and Drinfeld modular curves as well as Shimura curves. This will enable us to construct a Kolyvagin system for these instances.

7.1 Kolyvagin Systems

(7.1.1) A Kolyvagin system is a framework for all the derived cohomology classes attached to an Euler system. The derived cohomology classes of Euler systems can provide finiteness theorems and structure theorems for Selmer groups. Mazur and Rubin [54] showed that these theorems were consequences of the Kolyvagin system itself and furthermore that there are Kolyvagin systems of Galois representations for which no corresponding Euler system is known.

Kolyvagin systems were originated by Mazur and Rubin and are expounded in their monograph [54]. The theory of the monograph [54] applies to many Kolyvagin systems, but one case it does not cover is that of Heegner points on elliptic curves over the rational number field. Howard [30] extended the Kolyvagin systems of Mazur and Rubin to include this latter case.

We shall extend the Kolyvagin systems of [30] and [54] in order to include Heegner points on classical modular curves as well as Drinfeld modular curves and CM points on Shimura curves.

(7.1.2) For this section, we let

k be a global field;
F be a finite Galois field extension of k;
R be a coefficient ring where the residue characteristic l of R is different from the characteristic of k (see (4.1.2));
$\mathfrak{m}$ be the maximal ideal of R;
T be an R-representation of G_F.

Selmer triples, the integers $d_\mathfrak{q}$, ideals I_n and groups Γ_n

(7.1.3) A *Selmer triple* $(T, \mathcal{F}, \mathcal{L})$ on F consists of an R-representation T of G_F which is a finite free R-module which is unramified outside a finite set of primes of F, a choice of Selmer structure $\mathcal{F}$ on T (see (5.1.1)) and a set $\mathcal{L}$ of places of F disjoint from $S(\mathcal{F})$.

Definition 7.1.4 Suppose that $(T, \mathcal{F}, \mathcal{L})$ is a Selmer triple, and suppose that T extends to a G_k-representation.

- For each place $\mathfrak{q}$ of F in $\mathcal{L}$, let $d_\mathfrak{q}$ be a positive integer dividing $|\kappa(\mathfrak{q})^*|$.
- For $\mathfrak{q} \in \mathcal{L}$, the ideal $I_\mathfrak{q}$ of R is that generated by $d_\mathfrak{q}$ and $P_\mathfrak{q}(1)$ where $\mathfrak{p}$ is the prime of k lying below $\mathfrak{q}$ of F and

$$P_\mathfrak{q}(x) = \det(1 - \mathrm{Fr}_\mathfrak{p}.x|T) \in R[x]$$

is the characteristic polynomial of a Frobenius element $\mathrm{Fr}_\mathfrak{p} \in G_k$ at $\mathfrak{p}$ acting on T [NB. Note that $P_\mathfrak{q}(x)$ is the characteristic polynomial of $\mathrm{Fr}_\mathfrak{p}$ and not of $\mathrm{Fr}_\mathfrak{q}$.]
- The sets of primes $\mathcal{L}_i$ of F for positive integers i are subsets of $\mathcal{L}$ defined by

$$\mathcal{L}_i = \left\{ \mathfrak{q} \in \mathcal{L} \;\middle|\; \begin{array}{c} I_\mathfrak{q} \subseteq \mathfrak{m}^i \text{ and} \\ T/(\mathfrak{m}^i T + (\mathrm{Fr}_\mathfrak{q} - 1)T) \text{ is free} \\ \text{of finite rank over } R/\mathfrak{m}^i. \end{array} \right\} \qquad (i \geq 1).$$

7.1 Kolyvagin Systems

We put conventionally $\mathcal{L}_0 = \mathcal{L}$.
- For $\mathfrak{q} \in \mathcal{L}$, $\Gamma_\mathfrak{q}$ is defined to be the multiplicative group

$$\Gamma_\mathfrak{q} = \kappa(\mathfrak{q})^*.$$

Denote by $\sigma_\mathfrak{q} \in \Gamma_\mathfrak{q}$ a fixed generator of the finite cyclic group $\Gamma_\mathfrak{q}$.
- The set $\mathcal{M}_{i,F}$, for $i \in \mathbb{N}$, is that of squarefree cycles of F which are products of distinct primes $\mathfrak{q}$ of $\mathcal{L}_i$ lying over primes $\mathfrak{p}$ of k which are inert and unramified in F/k. We assume by convention the unit cycle 1 belongs to $\mathcal{M}_{i,F}$ for all $i \in \mathbb{N}$.
- For a cycle $n \in \mathcal{M}_{i,F}$, define

$$I_n = \sum_{\mathfrak{q}|n} I_\mathfrak{q}, \qquad \Gamma_n = \bigotimes_{\mathfrak{q}|n} \Gamma_\mathfrak{q}$$

where the sums and products run over all primes $\mathfrak{q}$ of $\mathcal{M}_{i,F}$ dividing n and the tensor products are over $\mathbb{Z}$ and where I_n is an ideal of R. We write by convention $I_1 = 0$, $\Gamma_1 = \mathbb{Z}$.

(7.1.5) Let $(T, \mathcal{F}, \mathcal{L})$ be a Selmer triple. Suppose that a, b, c are cycles of F without Archimedean components and whose supports are relatively coprime and that c is not divisible by any prime in $S(\mathcal{F})$. Then we may define a Selmer triple $(T, \mathcal{F}_a^b(c), \mathcal{L}(abc))$ by taking $\mathcal{F}_a^b(c)$ to be the modified Selmer structure of Sect. 5.2 and putting
$S(\mathcal{F}_a^b(c)) = S(\mathcal{F}) \cup \mathrm{Supp}(abc)$
$\mathcal{L}(abc) = \mathcal{L} \setminus \mathrm{Supp}(abc)$
that is to say $S(\mathcal{F}_a^b(c))$ is equal to $S(\mathcal{F})$ augmented by the primes dividing abc and $\mathcal{L}(abc)$ is equal to $\mathcal{L}$ with the primes dividing abc removed.

For the Selmer structure $\mathcal{F}_a^b(c)$, one takes the Selmer structure $\mathcal{F}$ for primes not dividing abc, the strict condition for primes dividing a, the relaxed condition for primes dividing b, and the transverse condition for primes dividing c.

For the case of places v dividing c, a choice of field extension L/F_v is made for each $v|c$ where L is a maximal tamely ramified abelian extension of F_v depending on v; this is in order to define the L-transverse condition at v.

If one of a, b, c is equal to 1, then it is omitted from the notation $(T, \mathcal{F}_a^b(c), \mathcal{L}(abc))$.

Comparing Selmer modules

(7.1.6) Let $(T, \mathcal{F}, \mathcal{L})$ be a Selmer triple, and suppose that T extends to a G_k-representation.

Let i be a positive integer. Suppose that $m \in \mathcal{M}_{i,F}$ and $\mathfrak{q}$ is a prime of F dividing m. Let $\mathfrak{p}$ be the prime of k lying under $\mathfrak{q}$. Then we have the following, from the definitions of this section:

(i) $T/I_m T$ is a finite free R/I_m-module and $|\kappa(\mathfrak{q})^*|.(T/I_m T) = 0$;

(ii) If $\text{Fr}_\mathfrak{q}$ is a Frobenius element of G_F at $\mathfrak{q}$, then we have

$$\det(1 - \text{Fr}_\mathfrak{q} \mid T/I_m T) = 0.$$

[To see this, let $\text{Fr}_\mathfrak{p}$ be a Frobenius element of $G_{k_\mathfrak{p}}$ at $\mathfrak{p}$. Then we have $\text{Fr}_\mathfrak{q} = \text{Fr}_\mathfrak{p}^d$, where $d = [F : k]$, as $\mathfrak{p}$ is inert and unramified in F/k. We then obtain

$$\det(1 - \text{Fr}_\mathfrak{q} \mid T/I_m T) = \det(1 - \text{Fr}_\mathfrak{p}^d \mid T/I_m T) =$$

$$= \det((1 - \text{Fr}_\mathfrak{p})(\sum_{i=0}^{d-1} \text{Fr}_\mathfrak{p}^i) \mid T/I_m T) =$$

$$= \det(1 - \text{Fr}_\mathfrak{p} \mid T/I_m T)\det(\sum_{i=0}^{d-1} \text{Fr}_\mathfrak{p}^i \mid T/I_m T) = 0];$$

(iii) $T/(I_m T + (\text{Fr}_\mathfrak{q} - 1)T)$ is a finite free R/I_m-module.

Therefore the hypotheses for the existence of a finite-singular comparison isomorphism $\phi_\mathfrak{q}^{\text{fs}}$, Definition 4.5.5, are satisfied at $\mathfrak{q}$ for the field F, and we obtain an isomorphism, which depends on the corresponding block decomposition,

$$\phi_\mathfrak{q}^{\text{fs}} : H_f^1(F_\mathfrak{q}, T/I_m T) \longrightarrow H_s^1(F_\mathfrak{q}, T/I_m T) \otimes_\mathbb{Z} \Gamma_\mathfrak{q}.$$

(7.1.7) Suppose now that i is a positive integer and $n\mathfrak{q} \in \mathcal{M}_{i,F}$ where $\mathfrak{q}$ is a prime place of F, so that n and $\mathfrak{q}$ are coprime. Then we may compare $H^1_{\mathcal{F}(n)}(F, T/I_n T)$ and $H^1_{\mathcal{F}(n\mathfrak{q})}(F, T/I_{n\mathfrak{q}} T)$ at the prime $\mathfrak{q}$ as follows.

We have the diagram of homomorphisms where the two homomorphisms labelled $\text{res}_{F_\mathfrak{q}/F}$ are localizations at the prime $\mathfrak{q}$ of F where the topmost $\text{res}_{F_\mathfrak{q}/F}$ is the composition of localization at $\mathfrak{q}$ with the change of representation $T/I_n T \to T/I_{n\mathfrak{q}} T$ and $\phi_\mathfrak{q}^{\text{fs}}$ is a finite-singular comparison map $H_f^1(F_\mathfrak{q}, T/I_{n\mathfrak{q}} T) \to H_s^1(F_\mathfrak{q}, T/I_{n\mathfrak{q}} T) \otimes_\mathbb{Z} \Gamma_\mathfrak{q}$ tensored over $\mathbb{Z}$ with Γ_n

$$\begin{array}{c} H^1_{\mathcal{F}(n)}(F, T/I_n T) \otimes_\mathbb{Z} \Gamma_n \\ \downarrow \text{res}_{F_\mathfrak{q}/F} \\ H_f^1(F_\mathfrak{q}, T/I_{n\mathfrak{q}} T) \otimes_\mathbb{Z} \Gamma_n \\ \downarrow \phi_\mathfrak{q}^{\text{fs}} \\ H^1_{\mathcal{F}(n\mathfrak{q})}(F, T/I_{n\mathfrak{q}} T) \otimes_\mathbb{Z} \Gamma_{n\mathfrak{q}} \xrightarrow{\text{res}_{F_\mathfrak{q}/F}} H_s^1(F_\mathfrak{q}, T/I_{n\mathfrak{q}} T) \otimes_\mathbb{Z} \Gamma_{n\mathfrak{q}} \end{array} \quad (7.1.8)$$

7.1 Kolyvagin Systems

Definition of Kolyvagin Systems

Definition 7.1.9 Suppose that $(T, \mathcal{F}, \mathcal{L})$ is a Selmer triple and that T extends to a G_k-representation and i is a positive integer. A *Kolyvagin system* for $(T, \mathcal{F}, \mathcal{L})$ is a collection κ of cohomology classes

$$\kappa_n \in H^1_{\mathcal{F}(n)}(F, T/I_n T) \otimes_{\mathbb{Z}} \Gamma_n$$

for every $n \in \mathcal{M}_{i,F}$ such that for any $n\mathfrak{q} \in \mathcal{M}_{i,F}$ where $\mathfrak{q}$ is a prime divisor of F then the images of κ_n and $\kappa_{n\mathfrak{q}}$ in $H^1_s(F_\mathfrak{q}, T/I_{n\mathfrak{q}} T) \otimes_{\mathbb{Z}} \Gamma_{n\mathfrak{q}}$ agree under the maps of diagram (7.1.8), that is to say they satisfy the equation

$$\phi^{\text{fs}}_\mathfrak{q}(\text{res}_{F_\mathfrak{q}/F}(\kappa_n)) = \text{res}_{F_\mathfrak{q}/F}(\kappa_{n\mathfrak{q}}) \in H^1_s(F_\mathfrak{q}, T/I_{n\mathfrak{q}} T) \otimes_{\mathbb{Z}} \Gamma_{n\mathfrak{q}}.$$

For a fixed positive integer i and a fixed choice of finite-singular comparison maps $\phi^{\text{fs}}_\mathfrak{q}$, the Kolyvagin systems for $(T, \mathcal{F}, \mathcal{L})$ form an R-module with the evident operations of addition and multiplication by elements of R; this module is denoted $\mathbf{KS}(T, \mathcal{F}, \mathcal{L})$. Note that the tensor products $- \otimes_{\mathbb{Z}} \Gamma_n$ in the above definition are over $\mathbb{Z}$ (and not R).

Remarks 7.1.10

(i) The group $\Gamma_\mathfrak{q}$ is cyclic of order $|\kappa(\mathfrak{q})^*|$ which is divisible by $d_\mathfrak{q}$. It then follows from the definitions of 7.1.4 that $\Gamma_n \otimes_{\mathbb{Z}} R/I_n$ is a free R/I_n-module of rank 1 for all $n \in \mathcal{M}_{0,F}$.

(ii) (*L-transverse condition.*) For places v dividing c where the transverse condition is imposed for $\mathcal{F}^b_a(c)$, a choice of field extension L/F_v is made, so that L/F_v is a maximal totally tamely ramified and separable abelian extension (see paragraph (4.3.4) for the definition of the *L*-transverse condition). In general, there is no canonical choice for these fields L, but in the special cases where k is $\mathbb{Q}$ or F is an imaginary quadratic extension of $\mathbb{Q}$, then there is a canonical choice of such fields L; see (iii).

By local class field theory, a tamely totally ramified abelian extension of a non-Archimedean local field K has degree at most $|\kappa^*|$ where κ is the residue field of K. Furthermore, such extensions of degree $|\kappa^*|$ always exist but need not be unique. Thus for the L-transverse local condition, the field L is necessarily maximal with the properties that it is a tamely totally ramified abelian extension.

(iii) (*L-transverse condition continued.*) For the field $\mathbb{Q}_l$, where l is a prime number, then $\mathbb{Q}_l(\mu_l)/\mathbb{Q}_l$ is a maximal totally tamely ramified abelian extension. Therefore, for $\mathbb{Q}_l$, this can be a canonical choice of L-transverse condition.

Suppose that the ground field k is $\mathbb{Q}$ or more generally is a totally real algebraic number field with class number 1. Suppose that F/k is a totally imaginary quadratic field extension, $\mathfrak{p}$ is a non-Archimedean prime of k which remains inert in F, and $\mathfrak{q}$ is the prime of F over $\mathfrak{p}$. Let $F[\mathfrak{p}]$ be the ring class field of conductor $\mathfrak{p}$. Then $\mathfrak{q}$ splits completely in the Hilbert class field extension $F[1]/F$ as $\mathfrak{p}$ and q are principal ideals of their corresponding rings of integers.

Then the local extension $F[\mathfrak{p}]_q/F_q$ is a maximal totally tamely ramified abelian extension of F_q whose galois group $\operatorname{Gal}(L/F_q)$ is canonically identified with $\kappa(q)^*/\kappa(\mathfrak{p})^*$ by class field theory. Therefore, for such a q, there is a canonical choice of L-transverse condition.

(iv) The module $\mathbf{KS}(T, \mathcal{F}, \mathcal{L})$, as well as a Kolyvagin system itself, depends on the choice of the positive integer i and the choice of finite singular comparison isomorphisms ϕ_q^{fs}, where these latter depend in turn upon the block decomposition of the cohomology groups. So this module should really be written $\mathbf{KS}(T, \mathcal{F}, \mathcal{L}, i, \{\phi_q^{\text{fs}}\})$.

Functoriality of Kolyvagin systems

(7.1.11) The module $\mathbf{KS}(T, \mathcal{F}, \mathcal{L})$, with a fixed choice of the positive integer i and a fixed choice of finite singular comparison isomorphisms ϕ_q^{fs}, has evident functorial properties.

- If $\mathcal{K} \subseteq \mathcal{L}$, there is a R-module
  homomorphism $\mathbf{KS}(T, \mathcal{F}, \mathcal{L}) \to \mathbf{KS}(T, \mathcal{F}, \mathcal{K})$;
- If $\mathcal{F} \leq \mathcal{G}$ are Selmer structures on T and $\mathcal{L}$ is disjoint from $S(\mathcal{F}) \cup S(\mathcal{G})$ then there is an inclusion of R-modules $\mathbf{KS}(T, \mathcal{F}, \mathcal{L}) \subseteq \mathbf{KS}(T, \mathcal{G}, \mathcal{L})$;
- If $R \to R'$ is a local homomorphism of coefficient rings, then there is an R'-module homomorphism $\mathbf{KS}(T, \mathcal{F}, \mathcal{L}) \otimes_R R' \to \mathbf{KS}(T \otimes_R R', \mathcal{F} \otimes_R R', \mathcal{L})$ where the local condition $\mathcal{F} \otimes_R R'$ is defined to be the image of

$$H^1_{\mathcal{F}}(F_q, T) \otimes_R R' \to H^1(F_q, T \otimes_R R')$$

  for q a place of F and $S(\mathcal{F} \otimes_R R') = S(\mathcal{F})$;
- Let $T \to T'$ be an R-homomorphism of G_k-representations, $\mathcal{F}$ be a Selmer structure on T, and $\mathcal{F}'$ be the Selmer on T' where the local condition $\mathcal{F}'$ on T' at a place q of F is defined to be the image of

$$H^1_{\mathcal{F}}(F_q, T) \to H^1(F_q, T')$$

  for q a place of F and $S(\mathcal{F}') = S(\mathcal{F})$. Then there is an induced homomorphism

$$\mathbf{KS}(T, \mathcal{F}, \mathcal{L}) \longrightarrow \mathbf{KS}(T', \mathcal{F}', \mathcal{L});$$

- If $(T_1, \mathcal{F}_1, \mathcal{L})$, $(T_2, \mathcal{F}_2, \mathcal{L})$ are Selmer triples over R with the same $\mathcal{L}$, then there is an isomorphism
$$\mathbf{KS}(T_1 \oplus T_2, \mathcal{F}_1 \oplus \mathcal{F}_2, \mathcal{L}) \cong \mathbf{KS}(T_1, \mathcal{F}_1, \mathcal{L}) \oplus \mathbf{KS}(T_2, \mathcal{F}_2, \mathcal{L}).$$

7.1 Kolyvagin Systems

(7.1.12) (*Change a finite set of places to the transversal condition.*) Let i be a positive integer, and let $m \in \mathcal{M}_{i,F}$. Fix a choice of finite singular comparison isomorphisms $\phi_{\mathfrak{q}}^{\text{fs}}$ for all prime places $\mathfrak{q} \in \mathcal{M}_{i,F}$.

Let $\mathcal{L}(m)$ be the set of prime places of $\mathcal{L}$ not dividing m. Let $\mathcal{F}(m)$ be the Selmer structure $\mathcal{F}$ modified at m so that the local condition of $\mathcal{F}(m)$ at every prime dividing m is the transverse condition and at every other place the local condition is that of $\mathcal{F}$ (see Sect. 5.2). Let $h \in \text{Hom}(\Gamma_m, R/I_m)$; consider $\otimes h$ as a map $\Gamma_{mr} \otimes (R/I_m) = \Gamma_m \otimes \Gamma_r \otimes (R/I_m) \to \Gamma_r \otimes (R/I_m)$ for all r.

Then there is an R-module homomorphism

$$\mathbf{KS}(T, \mathcal{F}, \mathcal{L}) \longrightarrow \mathbf{KS}(T/I_m T, \mathcal{F}(m), \mathcal{L}(m))$$

given by,

$$\kappa \mapsto (\kappa_r^{(m)})$$

where

$$\kappa_r^{(m)} = \kappa_{rm} \otimes h \in H^1_{\mathcal{F}(mr)}(F, T/I_{mr}T) \otimes_{\mathbb{Z}} \Gamma_r \quad \text{(for all } r \in \mathcal{M}_{i,F} \text{ coprime to } m\text{)}.$$

Remarks 7.1.13

(i) Euler systems give rise to Kolyvagin systems (see Sect. 7.2). The derived cohomology classes of an Euler system in many cases form a relaxed Kolyvagin system (see Definition 7.1.15); this can then be refined to a Kolyvagin system by taking linear combinations of these derived classes.

(ii) The Kolyvagin systems $\mathbf{KS}(T, \mathcal{F}, \mathcal{L})$ can be interpreted as the global sections of a simplicial sheaf on a graph. See (7.1.17)–(7.1.18).

(iii) The Kolyvagin systems of [54] are all specializations of that above by taking $k = F = \mathbb{Q}$, i a positive integer, and $d_p = p - 1$ for every prime number p of $\mathbb{Z}$ different from the residue characteristic of R.

The Heegner point Kolyvagin system of [30] for elliptic curves over $\mathbb{Q}$ is also a specialization of that above. For this, one takes $k = \mathbb{Q}$, F an imaginary quadratic field extension of $\mathbb{Q}$ and if p is a prime number, different from the residue characteristic of R, which is inert and unramified in F and $\mathfrak{q}$ is the prime place of F overlying p then $d_{\mathfrak{q}} = p + 1$; the integers $d_{\mathfrak{q}}$ for other places $\mathfrak{q}$ of F are not defined.

Heegner point Kolyvagin systems are considered in greater generality in the rest of this chapter.

[See Auxiliary Remarks 7.1.19 for more details on the connection with [54] and [30].]

Relaxed Kolvagin systems

(7.1.14) Kolyvagin systems may be constructed from Euler systems by taking the derived cohomology classes of those of the Euler system. Frequently, these derived classes do not form a Kolyvagin system in that they may not satisfy $\kappa_n \in H^1_{\mathcal{F}(n)}(F, T/I_nT) \otimes_{\mathbb{Z}} \Gamma_n$ but only satisfy the weaker constraint obtained by relaxing the local conditions at all places dividing n.

Definition 7.1.15 Suppose that $(T, \mathcal{F}, \mathcal{L})$ is a Selmer triple and that T extends to a G_k-representation. Fix a positive integer i. A *relaxed Kolyvagin system* for $(T, \mathcal{F}, \mathcal{L})$ is a collection κ of cohomology classes

$$\kappa_n \in H^1_{\mathcal{F}^n}(F, T/I_nT) \otimes_{\mathbb{Z}} \Gamma_n$$

for every $n \in \mathcal{M}_{i,F}$ such that for any $nq \in \mathcal{M}_{i,F}$ where q is a prime divisor of F coprime to n, then the images of κ_n and κ_{nq} in $H^1_s(F_q, T/I_{nq}T) \otimes_{\mathbb{Z}} \Gamma_{nq}$ satisfy the equation, where the maps $\mathrm{res}_{F_q/F}$ have the same meaning as in the equation of Definition 7.1.9 and diagram (7.1.8),

$$\phi_q^{\mathrm{fs}}(\mathrm{res}_{F_q/F}(\kappa_n)) = \mathrm{res}_{F_q/F}(\kappa_{nq}) \in H^1_s(F_q, T/I_{nq}T) \otimes_{\mathbb{Z}} \Gamma_{nq}.$$

It is assumed here that choices have been fixed for the finite-singular comparison isomorphisms ϕ_q^{fs}.

The difference from a Kolyvagin system is that the cohomology classes κ_n are only required to belong to $H^1_{\mathcal{F}^n}(F, T/I_nT) \otimes_{\mathbb{Z}} \Gamma_n$ obtained by relaxing at n the local conditions of the group $H^1_{\mathcal{F}(n)}(F, T/I_nT) \otimes_{\mathbb{Z}} \Gamma_n$.

Remarks 7.1.16

(i) A relaxed Kolyvagin system is called a "weak Kolyvagin system" in [54, Definition 3.1.8]. Relaxed Kolyvagin systems have a lack of rigidity compared to Kolyvagin systems (see [54, Remark 3.1.9, Example 3.1.10] for more details).
(ii) For the applications of Euler systems to show the finiteness and determine the structure of Selmer groups, it is usually sufficient to have only a relaxed Kolyvagin system. However for a relaxed system arising from an Euler system, under hypotheses, this relaxed Kolyvagin system can be refined to an unique Kolyvagin system by a process of elimination similar to that of Gaussian elimination in linear algebra (see Sect. 7.2 below and [54, Appendix A]). In this way, a Kolyvagin system is a canonical form of the derived classes of an Euler system.

7.1 Kolyvagin Systems

Simplicial Sheaves, Selmer Sheaves

Definition 7.1.17 Let G be a graph with set of vertices V and set of edges E. A *simplicial sheaf* $\mathcal{S}$ on G with values in the category of R-modules consists of

(a) R-modules $\mathcal{S}(e)$, $\mathcal{S}(v)$ for all $e \in E$ and all $v \in V$;
(b) an R-module homomorphism $f_{v,e} : \mathcal{S}(v) \to \mathcal{S}(e)$ whenever e is an edge which has the vertex v as an endpoint.

The R-module $\mathcal{S}(v)$ is the *stalk* of the sheaf $\mathcal{S}$ at the vertex $v \in V$. A *global section* of $\mathcal{S}$, that is to say an element of $\Gamma(G, \mathcal{S})$, is an ensemble of elements s_v for all $v \in V$ such that for every edge $e \in E$ then

$$f_{v_1,e}(s_{v_1}) = f_{v_2,e}(s_{v_2})$$

where the endpoints of e are the vertices v_1, v_2.

Simplicial sheaves may be defined similarly for any simplicial complex.

(7.1.18) Suppose that $(T, \mathcal{F}, \mathcal{L})$ is a Selmer triple on the global field F. Define a graph G where the set of vertices of G is the set of squarefree cycles $\mathcal{M}_{0,F}$ of F formed from $\mathcal{L}$ (Definition 7.1.4) and where the cycles $n, nq \in \mathcal{M}_{0,F}$ are joined by an edge whenever $q \in \mathcal{L}$ is a prime cycle.

The *Selmer sheaf* $\mathcal{H}$ corresponding to $(T, \mathcal{F}, \mathcal{L})$ is the simplicial sheaf on G given by

(a) $\mathcal{H}(n) = H^1_{\mathcal{F}(n)}(F, T/I_n T) \otimes \Gamma_n$ for all $n \in \mathcal{M}_{0,F}$;
(b) if e is the edge joining $n, nq \in \mathcal{M}_{0,F}$ then $\mathcal{H}(e) = H^1_s(F_q, T/I_{nq}T) \otimes \Gamma_{nq}$;
(c) the map $f_{nq,e}$ is $H^1_{\mathcal{F}(nq)}(F, T/I_{nq}T) \otimes \Gamma_{nq} \to H^1_s(F_q, T/I_{nq}T) \otimes \Gamma_{nq}$ and is the localization at q followed by the projection onto the singular quotient H^1_s;
(d) the map $f_{n,e}$ is $H^1_{\mathcal{F}(n)}(F, T/I_n T) \otimes \Gamma_n \to H^1_s(F_q, T/I_{nq}T) \otimes \Gamma_{nq}$ and is the composite map of localization at q with a finite-singular comparison isomorphism ϕ^{fs}_q (see (7.1.6) and Definition 4.5.5)

$$H^1_{\mathcal{F}(n)}(F, T/I_n T) \otimes \Gamma_n \to H^1_f(F_q, T/I_{nq}T) \otimes \Gamma_n \to H^1_s(F_q, T/I_{nq}T) \otimes \Gamma_{nq}.$$

The finite-singular comparison isomorphisms ϕ^{fs}_q are not uniquely determined in general and depend on the block decompositions. In this way, the Selmer sheaf $\mathcal{H}$ depends on a choice of these isomorphisms ϕ^{fs}_q, a change of such a choice of isomorphisms changing $\mathcal{H}$ by an automorphism.

A global section of the Selmer sheaf $\mathcal{H}$, an element of $\Gamma(G, \mathcal{H})$, is then exactly the same as a Kolyvagin system for $(T, \mathcal{F}, \mathcal{L})$, so there is a canonical isomorphism $\mathbf{KS}(T, \mathcal{F}, \mathcal{L}) \cong \Gamma(G, \mathcal{H})$.

(7.1.19) Auxiliary remarks on the Kolyvagin systems of [54] and [30].

The two references [54] and [30] define Kolyvagin systems for, respectively, (a) R-representations of $G_\mathbb{Q}$ where $k = F = \mathbb{Q}$, and (b) the $\mathbb{Z}_l$-representation of $G_\mathbb{Q}$ given by the Tate module of an elliptic curve over $k = \mathbb{Q}$ and F is an imaginary quadratic extension of $\mathbb{Q}$. The definitions there of a Kolyvagin system are largely similar to that of Definition 7.1.4 above, but there are some significant differences, because our construction of the finite-singular comparison map in Sect. 4.5 is more general than that of either of these two references. We give some comments here on these differences.

(i) *(Definition of the integers $d_\mathfrak{q}$.)* As already mentioned in Remark 7.1.13(iii), in Definition 7.1.4, if one takes $k = F = \mathbb{Q}$, i is a fixed positive integer, and $d_\mathfrak{p} = l - 1$ for every non-zero prime ideal $\mathfrak{p} = l\mathbb{Z}$, where $l > 0$, of $\mathbb{Z}$, then one recovers the Kolyvagin systems of Mazur and Rubin [54].

In Definition 7.1.4, if one takes R to be $\mathbb{Z}_p$, T to be the p-adic Tate module of an elliptic curve over $\mathbb{Q}$, $k = \mathbb{Q}$ and F to be an imaginary quadratic extension field of $\mathbb{Q}$ and $d_\mathfrak{q} = l + 1$, for every prime number $l \in \mathbb{N}$ inert and unramified in $K/\mathbb{Q}$, where $\mathfrak{q}$ is the unique prime place of K lying over l, then one recovers the Heegner point Kolyvagin system of Howard [30] for elliptic curves over the rational numbers.

(ii) *(Definition of the ideals I_n.)* In [54], the ideal I_l, for a prime number l, is defined to be that generated by $l - 1$ and $P_l(1)$ where $P_l(x) = \det(1 - \mathrm{Fr}_l.x|T)$. This is the same as that of $I_\mathfrak{q}$ in Definition 7.1.4 taking $d_\mathfrak{q} = l - 1$ for the prime place $\mathfrak{p}$ of $\mathbb{Q}$ corresponding to l. Therefore, the ideals I_n of [54] coincide with the ideals I_n we have defined in Definition 7.1.4 with this specialization $d_\mathfrak{q} = l - 1$.

In [30], where $k = \mathbb{Q}$ and F is an imaginary quadratic field, in Definition 7.1.4, the ideal $I_\mathfrak{q}$, for a prime number l inert and unramified in $K/\mathbb{Q}$ lying under $\mathfrak{q}$, we take to be that ideal generated by $l + 1$ and $P_l(1)$ where $P_l(x) = \det(1 - \mathrm{Fr}_l.x|T)$. This coincides with Howards definition [30, Definition 1.2.1] of $I_\mathfrak{q}$, but it is there stated differently (for the equivalence between the two definitions, see Remark 7.3.6(ii)). Therefore, the ideals I_n of [30] coincide with the ideals I_n we have defined in (7.1.4) with this specialization $d_\mathfrak{q} = l + 1$ for all prime numbers l unramified and inert in $F/\mathbb{Q}$.

(iii) *(Definition of the sets of prime places $\mathcal{L}_i$.)* The set $\mathcal{L} = \mathcal{L}_0$ is that given in the Selmer triple $(T, \mathcal{F}, \mathcal{L})$. The sets of primes $\mathcal{L}_i$ of F for $i \geq 1$ are given in Definition 7.1.4 and are subsets of $\mathcal{L}$ defined by

$$\mathcal{L}_i = \left\{ \mathfrak{q} \in \mathcal{L} \;\middle|\; \begin{array}{c} I_\mathfrak{q} \subseteq \mathfrak{m}^i \text{ and} \\ T/(\mathfrak{m}^i T + (\mathrm{Fr}_\mathfrak{q} - 1)T) \text{ is free} \\ \text{of finite rank over } R/\mathfrak{m}^i. \end{array} \right\} \qquad (i \geq 1). \tag{7.1.20}$$

This definition of the $\mathcal{L}_i$ for $i \geq 1$ differs from the corresponding definition in [54, Definition 3.1.6] where the latter defines $\mathcal{L}_i$ to be the set of prime numbers $l \notin S(\mathcal{F})$ exactly as above in (7.1.20) except that $T/(\mathfrak{m}^i T + (\mathrm{Fr}_l - 1)T)$ is assumed to be free

7.1 Kolyvagin Systems

of rank 1 over $R/\mathfrak{m}^i$, instead of free of finite rank. Hence, the $\mathcal{L}_i$ of [54] is a subset of the set $\mathcal{L}_i$ in Definition 7.1.4.

For the case of elliptic curves over $\mathbb{Q}$, the sets $\mathcal{L}_i$ for $i \geq 1$ defined above (Definition 7.1.4) coincide with the sets of prime numbers $\mathcal{L}_i$ of [30, Definition 1.2.1] which are there restricted to the special case where $k = \mathbb{Q}$ and F is an imaginary quadratic extension of k. Furthermore, in this case of rational elliptic curves, $T/(\mathfrak{m}^i T + (\mathrm{Fr}_\mathfrak{q} - 1)T)$ is free of rank 2 over $R/\mathfrak{m}^i$ for $\mathfrak{q} \in \mathcal{L}_i$, $i > 0$ (see Remarks 7.3.6(ii) and (iii)).

(iv) *(Iwasawa theory.)* Suppose first that $\mathbb{Q}_\infty$ is the cyclotomic $\mathbb{Z}_p$-extension of $\mathbb{Q}$ where p is a prime number. Let Λ be the Iwasawa algebra

$$\Lambda = \mathbb{Z}_p[[\mathrm{Gal}(\mathbb{Q}_\infty/\mathbb{Q})]].$$

Let T be a $\mathbb{Z}_p$-representation of $G_\mathbb{Q}$ and which is a finite free $\mathbb{Z}_p$-module. Let S be a finite set of places of $\mathbb{Q}$ containing p, ∞ and the primes where T is ramified.

Set $\mathbf{T} = T \otimes_{\mathbb{Z}_p} \Lambda$ where $G_\mathbb{Q}$ acts on both factors of the tensor product.

Define a Selmer structure $\mathcal{F}_\Lambda$ on $\mathbf{T}$ by putting $S(\mathcal{F}_\Lambda) = S$ and $H^1_{\mathcal{F}_\Lambda}(\mathbb{Q}_v, \mathbf{T}) = H^1(\mathbb{Q}_v, \mathbf{T})$ for $v \in S$. Then we have $H^1_f(\mathbb{Q}_v, \mathbf{T}) = H^1(\mathbb{Q}_v, \mathbf{T})$ for $v \notin S$ (by [54, Lemma 5.3.1(ii)]); in particular, the Selmer structure $\mathcal{F}_\Lambda$ is independent of the choice of S. Take $\mathcal{L}$ to be a set of prime numbers disjoint from S. Then $(\mathbf{T}, \mathcal{F}_\Lambda, \mathcal{L})$ is a Selmer triple (7.1.3).

Kolyvagin systems are then defined for $(\mathbf{T}, \mathcal{F}_\Lambda, \mathcal{L})$ as in Definition 7.1.9 taking $k = F = \mathbb{Q}$ where the coefficient ring is $R = \Lambda$. The definitions of Kolyvagin systems for $\mathbf{T}$ in Definition 7.1.9 and [54, Definition 3.1.3 and §5.3] are essentially the same.

Suppose now that E is an elliptic curve over the rational number field $\mathbb{Q}$ and $T = T_p(E)$ is the p-adic Tate module of E. Then T is an $\mathbb{Z}_p$-representation of $G_\mathbb{Q}$, and we take $k = \mathbb{Q}$ and F to be an imaginary quadratic field over $\mathbb{Q}$. If Λ is the Iwasawa algebra for the anticyclotomic $\mathbb{Z}_p$ extension of F and taking the coefficient ring R to be Λ, then $T \otimes \Lambda$ has an action by G_F but which does not usually extend to an action by G_k, so that the definitions of 7.1.4 and Definition 7.1.9 of a Kolyvagin system do not immediately apply to the representation $T \otimes \Lambda$.

Let l be a prime number inert and unramified in $F/\mathbb{Q}$. For this elliptic curve case, let $\mathfrak{q}$ be the prime of F over l. It is then necessary to restrict to primes $\mathfrak{q}$ of K for which $\mathrm{Fr}_\mathfrak{q}$ acts trivially on T; the finite-singular comparison map at $\mathfrak{q}$ in this case is then obvious and induced from the identity isomorphism $T \to T$, and this allows a Kolyvagin system definition for $T \otimes \Lambda$ (see [30] for more details).

(v) The finite singular comparison map of Definition 4.5.5 is more general than that of [54, definition 1.2.2 and lemma 1.2.3] because of our decomposition Theorem 4.4.14; this is because (in the notation of Definition 4.5.5) $T/(\mathrm{Fr} - 1)T$ is restricted to be of free rank 1 in [54], whereas in Definition 4.5.5, $T/(\text{Fr}-1)T$ is taken to be of free of any finite rank. Furthermore, although [54, definition 1.2.2] allows a finite singular comparison homomorphism to be defined under weaker conditions than Definition 4.5.5, this [54, definition 1.2.2] does not give the right comparison map if, for example, $T/(\text{Fr}-1)T$ is free of rank ≥ 2. The two definitions 4.5.5 and [54, definition 1.2.2] give exactly the same finite singular comparison map only if the 1-blocks of T for Fr all have the same length, in particular when $T/(\text{Fr}-1)T$ is free of rank 1 over R (see Remarks 4.4.24).

7.2 Euler Systems Induce Kolyvagin Systems

(7.2.1) Euler systems give rise to Kolyvagin systems in that the derived cohomology classes of an Euler system often form a relaxed Kolyvagin system which can then be refined to a Kolyvagin system. An instance of this is Theorem 7.2.4 relating Euler systems for the rational field $\mathbb{Q}$ and Kolyvagin systems.

We shall see later in this chapter that the derived classes of Euler systems arising from Heegner systems can be refined to Kolyvagin systems.

(7.2.2) Let R be a coefficient ring with maximal ideal $\mathfrak{m}$ which is the ring of integers of a finite extension field of $\mathbb{Q}_l$ of residue characteristic l and with the notation of the last section Sect. 7.1; let

$k = F = \mathbb{Q}$;
$\mathcal{K}$ be a, possibly infinite, abelian algebraic extension field of $\mathbb{Q}$;
T be an R-representation of $G_\mathbb{Q}$ which is a free R-module of finite rank;
$\mathcal{F}_{\text{can}}$ be the canonical Selmer structure on T (Definition 5.1.4);
$\mathcal{L}$ be a set of prime numbers different from l and the places where T is ramified and let $(T, \mathcal{F}_{\text{can}}, \mathcal{L})$ be the corresponding Selmer triple;
$\mathcal{N}$ be a cycle of $\mathbb{Q}$ divisible by the Archimedean place of $\mathbb{Q}$, all places of $\mathbb{Q}$ where T is ramified and the place at the prime number l.

Let $\mathbf{ES}(T, \mathcal{K}, \mathcal{N})$ be the module of Euler systems of Definition 6.2.6.

(7.2.3) Let n be a positive integer, and let $\mathcal{L}_n$ be the set of prime numbers $p \in \mathcal{L}$ such that

(a) $T/(\mathfrak{m}^n T + (\text{Fr}_p - 1)T)$ is a free $R/\mathfrak{m}^n$-module of finite rank;
(b) $I_p \subseteq \mathfrak{m}^n$.

Define

$$\overline{\mathbf{KS}}(T) = \overline{\mathbf{KS}}(T, \mathcal{F}_{\text{can}}, \mathcal{L}) = \varprojlim_n \varinjlim_m \mathbf{KS}(T/\mathfrak{m}^n T, \mathcal{F}_{\text{can}}, \mathcal{L}_m)$$

7.2 Euler Systems Induce Kolyvagin Systems

where the transition maps of the limit and colimit are the evident maps of (7.1.11).

For the next theorem, we say that an R-module M is *pure* if M is a free R/I-module of finite rank where $I \subseteq R$ is the annihilator ideal of M. We also assume that all finite singular comparison isomorphisms have been chosen once and for all, depending on the block decompositions of the modules in question (see Definition 4.5.5).

Theorem 7.2.4 *Suppose that $\mathcal{K}$ contains the maximal abelian l-extension of $\mathbb{Q}$ which is unramified outside l and $\mathcal{L}$. Suppose that*

(a) $T/(\operatorname{Fr}_p - 1)T$ *is a pure R-module for every $p \in \mathcal{L}$;*
(b) *The endomorphism* $\operatorname{Fr}_p^{l^n} - 1$ *of T is injective for all $p \in \mathcal{L}$ and all $n \in \mathbb{N}$.*

Then there is a canonical $G_\mathbb{Q}$-equivariant homomorphism $\mathbf{ES}(T, \mathcal{K}, \mathcal{N}) \longrightarrow \overline{\mathbf{KS}}(T, \mathcal{F}_{\mathrm{can}}, \mathcal{L})$ *where $G_\mathbb{Q}$ acts trivially on $\overline{\mathbf{KS}}(T, \mathcal{F}_{\mathrm{can}}, \mathcal{L})$ such that if c maps to κ then $\kappa_1 = c_\mathbb{Q}$.*

If further that $H^0(\mathbb{Q}_l, T^*)$ is a divisible R-module, then there is a canonical $G_\mathbb{Q}$-equivariant homomorphism $\mathbf{ES}(T, \mathcal{K}, \mathcal{N}) \longrightarrow \mathbf{KS}(T, \mathcal{F}_{\mathrm{can}}, \mathcal{L})$ *such that if c maps to κ then $\kappa_1 = c_\mathbb{Q}$.*

[This theorem is stated and proved by Mazur and Rubin in [54, Theorem 3.2.4] under a stronger hypothesis than (a) above, namely, the hypothesis there is taken to be that $T/(\operatorname{Fr}_p - 1)T$ be a cyclic R-module for every $p \in \mathcal{L}$. With our more general construction of the finite-singular isomorphism in Sect. 4.5, this can be replaced by the hypothesis (a) above. The rest of the proof of the theorem is virtually unchanged from [54, Theorem 3.2.4 and Appendix A].] □

Remarks 7.2.5

(i) The proof of Theorem 7.2.4 involves taking the derived cohomology classes κ_n of an Euler system and first showing that they form a relaxed Kolyvagin system. One then computes the finite projection $(\kappa_n)_{p,\mathrm{f}}$ of the localization $(\kappa_n)_p$ of κ_n at a prime p dividing n of a derived cohomology class κ_n. By taking linear combinations κ'_n of the κ_n, the finite projections can be eliminated, so that only the singular projections remain and these κ'_n therefore form a Kolyvagin system.
(ii) A similar result when the coefficient ring R is an Iwasawa algebra is proved in [54, Theorem 5.3.3] and where again the hypothesis (a) there stated that $T/(\operatorname{Fr}_p - 1)T$ be a cyclic R-module for every $p \in \mathcal{L}$ can be weakened to the hypothesis that $T/(\operatorname{Fr}_p - 1)T$ be a pure R-module for every $p \in \mathcal{L}$ as above.

7.3 Preliminaries on Heegner Point and CM Point Kolyvagin Systems

(7.3.1) In the next four sections Sects. 7.3–7.6, we construct Kolyvagin systems from the Heegner point and CM point Euler systems, notably for the special case of elliptic curves. These all follow the pattern of Sect. 7.1, and we only require to specify the parameters and to construct the cohomology classes κ_m.

In this section, we define the Selmer triple $(T, \mathcal{F}, \mathcal{L})$, the ideals I_n, the integers $d_\mathfrak{q}$ the groups Γ_n, the sets of primes $\mathcal{L}_i$, and the sets of cycles $\mathcal{M}_{i,F}$ for this family of Kolyvagin systems for abelian varieties.

The only differences between the system of objects $(T, \mathcal{F}, \mathcal{L})$, $d_\mathfrak{q}$, $\mathcal{L}_i$, Γ_n, $\mathcal{M}_{i,F}$, I_n of Sect. 7.1 and that of this section Sect. 7.3 are the following. The system of this section is a specialization of that of Sect. 7.1 and more exactly here (i) T is the l-adic Tate module of an abelian variety J/k (ii) F/k is a quadratic field extension (iii) $\mathcal{F}$ is the classical Selmer structure for the abelian variety $J \times_k F$ (iii) $d_\mathfrak{q} = |\kappa(\mathfrak{p})| + 1$, if $\mathfrak{p}$ is inert and unramified in F/k and $\mathfrak{q}$ is the prime of F over $\mathfrak{p}$, and $d_\mathfrak{q} = 1$ otherwise.

The objects $\mathcal{L}_i$, Γ_n, $\mathcal{M}_{i,F}$, I_n of this Sect. 7.3 are then exactly those of Sect. 7.1 given by these specializations of T, $\mathcal{F}$, F/k, $d_\mathfrak{q}$.

The Selmer Triple $(T, \mathcal{F}, \mathcal{L})$.

(7.3.2) We assume that

k is a global field;
l is a prime number distinct from the characteristic of k;
R is the coefficient ring $\mathbb{Z}_l$;
J/k is an abelian variety;
$T = T_l(J)$ is the l-adic Tate module of J equipped with its action by the Galois group G_k;
F/k is a quadratic galois field extension.

Definition 7.3.3 Let $(T, \mathcal{F}, \mathcal{L})$ be the Selmer triple given by J and the quadratic field extension F/k; that is to say

- T is the l-adic Tate module of J/k as a G_k-module and hence by restriction as a G_F-module;
- $S(\mathcal{F})$ is the finite set of places of the field F where T is ramified, i.e. the places of bad reduction of $J \times_k F$, together with the places above l (if any) and the archimedean places (if any);
- $\mathcal{F}$ is the classical Selmer structure for the abelian variety $J \times_k F$ with exceptional set $S(\mathcal{F})$ (see Sect. 5.5);
- $\mathcal{L}$ is a subset of Σ_F disjoint from $S(\mathcal{F})$.

7.3 Preliminaries on Heegner Point and CM Point Kolyvagin Systems

The integers $d_\mathfrak{q}$, ideals I_n, groups Γ_n, sets of primes $\mathcal{L}_i$ and the sets of cycles $\mathcal{M}_{i,F}$

(7.3.4) Let $\mathfrak{q}$ be a place of $\mathcal{L}$, and let $\mathfrak{p}$ be the place of k underlying $\mathfrak{q}$. Define the integer $d_\mathfrak{q}$ to be

$$d_\mathfrak{q} = \begin{cases} |\kappa(\mathfrak{p})| + 1 & \text{if } \mathfrak{p} \text{ is inert and unramified in } F/k \\ 1 & \text{otherwise.} \end{cases}$$

Here $d_\mathfrak{q}$ divides $|\kappa(\mathfrak{q})^*|$ for all $\mathfrak{q}$ as F/k has degree 2.

From Definition 7.1.4, for this choice of the integers $d_\mathfrak{q}$, we then have $I_n, \Gamma_n, \mathcal{L}_i, \mathcal{M}_{i,F}$ there defined and which we recall below.

- For $\mathfrak{q} \in \mathcal{L}$, the ideal $I_\mathfrak{q}$ of R is generated by $d_\mathfrak{q}$ and $P_\mathfrak{q}(1)$ where $P_\mathfrak{q}(x) = \det(1 - \mathrm{Fr}_\mathfrak{p}.x|T) \in R[x]$ is the characteristic polynomial of a Frobenius element $\mathrm{Fr}_\mathfrak{p} \in G_k$ at $\mathfrak{p}$ acting on T where $\mathfrak{p}$ is the prime of k lying below $\mathfrak{q}$;
- The sets of primes $\mathcal{L}_i$, for $i \geq 1$, of F are subsets of $\mathcal{L}$ defined by

$$\mathcal{L}_i = \left\{ \mathfrak{q} \in \mathcal{L} \;\middle|\; \begin{array}{c} I_\mathfrak{q} \subseteq \mathfrak{m}^i \text{ and} \\ T/(\mathfrak{m}^i T + (\mathrm{Fr}_\mathfrak{q} - 1)T) \text{ is free} \\ \text{of finite rank over } R/\mathfrak{m}^i. \end{array} \right\} \qquad (i \geq 1).$$

We put conventionally $\mathcal{L}_0 = \mathcal{L}$;
- For $\mathfrak{q} \in \mathcal{L}$, $\Gamma_\mathfrak{q}$ is the multiplicative group

$$\Gamma_\mathfrak{q} = \kappa(\mathfrak{q})^*.$$

Denote by $\sigma_\mathfrak{q} \in \Gamma_\mathfrak{q}$ a fixed generator of the finite cyclic group $\Gamma_\mathfrak{q}$;
- The set $\mathcal{M}_{i,F}, i \in \mathbb{N}$ is that of squarefree cycles formed of distinct primes $\mathfrak{q}$ of $\mathcal{L}_i$ lying over primes $\mathfrak{p}$ of k which are inert and unramified in F/k and by convention the unit cycle 1 belongs to $\mathcal{M}_{i,F}, i \in \mathbb{N}$;
- For $n \in \mathcal{M}_{i,F}, i \geq 1$, define

$$I_n = \sum_{\mathfrak{q}|n} I_\mathfrak{q}, \qquad \Gamma_n = \bigotimes_{\mathfrak{q}|n} \Gamma_\mathfrak{q}.$$

Definition 7.3.5 A *Heegner point (or CM point) Kolyvagin system* for the Selmer triple $(T, \mathcal{F}, \mathcal{L})$ over F is a Kolyvagin system for $(T, \mathcal{F}, \mathcal{L})$ (Definition 7.1.9) where the field extension F/k, the integers $d_\mathfrak{q}$, ideals I_n, groups Γ_n, sets of primes $\mathcal{L}_i$, and the sets of cycles $\mathcal{M}_{i,F}$ are those defined in this section Sect. 7.3.

Remarks 7.3.6

(i) In Definition 7.3.4, the integer $d_\mathfrak{q}$ is different from 1 if and only if the place $\mathfrak{q}$ of $\mathcal{L}$ lies over a place $\mathfrak{p}$ of k which is inert and unramified in F/k. In particular, the sets $\mathcal{L}_i$ for $i \geq 1$ of (7.3.4) only contain places $\mathfrak{q}$ of $\mathcal{L}$ which lie over places $\mathfrak{p}$ of k which are inert and unramified in F/k.

(ii) *(Definition of $I_\mathfrak{q}$).* The definition of $I_\mathfrak{q}$ above in (7.3.4) specializes that of Definition 7.1.4. Howard's definition [30, Definition 1.2.1] of the ideal $I_\mathfrak{q}$ for Kolyvagin systems of Heegner points on rational elliptic curves is stated differently, but it is equivalent to that above in (7.3.4).

To see this equivalence, E be an elliptic curve over a the rational number field $\mathbb{Q}$ and $T = T_l(E)$ be the l-adic Tate module of E where the coefficient ring is $\mathbb{Z}_l$. Let $F/\mathbb{Q}$ be an imaginary quadratic field extension, and let p be a prime number inert and unramified in $F/\mathbb{Q}$. Let $\mathfrak{q}$ be the prime of F over p. Howard [30] defines $I_\mathfrak{q}$ to be the smallest ideal of $R = \mathbb{Z}_l$ containing $p + 1$ for which $\mathrm{Fr}_\mathfrak{q}$ acts trivially on $T/I_\mathfrak{q}T$.

We have $\mathrm{Fr}_\mathfrak{q} = \mathrm{Fr}_p^2$. Let $a_p = \mathrm{Tr}(\mathrm{Fr}_p|T)$ be the trace of the Frobenius at p on T. Suppose that the ideal I of $\mathbb{Z}_l$ contains $p + 1$. Then the ideal I is the smallest ideal of $\mathbb{Z}_l$ containing $p + 1$ and such that $\mathrm{Fr}_\mathfrak{q}$ acts trivially on T/IT if and only if $a_p \in I$; this shows that Howard's definition of $I_\mathfrak{q}$ is equivalent to that in (7.3.4) (and also that in Definition 7.1.4).

Howard's definition *loc. cit.* of $I_\mathfrak{q}$ makes sense when the representation does not have an action by $G_\mathbb{Q}$, but only an action by G_F; for example, if Λ is the Iwasawa algebra for the anticyclotomic $\mathbb{Z}_l$ extension of F, then $T \otimes \Lambda$ has an action by G_F but not by G_k and Howard's definition of $I_\mathfrak{q}$ for $T \otimes \Lambda$ applies and Howard [30] is then able to define Kolyvagin systems for $T \otimes \Lambda$ (see also Remark 7.1.19(iv)).

(iii) Suppose the abelian variety J/k is an elliptic curve. It follows from the equivalence argument in (ii) transposed to the ground field k instead of $\mathbb{Q}$, that for definition of the sets $\mathcal{L}_i$ given in (7.3.4) where $i \geq 1$, one has the following. If $\mathfrak{q} \in \mathcal{L}_i$ where $i \geq 1$ and $I_\mathfrak{q} \subseteq \mathfrak{m}^i$ then $\mathrm{Fr}_\mathfrak{q}$ acts trivially on $T/\mathfrak{m}^i T$ and hence that $T/(\mathfrak{m}^i T + (\mathrm{Fr}_\mathfrak{q} - 1)T)$ is a free $R/\mathfrak{m}^i$-module of rank 2.

In particular, the definition of the sets $\mathcal{L}_i$ in (7.3.4) above coincides with definition of the sets $\mathcal{L}_i$ of [30] for the case of rational elliptic curves.

7.4 Derived Cohomology Classes

(7.4.1) In the rest of this chapter in Sects. 7.4–7.6, we construct a Kolyvagin system on an elliptic curve of the Heegner point and CM point type defined in Sect. 7.3 starting from an axiomatic Heegner system of Sect. 6.5.

As these axioms of Sect. 6.5 have already been shown to hold for Heegner points on classical/Drinfeld modular curves and also for CM points on Shimura curves (Sect. 6.6, Sect. 6.7), this constructs a Kolyvagin system starting from these Heegner points and CM points.

7.4 Derived Cohomology Classes

In this section, we define the derived cohomology classes starting from the axiomatic Heegner system of Sect. 6.5. In the next section Sect. 7.5, we give some properties of the derived cohomology classes, and in Sect. 7.6, we obtain the Kolyvagin system from these derived classes.

Note that for the rest of this Chap. 7, we only consider elliptic curves in contrast to the abelian varieties of Sect. 7.3.

Notation and Hypotheses

(7.4.2) The notation of the previous section Sect. 7.3 holds in this section with the following modifications. Let

k be a global field equipped with a special set of places ∞_k;

F be a finite Galois field extension of k which is imaginary quadratic with respect to the special set of places ∞_k; F is equipped with a special set of places ∞_F consisting of those places lying above ∞_k;

O_k, resp. O_F, be the ring of integers of k with respect to ∞_k, resp., the ring of integers of F with respect to ∞_F;

l be a prime number distinct from the characteristic of k;

R be the coefficient ring $\mathbb{Z}_l$;

E/k be an elliptic curve;

$T = T_l(E)$ be the l-adic Tate module of J equipped with its action by the Galois group G_k.

(7.4.3) (Selmer triple). Let $(T, \mathcal{F}, \mathcal{L})$ be the Selmer triple given by E and the imaginary quadratic field extension F/k as in Definition 7.3.3.

The Selmer triple $(T, \mathcal{F}, \mathcal{L})$, given by E and the imaginary quadratic field extension F/k, is more precisely given as follows. The set $S(\mathcal{F})$ is the finite set of places of the field F where T is ramified, together with the places above l and the Archimedean places; $\mathcal{F}$ is the classical Selmer structure for the elliptic curve $E \times_k F/F$ over F with exceptional set $S(\mathcal{F})$ (see Sect. 5.5); $\mathcal{L}$ is a subset of Σ_F disjoint from $S(\mathcal{F})$.

(7.4.4) (Integers d_q, ideals I_n, groups Γ_n, sets of primes $\mathcal{L}_i$, sets of cycles $\mathcal{M}_{i,F}$). Associated with the Selmer triple $(T, \mathcal{F}, \mathcal{L})$, we then have the integers d_q, ideals I_n, groups Γ_n, sets of primes $\mathcal{L}_i$, and the sets of cycles $\mathcal{M}_{i,F}$, for $i \in \mathbb{N}$, defined as in (7.3.4).

We recall from (7.3.4) that the set $\mathcal{M}_{i,F}$ for $i \in \mathbb{N}$ is that of squarefree cycles formed of distinct primes $\mathfrak{q}$ of $\mathcal{L}_i$ lying over primes $\mathfrak{p}$ of k which are inert and unramified in F/k and by convention the unit cycle 1 belongs to $\mathcal{M}_{i,F}$ for $i \in \mathbb{N}$.

(7.4.5) (Sets of cycles $\mathcal{M}_{i,F}$, sets of cycles $\mathcal{M}_{i,k}$). For any $i \in \mathbb{N}$, we define $\mathcal{M}_{i,k}$ to be the all the cycles on k which lie below the cycles of $\mathcal{M}_{i,F}$; that is to say if $c = \prod_j \mathfrak{q}_j \in \mathcal{M}_{i,F}$ where $\mathfrak{q}_j \in \mathcal{M}_{i,F}$ are prime cycles and if $\mathfrak{p}_j$ is the prime of k below $\mathfrak{q}_j$, then the cycle $\prod_j \mathfrak{p}_j \in \mathcal{M}_{i,k}$ and all elements of $\mathcal{M}_{i,k}$ are of this form.

There is a set bijection, for any $i \in \mathbb{N}$,

$$\sharp : \mathcal{M}_{i,k} \longrightarrow \mathcal{M}_{i,F}$$

taking a cycle $c = \prod \mathfrak{p}$ in $\mathcal{M}_{i,k}$ to the corresponding cycle $c^\sharp = \prod \mathfrak{q}$ of F where $\mathfrak{q}$ runs over the primes overlying primes $\mathfrak{p}$ of k dividing c, where this is well defined as the primes of k below primes dividing cycles in $\mathcal{M}_{i,F}$ are inert and unramified in F/k. The inverse bijection is denoted

$$\flat : \mathcal{M}_{i,F} \longrightarrow \mathcal{M}_{i,k}, n \mapsto n^\flat.$$

(7.4.6) (Heegner system). We assume there is a Heegner system on the elliptic curve E/k with l-adic Tate module T so that the axioms Ax.1–Ax.7 of Sect. 6.5 hold. We then have, where these objects form the Heegner system and are assumed to satisfy the axioms,

$\mathcal{C}$ is a set of cycles on k;
F_c is a finite abelian separable field extension of F for all $c \in \mathcal{C}$;
$a_z = \mathrm{Tr}(\mathrm{Fr}_z | T)$ is the trace of the Frobenius $\mathrm{Fr}_z \in \mathrm{Gal}(k^{\mathrm{sep}}/k)$ on T at a non-Archimedean place z of k where T is unramified;
$x_c \in E(F_c)$ is a point of E defined over F_c for all $c \in \mathcal{C}$;
$u(c)$ is a positive integer for all $c \in \mathcal{C}$.

In order to connect the Heegner system with the Selmer triple $(T, \mathcal{F}, \mathcal{L})$ and its sets $\mathcal{M}_{i,k}$ of (7.4.4), we now impose the condition

$$\mathcal{M}_{0,k} = \left\{ c \in \mathcal{C} \;\middle|\; \begin{array}{l} c \text{ is squarefree and all prime components} \\ \text{of } c \text{ are unramified and inert in } F/k \end{array} \right\}.$$

This means that $\mathcal{M}_{0,k}$ consists of squarefree cycles of $\mathcal{C}$ whose prime components are inert and unramified in F/k.

(7.4.7) (Groups $\mathfrak{G}_n$). For $\mathfrak{q}$ a prime cycle of $\mathcal{M}_{0,F}$, put

$$\mathfrak{G}_\mathfrak{q} = \kappa(\mathfrak{q})^* / \kappa(\mathfrak{p})^*$$

where $\mathfrak{q}$ is a prime place of F overlying the prime $\mathfrak{p}$ of k. The group $\mathfrak{G}_\mathfrak{q}$ is cyclic and has order

$$|\mathfrak{G}_\mathfrak{q}| = |\kappa(\mathfrak{p})| + 1 = d_\mathfrak{q}.$$

7.4 Derived Cohomology Classes

For a cycle $n \in \mathcal{M}_{0,F}$, we also define the group

$$\mathfrak{G}_n = \bigotimes_{q|n} \mathfrak{G}_q.$$

where q runs over all primes of F dividing n. The group $\mathfrak{G}_n$ is also cyclic.

For $c \in \mathcal{M}_{0,k}$, a cycle on k, we have the field F_c by the Heegner system which is a finite separable abelian extension of F (Ax.2). We write the associated galois groups as, where $b, c \in \mathcal{M}_{0,k}$ are cycles such that b divides c and where the decomposition follows from Ax.3,

$$\mathcal{G}(c) = \mathrm{Gal}(F_c/F), \qquad G(c,b) = \mathrm{Gal}(F_c/F_b) \cong \begin{cases} \left(\prod_{q|c/b} \mathfrak{G}_q\right)/\Delta_b & \text{if } b = 1 \\ \prod_{q|c/b} \mathfrak{G}_q & \text{if } b \neq 1. \end{cases}$$
(7.4.8)

Here the products run over all prime places q of F dividing the cycle c/b on k and Δ_b is a subgroup of the product of order $u(b)$ by Ax.3. Furthermore, there is the exact sequence $0 \to G(c,1) \to \mathcal{G}(c) \to \mathrm{Gal}(F_1/F) \to 0$.

Lemma 7.4.9 *For a cycle $n \in \mathcal{M}_{0,F}$, we have isomorphisms of R-modules*

$$\mathfrak{G}_n \otimes_{\mathbb{Z}} R/I_n R \cong \Gamma_n \otimes_{\mathbb{Z}} R/I_n R.$$

Proof This follows immediately from the definitions of I_n and Γ_n in (7.3.4). □

The Kolyvagin Operators D_n

(7.4.10) Let q be a prime place of F belonging to $\mathcal{M}_{0,F}$. Let ϕ_q be a generator of the cyclic group $\mathfrak{G}_q$. Define the operator

$$D_q \in \mathbb{Z}_l[\mathfrak{G}_q], \qquad D_q = \sum_{i=1}^{|\mathfrak{G}_q|-1} i\phi_q^i.$$

For $n \in \mathcal{M}_{0,F}$, define the corresponding operator $D_n \in \mathbb{Z}_l[\mathfrak{G}_n]$ by

$$D_n = \prod_{q|n} D_q.$$

where the product runs over prime places q of F dividing n.

For $c \in \mathcal{M}_{0,k}$, select a set $\mathcal{S}_c$ of coset representatives of the subgroup $G(c, 1) = \mathrm{Gal}(F_c/F_1)$ of $\mathcal{G}(c) = \mathrm{Gal}(F_c/F)$.

(7.4.11) Define for $c \in \mathcal{M}_{0,k}$

$$\gamma_c = \sum_{s \in \mathcal{S}_c} s D_{c^\sharp} x_c \in E(F_c) \qquad (7.4.12)$$

where $x_c \in E(F_c)$ is the given point of the Heegner system.

Lemma 7.4.13 *Let $c \in \mathcal{M}_{1,k}$ and suppose that $\mathfrak{p} \in \mathcal{M}_{1,k}$ divides c. Let $\mathfrak{q}$ be the prime of F overlying $\mathfrak{p}$.*

The point x_c belongs to $E(F_c)$ and $\mathfrak{G}_\mathfrak{q}$ has $\mathrm{Gal}(F_c/F_{c/\mathfrak{q}})$ as a homomorphic image (see (7.4.8)), so that $\mathfrak{G}_\mathfrak{q}$ acts via galois on $E(F_c)$.

We then have the norm relation

$$N_{\mathfrak{G}_\mathfrak{q}} x_c = a_\mathfrak{p} x_{c/\mathfrak{p}}.$$

Proof By the norm relations for Heegner points Ax.4, we obtain

$$u(c/\mathfrak{p}) N_{F_c/F_{c/\mathfrak{p}}} x_c = a_\mathfrak{p} x_{c/\mathfrak{p}}$$

We have from (7.4.8)

$$\mathrm{Gal}(F_c/F_{c/\mathfrak{p}}) \cong \begin{cases} \mathfrak{G}_\mathfrak{q}/G & \text{if } c = \mathfrak{p} \\ \mathfrak{G}_\mathfrak{q} & \text{if } c \neq \mathfrak{p} \end{cases}$$

where G is a subgroup of order $u(1)$. The kernel of the homomorphism $\mathfrak{G}_\mathfrak{q} \to \mathrm{Gal}(F_c/F_{c/\mathfrak{q}})$ acts trivially on $E(F_c)$. The norm $N_{\mathfrak{G}_\mathfrak{q}} x_c$ is then equal to $a_\mathfrak{p} x_{c/\mathfrak{p}}$. $\square$

Proposition 7.4.14 *Let $c \in \mathcal{M}_{1,k}$ and L be the subgroup $(I_{c^\sharp} \cap \mathbb{Z}) E(F_c)$ of $E(F_c)$. Then the image of γ_c in $E(F_c)/L$ belongs to $(E(F_c)/L)^{\mathrm{Gal}(F_c/F)} = (E(F_c)/L)^{\mathcal{G}(c)}$.*

Proof First, we have $I_{c^\sharp} \otimes_\mathbb{Z} E(F_c) = L \otimes_\mathbb{Z} \mathbb{Z}_l$, because the ideal $I_\mathfrak{q}$ for $\mathfrak{q}$ a place of $\mathcal{L}$ is generated by the natural integers $d_\mathfrak{q}$, $P_\mathfrak{q}(1)$ (Definition 7.1.4 and (7.3.4)). Put

$$\tilde{\gamma}_c = D_{c^\sharp} x_c \in E(F_c)$$

so that $\gamma_c = \sum_{s \in \mathcal{S}_c} s \tilde{\gamma}_c$.

We have the identity

$$(\phi_\mathfrak{q} - 1) D_\mathfrak{q} = |\mathfrak{G}_\mathfrak{q}| - N_{\mathfrak{G}_\mathfrak{q}}.$$

7.4 Derived Cohomology Classes

We then have if $\mathfrak{p}$ is a place of k which divides c and $\mathfrak{q}$ is the prime of F lying over $\mathfrak{p}$

$$(\phi_\mathfrak{q} - 1)\tilde{\gamma}_c = (\phi_\mathfrak{q} - 1)D_{c\sharp}x_c = D_{c\sharp/\mathfrak{q}}(\phi_\mathfrak{q} - 1)D_\mathfrak{q} x_c$$

$$= D_{c\sharp/\mathfrak{q}}(|\mathfrak{G}_\mathfrak{q}| - N_{\mathfrak{G}_\mathfrak{q}})x_c.$$

The term $N_{\mathfrak{G}_\mathfrak{q}} x_c$ is given by Lemma 7.4.13, and we obtain

$$(\phi_\mathfrak{q} - 1)\tilde{\gamma}_c = |\mathfrak{G}_\mathfrak{q}| D_{c\sharp/\mathfrak{q}} x_c - a_\mathfrak{p} D_{c\sharp/\mathfrak{q}} x_{c/\mathfrak{p}}.$$

As the integers $a_\mathfrak{p}$, $|\mathfrak{G}_\mathfrak{q}|$ belong to $I_\mathfrak{q}$ and hence *a fortiori* belong to $I_{c\sharp}$ for all $\mathfrak{p}$ dividing c, we obtain

$$(\phi_\mathfrak{q} - 1)\tilde{\gamma}_c \equiv 0 \mod I_{c\sharp}.$$

It follows that the image of $\tilde{\gamma}_c$ in $E(F_c)/L$ is invariant under the group $\mathfrak{G}_{c\sharp}$, and hence the image is invariant under $\mathrm{Gal}(F_c/F_1)$ by (7.4.8). It follows that the image of $\gamma_c = \sum_{s \in \mathcal{S}_c} s\tilde{\gamma}_c$ in $E(F_c)/L$ is invariant under $\mathrm{Gal}(F_c/F)$. □

Definition of the Derived Classes

(7.4.15) Suppose that $c \in \mathcal{M}_{1,k}$. Write L for the subgroup $(I_{c\sharp} \cap \mathbb{Z})E(F_c)$ of $E(F_c)$. Then the Kummer map is the injective homomorphism

$$\mathrm{Kumm} : E(F_c)/L \longrightarrow H^1(F_c, T/I_{c\sharp}T).$$

Let δ_c be the image of γ_c under the Kummer map in $H^1(F_c, T/I_{c\sharp}T)$ so that

$$\delta_c = \mathrm{Kumm}(\gamma_c \bmod L) \in H^1(F_c, T/I_{c\sharp}T).$$

As $\gamma_c \bmod L$ belongs to $(E(F_c)/L)^{\mathrm{Gal}(F_c/F)}$ (Proposition 7.4.14) and the Kummer map is injective, we obtain that

$$\delta_c \in H^1(F_c, T/I_{c\sharp}T)^{\mathrm{Gal}(F_c/F)}.$$

We have the inflation-restriction sequence where the map labelled res here is the restriction map

$$0 \longrightarrow H^1(\mathrm{Gal}(F_c/F), (T/I_{c\sharp}T)^{G_{F_c}}) \longrightarrow H^1(F, T/I_{c\sharp}T) \overset{\mathrm{res}}{\longrightarrow}$$

$$H^1(F_c, T/I_{c\sharp}T)^{\mathrm{Gal}(F_c/F)} \longrightarrow H^2(\mathrm{Gal}(F_c/F), (T/I_{c\sharp}T)^{G_{F_c}}).$$

If $(T/I_{c^\sharp}T)^{G_{F_c}} = 0$, then this exact sequence reduces to the restriction isomorphism

$$H^1(F, T/I_{c^\sharp}T) \stackrel{\text{res}}{\cong} H^1(F_c, T/I_{c^\sharp}T)^{\text{Gal}(F_c/F)}. \qquad (7.4.16)$$

Definition 7.4.17 (Derived Cohomology Classes) By the axiom Ax.7 of Sect. 6.5 that is satisfied by our Heegner system, there is a finite set of prime numbers $\mathcal{E}$ depending on E/k and F such that for any $c \in \mathcal{M}_{1,k}$, then $(T/I_{c^\sharp}T)^{G_{F_c}} = 0$ if the residue characteristic of R satisfies $l \notin \mathcal{E}$.

Suppose that $c \in \mathcal{M}_{1,k}$ and that $l \notin \mathcal{E}$. Then we define the *derived cohomology class*

$$\theta_c \in H^1(F, T/I_{c^\sharp}T) \otimes \Gamma_{c^\sharp} \qquad (7.4.18)$$

to be the inverse image of $\delta_c = \text{Kumm}(\gamma_c \bmod I_{c^\sharp}) \in H^1(F_c, T/I_{c^\sharp}T)$ under the restriction isomorphism $H^1(F, T/I_{c^\sharp}T) \cong H^1(F_c, T/I_{c^\sharp}T)^{\text{Gal}(F_c/F)}$ of (7.4.16) and where the component in $\Gamma_{c^\sharp}$ arises because of the choices of generators ϕ_q we have made for each prime q dividing $c^\sharp$.

We may write indifferently $\Gamma_{c^\sharp}$ or $\mathfrak{G}_{c^\sharp}$ in this last formula (7.4.18) by Lemma 7.4.9.

Remark 7.4.19 Rubin [62] has a different construction of the derived cohomology classes, for number fields, based on the "universal Euler system".

7.5 Properties of the Derived Classes

(7.5.1) In this section, we give some properties of the derived cohomology classes. We first give an explicit cocyle formula for a derived cohomology class (Lemma 7.5.2) and then the main result on their behaviour under localization (Lemma 7.5.3).

The notation and hypotheses (7.4.2)–(7.4.8) of the previous section Sect. 7.4 holds in this section, so that we have the Heegner system $x_c \in E(F_c)$ on the elliptic curve E/k satisfying the axioms Ax.1–Ax.7.

Let $c \in \mathcal{M}_{1,k}$, and let $I_{c^\sharp}$ be the corresponding ideal. Then as in the last section, we have the elements γ_c, θ_c given by

- $\gamma_c = \sum_{s \in \mathcal{S}_c} s D_{c^\sharp} x_c \in E(F_c)$ where $x_c \in E(F_c)$ (see (7.4.12));
- $\theta_c \in H^1(F, T/I_{c^\sharp}T) \otimes \Gamma_{c^\sharp}$ is the derived cohomology class obtained from γ_c assuming that $(T/I_{c^\sharp}T)^{G_{F_c}} = 0$ (see Definition 7.4.17).

Lemma 7.5.2 *Let $c \in \mathcal{M}_{1,k}$ and $\gamma_c \in E(F_c)$ be the corresponding point of (7.4.12) and define $n \in \mathbb{N}$ which is an integral power of l by $I_{c^\sharp} = n\mathbb{Z}_l$. Assume that $(T/lT)^{G_{F_c}} = 0$.*

7.5 Properties of the Derived Classes

Fix a n-division point $\frac{\gamma_c}{n} \in E(F^{\text{sep}})$ of γ_c. Then the derived cohomology class $\theta_c \in H^1(F, T/I_{c^\sharp}T)$ is represented by the cocycle

$$g \mapsto \xi_c(g) = (g-1)\frac{\gamma_c}{n} - \frac{(g-1)\gamma_c}{n}, \quad G_F \to T/I_{c^\sharp}T,$$

where $\frac{(g-1)\gamma_c}{n}$, for $g \in G_F$, is the unique n-division point of $(g-1)\gamma_c$ in $E(F_c)$.

Proof The image of the element $\gamma_c \in E(F_c)$ in $E(F_c)/nE(F_c)$ belongs to $(E(F_c)/nE(F_c))^{\text{Gal}(F_c/F)}$ by Proposition 7.4.14.

As $(T/lT)^{G_{F_c}} = 0$, the inflation restriction sequence as in (7.4.15) and (7.4.16) shows that there is a restriction isomorphism

$$H^1(F, E[n](F^{\text{sep}})) \to H^1(F_c, E[n](F^{\text{sep}}))^{\text{Gal}(F_c/F)}.$$

Let $\frac{\gamma_c}{n} \in E(F^{\text{sep}})$ be a fixed nth division point of γ_c, that is to say $\frac{\gamma_c}{n}$ is any point which satisfies $n(\frac{\gamma_c}{n}) = \gamma_c$. Then the cocycle

$$\phi : g \mapsto g\left(\frac{\gamma_c}{n}\right) - \frac{\gamma_c}{n}, \quad G_{F_c} \to E[n](F^{\text{sep}}),$$

represents a cohomology class in $H^1(F_c, E[n](F^{\text{sep}}))^{\text{Gal}(F_c/F)}$ which is the image of γ_c mod $nE(F_c)$ in $E(F_c)/nE(F_c)$ under the coboundary map

$$\partial : (E(F_c)/nE(F_c))^{\text{Gal}(F_c/F)} \to H^1(F_c, E[n](F^{\text{sep}}))^{\text{Gal}(F_c/F)}.$$

The inflation of ϕ to G_F is given by the cocycle

$$\inf(\phi) : G_F \to E(F^{\text{sep}}), \quad g \mapsto g\left(\frac{\gamma_c}{n}\right) - \frac{\gamma_c}{n}$$

which *a priori* is not annihilated by n.

For any element $g \in G_F$, denote by

$$\frac{(g-1)\gamma_c}{n}$$

the unique nth root of $(g-1)\gamma_c$ in $E(F_c)$. This root exists, because γ_c mod $nE(F_c)$ lies in $(E(F_c)/nE(F_c))^{\text{Gal}(F_c/F)}$; furthermore, it is unique by the hypothesis $(T/lT)^{G_{F_c}} = 0$.

The cochain

$$\psi : G_F \to E(F_c), \quad g \mapsto -\frac{(g-1)\gamma_c}{n},$$

is a cocycle whose restriction to the subgroup G_{F_c} is the zero cochain. The cochain

$$\inf(\phi) + \psi : G_F \to E(F_c),\, g \mapsto g\Big(\frac{\gamma_c}{n}\Big) - \frac{\gamma_c}{n} - \frac{(g-1)\gamma_c}{n},$$

is a cocycle which is annihilated by n and whose restriction to G_{F_c} is the cocycle $\phi : G_{F_c} \to E[n](F^{\mathrm{sep}})$. Hence, the cocycle

$$\inf(\phi) + \psi : G_F \to E[n](F^{\mathrm{sep}})$$

represents the cohomology class θ_c in $H^1(F, E[n])$. □

Lemma 7.5.3 *Let $\mathcal{F}$ denote the classical Selmer structure on T (see Sect. 5.5). Suppose that l is coprime to the order of $\mathrm{Pic}(O_k)$, where O_k is the ring of integers of k, and $l \neq 2$. Assume that $c \in \mathcal{M}_{1,k}$ and $(T/lT)^{G_{F_c}} = 0$ and that the primes in the support of c are coprime to l when k is a number field. Then the derived cohomology class $\theta_c \in H^1(F, T/I_{c^\sharp}T) \otimes \Gamma_{c^\sharp}$ satisfies, for some choice of transverse condition subgroups,*

$$\theta_c \in H^1_{\mathcal{F}(c^\sharp)}(F, T/I_{c^\sharp}T) \otimes \Gamma_{c^\sharp}.$$

Proof of Lemma 7.5.3 In this proof, for a non-Archimedean local field K, we write K^{sh} for the field of fractions of the strict henselization of the valuation ring of K.

Step 1. *Suppose that z is a place of F not dividing the cycle c on k. Then the restriction $\mathrm{res}_{F_z/F}(\theta_c)$ belongs to $H^1_{\mathcal{F}}(F_z, T/I_{c^\sharp}T) \otimes \Gamma_{c^\sharp}$.*

If z is an Archimedean place of F, then k is a totally real algebraic number field and F is a totally imaginary quadratic extension of a totally real field (this follows from the definition in (1.1)), and we have that $F_z \cong \mathbb{C}$ and the Galois cohomology group $H^1(F_z, T/I_{c^\sharp}T) = 0$, so there is nothing to prove.

Suppose now that z is a non-Archimedean place of F and z does not divide the cycle $c \in \mathcal{C}$. Then F_c/F is unramified at z by Ax.3(i). So if w is any place of F_c over z, the maximal unramified extensions $(F_c)_w^{\mathrm{sh}}$ and F_z^{sh} are isomorphic. We then obtain the commutative diagram where the top row is a complex and where the first vertical arrow is restriction $\mathrm{res}_{F_c/F}$, the right-hand vertical arrow is an isomorphism, and the two right-hand horizontal arrows are obtained from localization

$$\begin{array}{ccccc} E(F_c)/(I_{c^\sharp} \cap \mathbb{Z})E(F_c) & \hookrightarrow & H^1(F_c, T/I_{c^\sharp}T) & \to & H^1((F_c)_w^{\mathrm{sh}}, T/I_{c^\sharp}T) \\ & & \uparrow \mathrm{res}_{F_c/F} & & \uparrow \cong \\ & & H^1(F, T/I_{c^\sharp}T) & \to & H^1(F_z^{\mathrm{sh}}, T/I_{c^\sharp}T) \end{array}$$

The element θ_c belongs to $H^1(F, T/I_{c^\sharp}T) \otimes_{\mathbb{Z}} \Gamma_{c^\sharp}$, and, ignoring the component in $\Gamma_{c^\sharp}$, it arises from the element γ_c of $E(F_c)$ where this element and θ_c have the same image in $H^1(F_c, T/I_{c^\sharp}T)$ in the diagram. Hence in the diagram, the image of θ_c in

7.5 Properties of the Derived Classes

$H^1((F_c)_w^{sh}, T/I_{c^{\sharp}}T)$ is zero. Therefore, the image of θ_c in $H^1(F_z^{sh}, T/I_{c^{\sharp}}T)$ is zero. That is to say, the restriction $\mathrm{res}_{F_z/F}(\theta_c)$ belongs to $H_{\mathcal{F}}^1(F_z, T/I_{c^{\sharp}}T)$ if z does not divide c.

This shows that we have (see also Remark 7.5.4)

$$\theta_c \in H^1_{\mathcal{F}^{c^{\sharp}}}(F, T/I_{c^{\sharp}}T).$$

Step 2. *Suppose that $d\mathfrak{p} \in \mathcal{M}_{1,k}$ where $\mathfrak{p} \in \mathcal{M}_{1,k}$ is a prime divisor of k coprime to $d \in \mathcal{M}_{1,k}$. Let O_k, O_F be the rings of ∞-integers of k, F, respectively. Let the order of the divisor class $[\mathfrak{p}]$ in $\mathrm{Pic}(O_k)$ of $\mathfrak{p}$ be n. Then for any prime $\mathfrak{r}$ of F_d lying over $\mathfrak{p}$, the degree $[F_{d,\mathfrak{r}} : k_{\mathfrak{p}}]$ divides n.*

Assume first that this Step 2 holds when F_d is a ring class field of the form $F[a]$ where a is a cycle on k coprime to ∞_k and coprime to $\mathfrak{p}$.

By Ax.2, there is a cycle d_0 on k coprime to ∞_k and coprime to all elements of $\mathcal{C}$ and a cycle $c \in \mathcal{C}$ on k such that ring class field $F[d_0c]$ for F/k satisfies $F_d \subseteq F[d_0c]$ and where the only primes of $\mathcal{C}$ dividing c are those dividing d, in particular d_0c is coprime to $\mathfrak{p}$.

By the assumption already made, for any prime $\mathfrak{r}$ of $F[d_0c]$ lying over $\mathfrak{p}$, we have that $[F[d_0c]_{\mathfrak{r}} : k_{\mathfrak{p}}]$ divides n. Then as $F \subseteq F_d \subseteq F[d_0c]$, for any prime $\mathfrak{s}$ of F_d lying over $\mathfrak{p}$, there is a prime $\mathfrak{r}$ of $F[d_0c]$ lying over $\mathfrak{s}$. It follows that the degree $[F_{d,\mathfrak{s}} : k_{\mathfrak{p}}]$ divides $[F[d_0c]_{\mathfrak{r}} : k_{\mathfrak{p}}]$ which then divides n by the case assumed for ring class fields. Hence $[F_{d,\mathfrak{s}} : k_{\mathfrak{p}}]$ divides n.

We are then reduced to proving Step 2 in the case where F_d is a ring class field of the form $F[a]$, for the field extension F/k where this is imaginary quadratic with respect to ∞_k, and for a cycle a on k coprime to ∞_k and to $\mathfrak{p}$ (see (6.5.1)).

Let O_a be the order of F relative to O_k, O_F, with conductor a. The reciprocity map gives a canonical isomorphism

$$\mathrm{Pic}(O_a) \cong \mathrm{Gal}(F[a]/F)$$

where the Frobenius element in $\mathrm{Gal}(F[a]/F)$ at a prime $\mathfrak{s}$ of F, which is coprime to a, maps to the divisor class $[\mathfrak{s} \cap O_a]$ of the locally free O_a-module $\mathfrak{s} \cap O_a$ of rank 1.

As $\mathfrak{p}$ has order n in $\mathrm{Pic}(O_k)$, we have that $\mathfrak{p}^n$ is a principal ideal equal to fO_k for some $f \in O_k$. Let I be the ideal of O_k cutting out the cycle a. It follows that the order O_a equals

$$O_a = O_k + IO_F.$$

Let $\mathfrak{q}$ be the prime of O_F lying over $\mathfrak{p}$, where the latter is inert in F/k as $\mathfrak{p} \in \mathcal{M}_{1,k}$.

Let M be an O_k-submodule of O_F. We have that

$$M = \bigcap_{\mathfrak{m}} M_{\mathfrak{m}}$$

where $\mathfrak{m}$ runs over all maximal ideals of O_k.

As the ideal $\mathfrak{p}^n$ is principal and equal to fO_k, it follows that $\mathfrak{q}^n$ is a principal ideal of O_F generated by f. Put

$$J = fO_F \cap O_a = fO_F \cap (O_k + IO_F).$$

Let $\mathfrak{m}$ be a maximal ideal of O_k. We then have

$$J_\mathfrak{m} = (fO_F \cap (O_k + IO_F))_\mathfrak{m} = (fO_F)_\mathfrak{m} \cap (O_k + IO_F)_\mathfrak{m}.$$

If $\mathfrak{m}$ is coprime to $\mathfrak{p}$, then $(fO_F)_\mathfrak{m} = O_{F,\mathfrak{m}}$. Hence, we have

$$J_\mathfrak{m} = O_{F,\mathfrak{m}} \cap (O_k + IO_F)_\mathfrak{m} = (f(O_k + IO_F))_\mathfrak{m}.$$

On the other hand if $\mathfrak{m}$ is coprime to I, then we have $(O_k + IO_F)_\mathfrak{m} = O_{F,\mathfrak{m}}$ and so

$$J_\mathfrak{m} = (fO_F)_\mathfrak{m} \cap (O_k + IO_F)_\mathfrak{m} = (fO_F)_\mathfrak{m} = (f(O_k + IO_F))_\mathfrak{m}.$$

As fO_k and I are coprime ideals, it follows that, where $\mathfrak{m}$ runs over all maximal ideals of O_k,

$$J = \bigcap_\mathfrak{m} J_\mathfrak{m} = \bigcap_\mathfrak{m} (f(O_k + IO_F))_\mathfrak{m} = f(O_k + IO_F) = fO_a.$$

That is to say J is a principal ideal of O_a generated by $f \in O_k$.

Now by definition $J = \mathfrak{q}^n \cap O_a$ and we have that $J = (\mathfrak{q} \cap O_a)^n$ because of the unique factorization into prime ideals of the order O_a of non-zero ideals of O_a coprime to its conductor. Hence we have that $\mathfrak{q} \cap O_a$ is a prime ideal of O_a coprime to the conductor and whose nth power is principal. This shows that the order of the Frobenius element $\mathrm{Fr}_\mathfrak{q} \in \mathrm{Gal}(F[a]/F)$, where $\mathfrak{q}$ is the prime of F lying over $\mathfrak{p}$, divides n. Hence, the degree of the unramified field extension $F[a]_\mathfrak{r}/k_\mathfrak{p}$ divides n for any prime $\mathfrak{s}$ of $F[a]$ lying over $\mathfrak{p}$, which proves Step 2.

Step 3. *Suppose that the non-Archimedean place $\mathfrak{p}$ of k divides c and that $l \neq 2$ and let λ be the place of F above $\mathfrak{p}$. Fix a prime λ_2 of F_c above λ. Then the restriction of θ_c to $H^1(F_{c,\lambda_2}, T/I_{c^\sharp} T)$ is zero.*

Assume that $\mathfrak{p}$ is a place of k where $\mathfrak{p}|c$ where $c, \mathfrak{p} \in \mathcal{M}_{1,k}$. Let λ be the unique place of F lying over $\mathfrak{p}$. Let λ_1 be a prime of $F_{c/\mathfrak{p}}$ over λ, where this prime is not usually unique. Fix a prime λ_2 of F_c over λ_1. We have the following inclusions of fields in the top row and primes in the second row where each prime is associated with the field immediately above it and lies over the immediately preceding prime in the row

7.5 Properties of the Derived Classes

$$
\begin{array}{ccccccc}
k & \subseteq & F & \subseteq & F_{c/\mathfrak{p}^m} & \subseteq & F_c \\
| & & | & & | & & | \\
\mathfrak{p} & & \lambda & & \lambda_1 & & \lambda_2
\end{array}
$$

Let

$$\mathrm{Kumm}_c : E(F_c)/I_{c\sharp} E(F_c) \to H^1_{\mathcal{F}}(F_c, T/I_{c\sharp}T)$$

be the Kummer homomorphism. The group $\mathrm{Gal}(F_c/F)$ acts on $E(F_c)/I_{c\sharp} E(F_c)$ and $H^1_{\mathcal{F}}(F_c, T/I_{c\sharp}T)$, and this action commutes with Kumm_c. Furthermore, we have the group algebra element (as in (7.4.12))

$$\sum_{s \in \mathcal{S}_c} s D_{c\sharp} \in \mathbb{Z}[\mathrm{Gal}(F_c/F)]$$

and we put

$$\gamma_c = \sum_{s \in \mathcal{S}_c} s D_{c\sharp} x_c \in E(F_c).$$

By Definition 7.4.17, we have the derived class $\theta_c \in H^1(F, T/I_{c\sharp}T) \otimes \Gamma_{c\sharp}$ which is defined as the inverse image of $\mathrm{Kumm}_c(\gamma_c \mod I_{c\sharp}) \in H^1(F_c, T/I_{c\sharp}T)^{\mathrm{Gal}(F_c/F)}$ under the restriction isomorphism

$$H^1(F, T/I_{c\sharp}T) \stackrel{\mathrm{res}}{\cong} H^1(F_c, T/I_{c\sharp}T)^{\mathrm{Gal}(F_c/F)}.$$

Then from the last two paragraphs, it suffices to show that, where the class $x_c \mod I_{c\sharp} E(F_c)$ belongs to $E(F_c)/I_{c\sharp} E(F_c)$,

$$\sum_{s \in \mathcal{S}_c} s D_{c\sharp} \mathrm{Kumm}_c(x_c \mod I_{c\sharp} E(F_c))$$

has trivial restriction to $H^1_f(F_{c,\lambda_2}, T/I_{c\sharp}T)$.

Let ϕ_λ be the generator of the group $\mathfrak{G}_\lambda = \kappa(\lambda)^*/\kappa(\mathfrak{p})^*$ fixed in (7.4.10) (see also (7.4.7)).

Let

$$\Theta = \mathrm{Kumm}_c(x_c \mod I_{c\sharp} E(F_c)) \in H^1_f(F_{c,\lambda_2}, T/I_{c\sharp}T)$$

and extend ϕ_λ to a generator of $\mathrm{Gal}(F^{\mathrm{sh}}_{c,\lambda_2}/K^{\mathrm{sh}}_\lambda)$. By definition of $I_{c\sharp}$, the Frobenius automorphism $\mathrm{Fr}_\lambda \in \mathrm{Gal}(F^{\mathrm{sh}}_\lambda/F_\lambda)$ acts trivially on $T/I_{c\sharp}T$, and so by Corollary 4.5.3(a), it suffices to show that

$$(D_{\mathfrak{p}^\sharp}\Theta)(\mathrm{Fr}_\lambda) \in T/I_{c^\sharp}T$$

is zero. Since ϕ_λ acts trivially on $T/I_{c^\sharp}T$ and hence acts trivially on $H^1(F_{c_{\lambda_2}}, T/I_{c^\sharp}T)$, we have

$$(D_{\mathfrak{p}^\sharp}\Theta)(\mathrm{Fr}_\lambda) = \sum_{i=1}^{|\kappa(\mathfrak{p})|} i\Theta(\mathrm{Fr}_\lambda) = \frac{|\kappa(\mathfrak{p})|(|\kappa(\mathfrak{p})|+1)}{2}\Theta(\mathrm{Fr}_\lambda) = 0$$

because $|\kappa(\mathfrak{p})|+1 \in I_{c^\sharp}$, which completes the proof of Step 3.

Step 4. *Let $N \subseteq G$ be profinite groups, where N is a normal subgroup of G of finite index, which act on the finite $\mathbb{Z}_l$-module M. If $|G/N|$ is coprime to l, then the restriction inflation sequence*

$$0 \to H^1(G/N, M^N) \to H^1(G, M) \to H^1(N, A)^{G/N} \to H^2(G/N, M^N) \to H^2(G, A)$$

becomes an isomorphism

$$H^1(G, M) \cong H^1(N, A)^{G/N}.$$

This follows because we have $H^i(G/N, M^N) = 0$ for all $i \geq 1$.

Step 5. *Suppose that $c \in \mathcal{M}_{1,k}$ and that $\mathfrak{p}$ is a prime divisor of k dividing c. Assume that $\mathfrak{p}$ is coprime to the prime number l if k is a number field. Suppose that l is coprime to the order of $\mathrm{Pic}(O_k)$. Let λ be the unique prime of F over $\mathfrak{p}$. Then there is a maximal totally and tamely ramified abelian extension L/F_λ of F_λ such that the restriction $\mathrm{res}_{L/F}(\theta_c)$ of the derived cohomology class $\theta_c \in H^1(F, T/I_{c^\sharp}T) \otimes \Gamma_{c^\sharp}$ to L is zero.*

Assume that $\mathfrak{p}$ is a place of k where $\mathfrak{p}|c$ where $c, \mathfrak{p} \in \mathcal{M}_{1,k}$. Let λ be the unique place of F lying over $\mathfrak{p}$. Let m be the greatest integer such that $\mathfrak{p}^m$ divides c. Let λ_3 be a prime of $F_{c/\mathfrak{p}^m}$ over λ. Fix a prime λ_2 of F_c over λ_3. We then have a similar diagram to that of Step 3 with the following inclusions of fields in the top row and primes in the second row where each prime is associated with the field immediately above it and lies over the immediately preceding prime in the row

$$\begin{array}{ccccccc}
k & \subseteq & F & \subseteq & F_{c/\mathfrak{p}} & \subseteq & F_c \\
| & & | & & | & & | \\
\mathfrak{p} & & \lambda & & \lambda_3 & & \lambda_2
\end{array}$$

From Step 3, we have that the restriction of θ_c to $H^1(F_{c,\lambda_2}, T/I_{c^\sharp}T)$ is zero.

7.5 Properties of the Derived Classes

We have by Ax.3 that as $\mathfrak{p}$ is coprime $c/\mathfrak{p}^m$, then $F_c/F_{c/\mathfrak{p}^m}$ is totally ramified at all places of $F_{c/\mathfrak{p}^m}$ lying over $\mathfrak{p}$ and $F_{c/\mathfrak{p}^m}/F$ is unramified at the place λ of F over $\mathfrak{p}$.

By Step 2, we have that the degree $[F_{c/\mathfrak{p}^m,\lambda_3} : F_\lambda]$ divides the order of the divisor class $[\mathfrak{p}]$ of $\mathfrak{p}$ in $\mathrm{Pic}(A)$, where the cycle $c/\mathfrak{p}^m$ is coprime to $\mathfrak{p}$. Therefore, the degree $[F_{c/\mathfrak{p}^m,\lambda_3} : F_\lambda]$ is coprime to l.

There is a maximal subextension

$$F_{c/\mathfrak{p}^m,\lambda_3} \subseteq K \subseteq F_{c,\lambda_2}$$

such that $[K : F_{c/\mathfrak{p}^m,\lambda_3}]$ is a power of l. As $F_{c,\lambda_2}/F_{c/\mathfrak{p}^m,\lambda_3}$ is an abelian extension, we then have that $[F_{c,\lambda_2} : K]$ is coprime to l. Applying Step 4 to the field extension $K \subseteq F_{c,\lambda_2}$ with $G = G_K, N = G_{F_{c,\lambda_2}}$, we obtain the restriction isomorphism

$$H^1(K, T) \cong H^1(F_{c,\lambda_2}, T)^{\mathrm{Gal}(F_{c,\lambda_2}/K)}.$$

From Step 3, we have that the restriction of $\theta_c \in H^1(F, T/I_{c\sharp}T)$ to $H^1(F_{c,\lambda_2}, T/I_{c\sharp}T)$ is zero. Hence by this last isomorphism, we have that the restriction of θ_c to K satisfies

$$\mathrm{res}_{K/F}(\theta_c) = 0.$$

We then have

- $K/F_{c/\mathfrak{p}^m,\lambda_3}$ has degree a power of l and is totally and tamely ramified as $\mathfrak{p}$ and l are coprime when k is a number field;
- $F_{c/\mathfrak{p}^m,\lambda_3}/F_\lambda$ has degree coprime to l and is unramified.

It follows as the group $G = \mathrm{Gal}(K/F_\lambda)$ is abelian that there are subgroups $G_1 \cong \mathrm{Gal}(K/F_{c/\mathfrak{p}^m,\lambda_3}), G_2 \cong \mathrm{Gal}(F_{c/\mathfrak{p}^m,\lambda_3}/F_\lambda])$ of G such that

$$G \cong G_1 \times G_2$$

and in particular that $|G_1|$ is power of l and $|G_2|$ is coprime to l.

Let K_1 be the subfield K^{G_2} of K, so that

$$F_\lambda \subseteq K_1 \subseteq K.$$

Then K_1/F_λ is totally and tamely ramified of degree a power of l, and K/K_1 has degree coprime to l. By Step 4 applied to K/K_1 with $G = G_{K_1}, N = G_K$, we obtain the restriction isomorphism

$$H^1(K_1, T) \cong H^1(K, T)^{\mathrm{Gal}(K/K_1)}.$$

We have already shown that $\mathrm{res}_{K/F}(\theta_c) = 0$ so we have $\mathrm{res}_{K_1/F}(\theta_c) = 0$. But K_1/F_λ is a totally and tamely ramified abelian extension of F_λ; hence, we must have $K_1 \subseteq L$, where L is a maximal totally tamely ramified abelian extension of F_λ. It follows from $\mathrm{res}_{K_1/F}(\theta_c) = 0$ that we must have $\mathrm{res}_{L/F}(\theta_c) = 0$, as required.

□

Remark 7.5.4 In Lemma 7.5.3 in the excluded case for $l = 2$, it follows from Step 1 that for $l = 2$, we only have the weaker conclusion that $\theta_c \in H^1_{\mathcal{F}^{c\sharp}}(F, T/I_{c\sharp}T) \otimes \Gamma_{c\sharp}$.

7.6 Heegner Point and CM Point Kolyvagin Systems

(7.6.1) The main result of this section, Theorem 7.6.14, constructs a Kolyvagin system from a Heegner system on an elliptic curve. This section is the conclusion of the sections Sect. 7.3–7.5. Lemma 7.5.3 shows that the derived classes θ_c satisfy

$$\theta_c \in H^1_{\mathcal{F}(c^\sharp)}(F, T/I_{c^\sharp}T) \otimes \Gamma_{c^\sharp}.$$

This is a principal step in showing that the classes θ_c can be formed into a Kolyvagin system. In general, these θ_c need to be modified by applying an automorphism, which is a product of Frobenius automorphisms, to the θ_c to obtain a Kolyvagin system. This is the essence of the proof of Theorem 7.6.14 which concludes this construction of a Kolyvagin system.

The technical Proposition 7.6.2 shows that the derived classes θ_c as c varies are related by Frobenius automorphisms, and these relations provide the automorphism used in the proof of Theorem 7.6.14.

The notation and hypotheses (7.4.2)–(7.4.8) of the section Sect. 7.4 holds in this section so that we have the Heegner system $x_c \in E(F_c)$ on the elliptic curve E/k satisfying the axioms Ax.1–Ax.7.

Let $c \in \mathcal{M}_{1,k}$ and let $I_{c^\sharp}$ be the corresponding ideal. Then as in the last section, we have the elements $\gamma_c, \theta_c, \phi_q$ given by

- $\gamma_c = \sum_{s \in \mathcal{S}_c} s D_{c^\sharp} x_c \in E(F_c)$ where $x_c \in E(F_c)$ (see (7.4.12));
- $\theta_c \in H^1(F, T/I_{c^\sharp}T) \otimes \Gamma_{c^\sharp}$ is the derived cohomology class obtained from γ_c assuming that $(T/I_{c^\sharp}T)^{G_{F_c}} = 0$ (see Definition 7.4.17);
- ϕ_q is a fixed generator of the cyclic group $\mathfrak{G}_q$ for any prime place q of F belonging to $\mathcal{M}_{1,F}$.

7.6 Heegner Point and CM Point Kolyvagin Systems

For any cycle $c \in \mathcal{M}_{1,k}$, write the ideal $I_{c^\sharp}$ of $\mathbb{Z}_l$ as $I_{c^\sharp} T = l^{M(c)} \mathbb{Z}_l$ where this defines the non-negative integer $M(c)$ (see (7.3.4) and Remark 7.3.6(ii)).

Proposition 7.6.2 *Let* $c\mathfrak{p} \in \mathcal{M}_{1,k}$ *where* $\mathfrak{p} \in \mathcal{M}_{1,k}$ *is a prime place of k and $c \in \mathcal{M}_{1,k}$. Let $\mathfrak{q}$ be the place of F over $\mathfrak{p}$. Assume that $l \neq 2$. Let $\Omega : T/I_{c^\sharp} T \to T/I_{c^\sharp \mathfrak{q}} T$ be the surjection obtained by multiplication by $\frac{l^{M(c)}}{l^{M(c\mathfrak{p})}}$ on $T/I_\mathfrak{q} T$. Let $\Phi_\mathfrak{p} \in G_F$ be any lifting of the generator $\phi_\mathfrak{q}$ of the Galois group $\mathfrak{G}_\mathfrak{q}/\Delta_c \cong \mathrm{Gal}(F_{c\mathfrak{p}}/F_c)$. A Frobenius element $\mathrm{Fr}_\mathfrak{p} \in G_k$ at $\mathfrak{p}$ induces a $\mathbb{Z}_l$-automorphism of $T/I_{c^\sharp \mathfrak{q}} T$ which is independent of the choice of $\mathrm{Fr}_\mathfrak{p}$ in its conjugacy class.*

Then we have the relation in $T/I_{c^\sharp \mathfrak{q}} T$

$$\theta_{c\mathfrak{p}}(\Phi_\mathfrak{p}) = \mathrm{Fr}_\mathfrak{p}(\Omega(\theta_c(\mathrm{Fr}_\mathfrak{q}))).$$

Proof The proof is in several steps. In this proof, for a non-Archimedean local field K, we write K^{sh} for the field of fractions of the strict henselization of the valuation ring of K.

Step 1. Let F/k be the given quadratic extension of global fields equipped with special sets of places. Let Δ_c be the finite abelian group given by Ax.3. Then for any squarefree cycle $c\mathfrak{p} \in \mathcal{M}_{1,k}$ where $\mathfrak{p}$ is a prime place of k, there is a canonical isomorphism $\mathrm{Gal}(F_{c\mathfrak{p}}/F_c) \cong \frac{\kappa(\mathfrak{q})^*}{\kappa(\mathfrak{p})^*}/\Delta_c$ where Δ_c depends only on c up to isomorphism. and where $\mathfrak{q}$ is the unique place of F over $\mathfrak{p}$.

This follows from Ax.3.

Step 2. We have by definition $\mathfrak{G}_\mathfrak{q} = \kappa(\mathfrak{q})^*/\kappa(\mathfrak{p})^*$ (see (7.4.7)). The generator $\phi_\mathfrak{q}$ mod Δ_c of the cyclic group $\mathfrak{G}_\mathfrak{q}/\Delta_c \cong \mathrm{Gal}(F_{c\mathfrak{p}}/F_c)$ can be identified with an element of $\mathrm{Gal}(F_{c\mathfrak{p}}/F_c)$ and lifted to an element

$$\Phi_\mathfrak{q} \in G_F.$$

In particular the restriction $\Phi_\mathfrak{q}|_{F_{c\mathfrak{p}}}$ is equal to $\phi_\mathfrak{q}$ mod $\Delta_c \in \mathrm{Gal}(F_{c\mathfrak{p}}/F_c)$. The set of all such liftings $\Phi_\mathfrak{q}$ of $\phi_\mathfrak{q}$ mod Δ_c is an unique coset of the closed subgroup $G_{F_{c\mathfrak{p}}}$ of G_F.

Let $\mathfrak{q}^{\mathrm{sep}}$ be a prime of F^{sep} above the place $\mathfrak{q}$ of F. The prime $\mathfrak{q}^{\mathrm{sep}}$ of F^{sep} restricts to a prime $\mathfrak{q}''$ of $F_{c\mathfrak{p}}$, and this latter prime restricts to a prime $\mathfrak{q}'$ of F_c.

There is this diagram of inclusions of fields, where the first column lists a prime place of the adjacent field in the table where the primes all lie successively above each other. The final column lists the fraction fields of the strict henselizations of the valuation rings of the fields

$$
\begin{array}{ccc}
\mathfrak{q}^{\text{sep}} & F^{\text{sep}} & \\
& \uparrow & \\
\mathfrak{q}'' & F_{c\mathfrak{p}} \longrightarrow & F^{\text{sh}}_{c\mathfrak{p},\mathfrak{q}''} \\
& \uparrow & \uparrow \\
\mathfrak{q}' & F_c \longrightarrow & F^{\text{sh}}_{\mathfrak{q}} \\
& | & \uparrow \\
\mathfrak{q} & F \longrightarrow & F^{\text{sh}}_{\mathfrak{q}} \\
& | & \\
\mathfrak{p} & k &
\end{array}
$$

Note that in this table $(F_{c,\mathfrak{q}'})^{\text{sh}} = F^{\text{sh}}_{\mathfrak{q}}$. The prime $\mathfrak{q}''$ of $F_{c\mathfrak{p}}$ is the unique prime of $F_{c\mathfrak{p}}$ lying over $\mathfrak{q}'$ of F_c by Ax.3.

The field extension $F_{c\mathfrak{p}}/F_c$ has Galois group $\mathfrak{G}_\mathfrak{q}/\Delta_c$ by Step 1 (see also (7.4.8)); furthermore, $F_{c\mathfrak{p}}/F_c$ is totally ramified at all primes of F_c over $\mathfrak{q}$ by Ax.3; therefore, $\text{Gal}(F_{c\mathfrak{p}}/F_c)$ is the inertia group of any prime of F_c lying over $\mathfrak{q}$; hence, the image of the generator $\phi_\mathfrak{q}$ of $\mathfrak{G}_\mathfrak{q} \cong \kappa(\mathfrak{q})^*/\kappa(\mathfrak{p})^*$ belongs to this inertia group $\text{Gal}(F_{c\mathfrak{p}}/F_c)$.

The isomorphism $\mathfrak{G}_\mathfrak{q}/\Delta_c \cong \text{Gal}(F_{c\mathfrak{p}}/F_c)$ enables $\phi_\mathfrak{q} \bmod \Delta_c$ to be identified with an element of $\text{Gal}(F_{c\mathfrak{p}}/F_c)$.

It follows that $\phi_\mathfrak{q} \bmod \Delta_c$ can be lifted to an element

$$\Phi_\mathfrak{q} \in G_F$$

where $\Phi_\mathfrak{q}|_{F_{c\mathfrak{p}}}$ coincides with the element $\phi_\mathfrak{q} \bmod \Delta_c \in \text{Gal}(F_{c\mathfrak{p}}/F_c)$; in particular, $\Phi_\mathfrak{q}$ restricts to the identity on F_c and its restriction to $F_{c\mathfrak{p}}$ generates the group $\text{Gal}(F_{c\mathfrak{p}}/F_c)$. The set of all such liftings $\Phi_\mathfrak{q}$ of $\phi_\mathfrak{q} \bmod \Delta_c$ is an unique coset of the closed subgroup $G_{F_{c\mathfrak{p}}}$ contained in G_F.

Step 3. Let $\mathcal{E}/\kappa(\mathfrak{p})$ be the reduction of the Néron model of E/k at $\mathfrak{p}$ and T be the l-adic Tate module of E. There are canonical isomorphisms

$$T/I_\mathfrak{q} T \cong E(F^{\text{sep}}_\mathfrak{q})[l^{M(\mathfrak{p})}] \cong \mathcal{E}(\kappa(\mathfrak{q})^{\text{sep}})[l^{M(\mathfrak{p})}] \cong \mathcal{E}(\kappa(\mathfrak{q}))[l^{M(\mathfrak{p})}]. \quad (7.6.3)$$

The elliptic curve E has good reduction at $\mathfrak{p}$, because $\mathfrak{p}$ is not in the set of places where T is ramified see (7.3.4). This identification of $\mathcal{E}(\kappa(\mathfrak{q})^{\text{sep}})[l^{M(\mathfrak{p})}]$ with $\mathcal{E}(\kappa(\mathfrak{q}))[l^{M(\mathfrak{p})}]$ holds, because the Frobenius $\text{Fr}_\mathfrak{q}$ acts trivially on $T/I_\mathfrak{q} T$ as well as on $\mathcal{E}(\kappa(\mathfrak{q}))$, where $I_\mathfrak{q} = l^{M(\mathfrak{p})}\mathbb{Z}_l$ (see (7.3.4) and Remark 7.3.6(ii)).

Step 4. Let $\mathcal{E}/\kappa(\mathfrak{p})$ be the reduction at $\mathfrak{p}$ of the Néron model of E/k. Then there is the relation which holds in the group $\mathcal{E}(\kappa(\mathfrak{q}^{\text{sep}}))$

$$\xi_{c\mathfrak{p}}(\Phi_\mathfrak{p}) \equiv \frac{a_\mathfrak{p} - (|\kappa(\mathfrak{p})| + 1)\text{Fr}_\mathfrak{p}^{-1}}{l^{M(c\mathfrak{p})}} \gamma_c \quad \bmod \mathfrak{q}^{\text{sep}}.$$

7.6 Heegner Point and CM Point Kolyvagin Systems

Let

$$\xi_{cp} : G_F \to T/I_{c^\sharp q} T \tag{7.6.4}$$

be the cocycle representing the derived cohomology class $\theta_{cp} \in H^1(F, T/I_{c^\sharp q}T)$ as in (7.4.18) and Lemma 7.5.2. Define $M(c\mathfrak{p}) \in \mathbb{N}$ by $I_{c^\sharp q} = l^{M(c)}\mathbb{Z}_l$ and fix a $l^{M(c\mathfrak{p})}$-division point $\frac{\gamma_c}{l^{M(c\mathfrak{p})}} \in E(F^{sep})$ of γ_{cp} where γ_{cp} is as in (7.6.1) and Lemma 7.5.2. From Lemma 7.5.2 and with the notation of that lemma, the class $\theta_{cp} \in H^1(F, T/I_{c^\sharp q}T)$ is represented by the cocycle

$$g \mapsto \xi_{cp}(g) = (g-1)\frac{\gamma_{cp}}{l^{M(c\mathfrak{p})}} - \frac{(g-1)\gamma_{cp}}{l^{M(c\mathfrak{p})}}, \quad G_F \to T/I_{(c\mathfrak{p})^\sharp}T, \tag{7.6.5}$$

where $\frac{(g-1)\gamma_{cp}}{l^{M(c\mathfrak{p})}}$, for $g \in G_F$, is the unique $l^{M(c\mathfrak{p})}$-division point of $(g-1)\gamma_{cp}$ in $E(F_{c\mathfrak{p}})$.

The generator $\phi_q \in \mathfrak{G}_q = \kappa(\mathfrak{q})^*/\kappa(\mathfrak{p})^*$ induces a generator of $\text{Gal}(F^{sh}_{c\mathfrak{p},\mathfrak{q}''}/F^{sh}_{c,\mathfrak{q}'})$, where $(F_{c,\mathfrak{q}'})^{sh} = F^{sh}_\mathfrak{q}$, and there is this canonical isomorphism

$$\text{Gal}(F^{sh}_{c\mathfrak{p},\mathfrak{q}''}/F^{sh}_\mathfrak{q}) \cong \mathfrak{G}_q/\Delta_c.$$

Furthermore, the field extension $F^{sh}_{c\mathfrak{p},\mathfrak{q}''}/F^{sh}_\mathfrak{q}$ is totally ramified by Ax.3. Therefore $\Phi_q \in G_F$, the element given by Step 2, acts trivially on $\frac{\gamma_{cp}}{l^{M(c\mathfrak{p})}}$ mod $\mathfrak{q}^{sep}$, the reduction at $\mathfrak{p}$ of $\frac{\gamma_{cp}}{l^{M(c\mathfrak{p})}}$, because the image of Φ_q belongs to the inertia group $\text{Gal}(F^{sh}_{c\mathfrak{p},\mathfrak{q}''}/F^{sh}_\mathfrak{q})$.

Putting $g = \Phi_q$ in the formula (7.6.5), we have that this cocycle ξ_{cp} satisfies, because by the previous paragraph, we have the congruence $(\Phi_q - 1)\frac{\gamma_{cp}}{l^{M(c\mathfrak{p})}} \equiv 0$ mod $\mathfrak{q}^{sep}$,

$$\xi_{cp}(\Phi_q) \equiv -\frac{(\Phi_q - 1)\gamma_{cp}}{l^{M(c\mathfrak{p})}} \mod \mathfrak{q}^{sep}. \tag{7.6.6}$$

From the definition of γ_c in (7.4.12) and the identity $(\phi_q - 1)D_q = |\mathfrak{G}_q| - N_{\mathfrak{G}_q}$, we may replace ϕ_q here by its lifting Φ_q, and we have the equality of points in $E(F_{c\mathfrak{p}})$, where we note that the restriction of Φ_q to F_{cp} is equal to ϕ_q mod Δ_c (Step 2), so the next formula makes sense,

$$(\Phi_q - 1)\gamma_{cp} = \sum_{s \in \mathcal{S}_{cp}} (\Phi_q - 1)sD_{c^\sharp q}x_{cp} = \sum_{s \in \mathcal{S}_{cp}} sD_{c^\sharp}(|\mathfrak{G}_q| - N_{\mathfrak{G}_q})x_{cp}.$$

From Lemma 7.4.13, we have the Euler system relation, where $N_{\mathfrak{G}_q}$ denotes the norm as in that Lemma 7.4.13 and $a_\mathfrak{p}$ is a Frobenius trace as in the definition of a Heegner system,

$$N_{\mathfrak{G}_q}x_{cp} = a_\mathfrak{p} x_c.$$

Hence, we obtain

$$(\Phi_q - 1)\gamma_{cp} = \sum_{s \in \mathcal{S}_{cp}} s D_{c^{\sharp}}\left(|\mathfrak{S}_q|x_{cp} - a_p x_c\right). \tag{7.6.7}$$

We have the congruence relation Ax.6 for the Heegner points $x_c \in E(F_c)$, where $\mathrm{Fr}_\mathfrak{p}$ denotes the arithmetic Frobenius element, namely, that for any prime $\mathfrak{q}^{\mathrm{sep}}$ of F^{sep} over $\mathfrak{q}$, we have

$$\mathrm{Fr}_\mathfrak{p} x_{cp} \equiv x_c \mod \mathfrak{q}^{\mathrm{sep}}. \tag{7.6.8}$$

We then obtain from this congruence relation and the congruence (7.6.6) and the formula (7.6.7) that

$$\xi_{cp}(\Phi_\mathfrak{p}) \equiv \frac{a_p - (|\kappa(\mathfrak{p})| + 1)\mathrm{Fr}_\mathfrak{p}^{-1}}{l^{M(cp)}} \gamma_c \mod \mathfrak{q}^{\mathrm{sep}}. \tag{7.6.9}$$

On the right-hand side here, the integers $a_\mathfrak{p}$ and $|\kappa(\mathfrak{p})| + 1$ are both divisible by $l^{M(\mathfrak{p})}$ where $I_\mathfrak{q} = l^{M(\mathfrak{p})}\mathbb{Z}_l$ and hence *a fortiori* the integers $a_\mathfrak{p}$ and $|\kappa(\mathfrak{p})| + 1$ are divisible by $l^{M(cp)}$ because $l^{M(cp)}$ divides $l^{M(\mathfrak{p})}$, which completes this Step 4.

Note that in the congruence (7.6.9), the element $\xi_{cp}(\Phi_\mathfrak{p}) \in T/I_{c^\sharp \mathfrak{q}}T$ depends on the choice of $\Phi_\mathfrak{p}$ made in Step 2, whereas the element $\frac{a_\mathfrak{p} - (|\kappa(\mathfrak{p})|+1)\mathrm{Fr}_\mathfrak{p}^{-1}}{l^{M(cp)}}\gamma_c \in E(F_c)$ is obviously independent of the choice of $\Phi_\mathfrak{p}$.

Step 5. Let $c\mathfrak{p} \in \mathcal{M}_{1,k}$ where $\mathfrak{p}$ is a prime cycle. Let $\mathcal{E}/\kappa(\mathfrak{p})$ be the reduction at $\mathfrak{p}$ of the Néron model of E/k. Then in the groups $\mathcal{E}(\kappa(\mathfrak{q}^{\mathrm{sep}})) \supseteq \mathcal{E}(\kappa(\mathfrak{q}))$ we have, where $\gamma_c \in E(F_c)$,

$$\gamma_c \mod \mathfrak{q}^{\mathrm{sep}} \in \mathcal{E}(\kappa(\mathfrak{q})).$$

By Step 4, we have the relation in the group $\mathcal{E}(\kappa(\mathfrak{q}^{\mathrm{sep}}))$

$$\xi_{cp}(\Phi_\mathfrak{p}) \equiv \frac{a_\mathfrak{p} - (|\kappa(\mathfrak{p})| + 1)\mathrm{Fr}_\mathfrak{p}^{-1}}{l^{M(cp)}}\gamma_c \mod \mathfrak{q}^{\mathrm{sep}}.$$

Here ξ_{cp} is a cocycle annihilated by $l^{M(cp)}$, and hence we have $l^{M(cp)}\xi_{cp}(\Phi_\mathfrak{p}) = 0$. Therefore, we have in the group $\mathcal{E}(\kappa(\mathfrak{q}^{\mathrm{sep}}))$

$$(a_\mathfrak{p} - (|\kappa(\mathfrak{p})| + 1)\mathrm{Fr}_\mathfrak{p}^{-1})\gamma_c \mod \mathfrak{q}^{\mathrm{sep}} \equiv 0.$$

Hence, we have

$$[(a_\mathfrak{p}\mathrm{Fr}_\mathfrak{p} - |\kappa(\mathfrak{p})| - \mathrm{Fr}_\mathfrak{p}^2)\mathrm{Fr}_\mathfrak{p}^{-1} + (\mathrm{Fr}_\mathfrak{p} - \mathrm{Fr}_\mathfrak{p}^{-1})]\gamma_c \equiv 0 \mod \mathfrak{q}^{\mathrm{sep}}.$$

7.6 Heegner Point and CM Point Kolyvagin Systems

As the Frobenius $\mathrm{Fr}_{\mathfrak{p}}$ acting on $\mathcal{E}(\kappa(\mathfrak{q}^{\mathrm{sep}}))$ satisfies $a_{\mathfrak{p}}\mathrm{Fr}_{\mathfrak{p}} - |\kappa(\mathfrak{p})| - \mathrm{Fr}_{\mathfrak{p}}^2 = 0$, we obtain

$$(\mathrm{Fr}_{\mathfrak{p}} - \mathrm{Fr}_{\mathfrak{p}}^{-1})\gamma_c \equiv 0.$$

It follows that $\gamma_c \mod \mathfrak{q}^{\mathrm{sep}}$ belongs to the subgroup $\mathcal{E}(\kappa(\mathfrak{q}))$ where $\kappa(\mathfrak{q})$ is the quadratic extension of $\kappa(\mathfrak{p})$.

[This result in the case of function fields also follows from [4, Theorem 4.6.19(ii)].]

Step 6 *End of proof: the term* $\Omega(\theta_c(\mathrm{Fr}_{\mathfrak{q}}))$.

From Step 4, we have the following congruence, where $\xi_{c\mathfrak{p}}(\Phi_{\mathfrak{p}}) \in T/I_{c^{\sharp}\mathfrak{q}}T$ and where we write $\mathfrak{q}$ in place of $\mathfrak{q}^{\mathrm{sep}}$ because $\gamma_c \mod \mathfrak{q} \in \mathcal{E}(\kappa(\mathfrak{q}))$ by Step 5,

$$\xi_{c\mathfrak{p}}(\Phi_{\mathfrak{p}}) \equiv \frac{a_{\mathfrak{p}} - (|\kappa(\mathfrak{p})| + 1)\mathrm{Fr}_{\mathfrak{p}}^{-1}}{l^{M(c\mathfrak{p})}} \gamma_c \mod \mathfrak{q}. \qquad (7.6.10)$$

There is a canonical surjective homomorphism obtained by multiplication by $\frac{l^{M(c)}}{l^{M(c\mathfrak{p})}}$ on $T/I_{\mathfrak{q}}T$ and where $T/I_{c^{\sharp}\mathfrak{q}}T$ can be considered to be a submodule of $T/I_{c^{\sharp}}T$

$$\Omega : T/I_{c^{\sharp}}T \to T/I_{c^{\sharp}\mathfrak{q}}T.$$

From Lemma 7.5.2 and with the notation of that lemma, the class $\theta_c \in H^1(F, T/I_{c^{\sharp}}T)$ is represented by the cocycle ξ_c

$$\xi_c : g \mapsto \xi_c(g) = (g-1)\frac{\gamma_c}{l^{M(c)}} - \frac{(g-1)\gamma_c}{l^{M(c)}}, \quad G_F \to T/I_{c^{\sharp}}T, \qquad (7.6.11)$$

where $\frac{(g-1)\gamma_c}{l^{M(c)}}$, for $g \in G_F$, is the unique $l^{M(c)}$-division point of $(g-1)\gamma_c$ in $E(F_c)$.

Evaluating the class θ_c at $\mathrm{Fr}_{\mathfrak{q}}$, we obtain from this last formula $\theta_c(\mathrm{Fr}_{\mathfrak{q}}) = (\mathrm{Fr}_{\mathfrak{q}} - 1)\frac{\gamma_c}{l^{M(c)}} - \frac{(\mathrm{Fr}_{\mathfrak{q}}-1)\gamma_c}{l^{M(c)}}$. Now $\gamma_c \mod \mathfrak{q}^{\mathrm{sep}}$ belongs to $\mathcal{E}(\kappa(\mathfrak{q}))$ by Step 5 and therefore $\frac{(\mathrm{Fr}_{\mathfrak{q}}-1)\gamma_c}{l^{M(c)}} \equiv 0 \mod \mathfrak{q}^{\mathrm{sep}}$. We then have

$$\theta_c(\mathrm{Fr}_{\mathfrak{q}}) \equiv (\mathrm{Fr}_{\mathfrak{q}} - 1)\frac{\gamma_c}{l^{M(c)}} \mod \mathfrak{q}^{\mathrm{sep}}.$$

It follows that

$$\Omega(\theta_c(\mathrm{Fr}_{\mathfrak{q}})) \equiv (\mathrm{Fr}_{\mathfrak{q}} - 1)\frac{\gamma_c}{l^{M(c\mathfrak{p})}} \mod \mathfrak{q}^{\mathrm{sep}}$$

As $\mathrm{Fr}_{\mathfrak{q}} = \mathrm{Fr}_{\mathfrak{p}}^2$ and as the Frobenius $\mathrm{Fr}_{\mathfrak{p}}$ acting on $\mathcal{E}(\kappa(\mathfrak{q}^{\mathrm{sep}}))$ satisfies $a_{\mathfrak{p}}\mathrm{Fr}_{\mathfrak{p}} - |\kappa(\mathfrak{p})| - \mathrm{Fr}_{\mathfrak{p}}^2 = 0$, we obtain

$$\Omega(\theta_c(\mathrm{Fr}_\mathfrak{q})) \equiv \frac{(a_\mathfrak{p} - (|\kappa(\mathfrak{p})|+1)\mathrm{Fr}_\mathfrak{p}^{-1})}{l^{M(c\mathfrak{p})}} \mathrm{Fr}_\mathfrak{p} \gamma_c \mod \mathfrak{q}^{\mathrm{sep}}$$

Here both integers $a_\mathfrak{p}$ and $|\kappa(\mathfrak{p})|+1$ are divisible by $l^{M(c\mathfrak{p})}$ and $\gamma_c \mod \mathfrak{q}^{\mathrm{sep}}$ belongs to $\mathcal{E}(\kappa(\mathfrak{q}))$ by Step 5; it follows that this congruence can be taken mod $\mathfrak{q}$ instead of mod $\mathfrak{q}^{\mathrm{sep}}$.

From (7.6.10), we have $\xi_{c\mathfrak{p}}(\Phi_\mathfrak{p}) \equiv \frac{a_\mathfrak{p} - (|\kappa(\mathfrak{p})|+1)\mathrm{Fr}_\mathfrak{p}^{-1}}{l^{M(c\mathfrak{p})}} \gamma_c \mod \mathfrak{q}$; hence, we obtain, where a priori this is a congruence mod $\mathfrak{q}^{\mathrm{sep}}$,

$$\xi_{c\mathfrak{p}}(\Phi_\mathfrak{p}) = \mathrm{Fr}_\mathfrak{p} \Omega(\theta_c(\mathrm{Fr}_\mathfrak{q})) \tag{7.6.12}$$

where the left-hand side $\xi_{c\mathfrak{p}}(\Phi_\mathfrak{p})$ belongs to $T/I_{c^\sharp \mathfrak{q}} T$ and the right-hand side $\mathrm{Fr}_\mathfrak{p} \Omega(\theta_c(\mathrm{Fr}_\mathfrak{q}))$ also belongs to $T/I_{c^\sharp \mathfrak{q}} T$; as these two elements are congruent mod $\mathfrak{q}^{\mathrm{sep}}$, by Step 3, these elements of $T/I_{c^\sharp \mathfrak{q}} T$ must be equal, and in Eq. (7.6.12), we have equality and not just a congruence mod $\mathfrak{q}^{\mathrm{sep}}$.

Therefore we have, where $\theta_{c\mathfrak{p}}(\Phi_\mathfrak{p}) \in T/I_{c^\sharp \mathfrak{q}} T$, $\theta_c(\mathrm{Fr}_\mathfrak{q}) \in T/I_{c^\sharp} T$ and $\Omega(\theta_c(\mathrm{Fr}_\mathfrak{q})) \in T/I_{c^\sharp \mathfrak{q}} T$,

$$\theta_{c\mathfrak{p}}(\Phi_\mathfrak{p}) = \mathrm{Fr}_\mathfrak{p} \Omega(\theta_c(\mathrm{Fr}_\mathfrak{q}))$$

and where the right-hand side here is independent of the choice of Frobenius $\mathrm{Fr}_\mathfrak{p}$ in its conjugacy class of G_k by the isomorphisms of Step 3. This proves the proposition. □

(7.6.13) As this is required in the next Theorem 7.6.14, we summarize the construction of the finite-singular comparison map of Definition 4.5.5 in the special case we have of a Kolyvagin system obtained from a Heegner system. Let L be a non-Archimedean local field with residue field κ of characteristic different from l and T is an R-representation of G_L where R has residue field of characteristic l.

It is assumed that G_L acts trivially on T and $|\kappa^*|.T = 0$. Then Corollary 4.5.3 gives the isomorphisms

$$H^1_f(L, T) \cong T/(\mathrm{Fr}-1)T, \quad H^1_s(L, T) \cong \mathrm{Hom}(I, T)^{\mathrm{Fr}=1}, \quad H^1_s(L, T) \otimes_\mathbb{Z} \kappa^* \cong T^{\mathrm{Fr}=1}.$$

Therefore as G_L acts trivially on T, we obtain the isomorphisms

$$H^1_f(L, T) \cong T/(\mathrm{Fr}-1)T \cong T \cong T^{\mathrm{Fr}=1} \cong H^1_s(L, T) \otimes_\mathbb{Z} \kappa^*.$$

The composite of these isomorphisms is

$$\phi^{fs} : H^1_f(L, T) \cong H^1_s(L, T) \otimes_\mathbb{Z} \kappa^*$$

7.6 Heegner Point and CM Point Kolyvagin Systems

which is the finite singular comparison isomorphism of Definition 4.5.5 and in the notation of that definition, the map $\psi : T/(\mathrm{Fr} - 1)T \to T^{\mathrm{Fr}=1}$ is taken to be the identity map on T.

In more detail, the canonical functorial isomorphisms of Corollary 4.5.3 are as follows. The map $H^1_f(L, T) \cong T/(\mathrm{Fr} - 1)T$ is given by the recipe: for $\eta \in H^1_f(L, T)$ represented by a cocycle ξ then

$$\eta \mapsto \xi(\mathrm{Fr}) \in T.$$

The map $H^1_s(L, T) \otimes_{\mathbb{Z}} \kappa^* \cong T^{\mathrm{Fr}=1}$ is given by the recipe:

$$c \otimes \alpha \mapsto c(\tau_\alpha)$$

where $\tau_\alpha \in \mathrm{Gal}(L^{\mathrm{ab}}/L^{\mathrm{unr}})$ is the Artin symbol of any lift of $\alpha \in \kappa^*$ to L, L^{ab} is the maximal abelian separable extension of L, and L^{unr} is the maximal abelian unramified separable extension of L.

The finite singular comparison map for $\eta \in H^1_f(L, T)$ represented by a cocycle ξ is then given by

$$\eta \mapsto \xi(\mathrm{Fr}) = c(\tau_\alpha) \mapsto c \otimes \alpha, \quad H^1_f(L, T) \to H^1_s(L, T) \otimes_{\mathbb{Z}} \kappa^*.$$

Theorem 7.6.14 *With the assumptions made in (7.6.1), assume that the residue characteristic l of the coefficient ring satisfies $l \neq 2$ and that l is coprime to the order of $\mathrm{Pic}(O_k)$ where O_k is the ring of integers of k. Let θ_c, for all $c \in \mathcal{M}_{1,k}$, be the derived cohomology classes obtained from the Heegner system x_c. Then there is a Kolyvagin system κ for $(T, \mathcal{F}, \mathcal{L})$ such that $\kappa_1 = \theta_1$.*

Proof The previous results, Lemma 7.5.3 and Proposition 7.6.2, show that the derived classes θ_c almost form a Kolyvagin system. It is only necessary to apply an automorphism to the groups $H^1(F, T/I_{c\sharp}T)$, and so to the classes θ_c, to obtain a Kolyvagin system.

Let $c \in \mathcal{M}_{1,k}$. For a prime divisor $\mathfrak{p} \in \mathcal{M}_{1,k}$ dividing c, let $\mathrm{Fr}_\mathfrak{p} \in G_k$ be a Frobenius automorphism at $\mathfrak{p}$, determined up to conjugation.

Let $\mathfrak{q}$ be the prime of F lying over $\mathfrak{p}$. The ideal $I_\mathfrak{q}$ for $\mathfrak{q} \in \mathcal{M}_{1,F}$ is such that $\mathrm{Fr}_\mathfrak{q}$ acts trivially on $T/I_\mathfrak{q}T$ (see (7.3.4) and also Remark 7.3.6(ii), Remarks 7.1.19(ii)). Therefore, $\mathrm{Fr}_\mathfrak{p}$ is such that $\mathrm{Fr}_\mathfrak{p}^2$ acts trivially on $T/I_{c\sharp}T$ because $I_{c\sharp} \supseteq I_\mathfrak{q}$; hence as $\mathrm{Fr}_\mathfrak{p}$ acts semi-simply on $T/I_{c\sharp}T$, it follows that $\mathrm{Fr}_\mathfrak{p}$ acts as a homothety on $T/I_{c\sharp}T$ (from the Weil conjectures) and so acts as a homothety on $H^1(F, T/I_{c\sharp}T)$. Therefore the elements $\mathrm{Fr}_\mathfrak{p}$, for all $\mathfrak{p} \in \mathcal{M}_{1,k}$ dividing c, act as homotheties on $H^1(F, T/I_{c\sharp}T)$ and therefore their actions commute.

For $c \in \mathcal{M}_{1,k}$, define the $\mathbb{Z}_l$-automorphism

$$\chi_c : H^1(F, T/I_{c\sharp}T) \to H^1(F, T/I_{c\sharp}T)$$

in the following way. For $\mathfrak{p} \in \mathcal{M}_{1,k}$ dividing c, the Frobenius $\mathrm{Fr}_\mathfrak{p} : T/I_\mathfrak{q} T \to T/I_\mathfrak{q} T$ induces an automorphism $\chi_\mathfrak{p}$ of $H^1(F, T/I_{c^\sharp} T)$. We will also write $\chi_\mathfrak{p}$ for this automorphism $\mathrm{Fr}_\mathfrak{p}$ acting on $T/I_\mathfrak{q} T$ and its quotients. These automorphisms $\chi_\mathfrak{p}$ of $H^1(F, T/I_{c^\sharp} T)$, for all prime divisors $\mathfrak{p}$ dividing c, commute from the previous paragraph and we then define $\chi_c : H^1(F, T/I_{c^\sharp} T) \to H^1(F, T/I_{c^\sharp} T)$ to be the composite automorphism

$$\chi_c = \prod_{\mathfrak{p}|c} \chi_\mathfrak{p}$$

as $\mathfrak{p}$ runs over all prime divisors in k in the support of c.

To construct the Kolyvagin system from the derived classes θ_c, we only need to apply the automorphisms χ_c. Define for $c \in \mathcal{M}_{1,k}$, where $\theta_c \in H^1(F, T/I_{c^\sharp} T) \otimes \Gamma_{c^\sharp}$ are the derived classes of (7.4.1) and Definition 7.4.17 obtained from the Heegner system,

$$\kappa_c = \chi_c^{-1}(\theta_c) \otimes (\bigotimes_{v|c} \phi_{v^\sharp}) \in H^1_{\mathcal{F}(c)}(F, T/I_{c^\sharp} T) \otimes \mathfrak{G}_{c^\sharp}.$$

We write here $\mathfrak{G}_{c^\sharp}$ in place of $\Gamma_{c^\sharp}$ because $\mathfrak{G}_n \otimes_{\mathbb{Z}} R/I_n R \cong \Gamma_n \otimes_{\mathbb{Z}} R/I_n R$ for $n \in \mathcal{M}_{0,F}$ by Lemma 7.4.9.

We now check the compatibility of these elements κ_c via the diagram (7.1.8) defining a Kolyvagin system, where $c, \mathfrak{p} \in \mathcal{M}_{1,k}$ are coprime and $\mathfrak{q}$ is the prime of F over $\mathfrak{p}$, and where, as already mentioned, the groups $\Gamma_{c^\sharp}$ are replaced by their homomorphic images $\mathfrak{G}_{c^\sharp}$,

$$\begin{array}{c} H^1_{\mathcal{F}(c^\sharp)}(F, T/I_{c^\sharp} T) \otimes_{\mathbb{Z}} \mathfrak{G}_{c^\sharp} \\ \downarrow \mathrm{res}_{F_\mathfrak{q}/F} \\ H^1_f(F_\mathfrak{q}, T/I_{c^\sharp \mathfrak{q}} T) \otimes_{\mathbb{Z}} \mathfrak{G}_{c^\sharp} \\ \downarrow \phi_\mathfrak{q}^{\mathrm{fs}} \\ H^1_{\mathcal{F}(c^\sharp \mathfrak{q})}(F, T/I_{c^\sharp \mathfrak{q}} T) \otimes_{\mathbb{Z}} \mathfrak{G}_{c^\sharp \mathfrak{q}} \xrightarrow{\mathrm{res}_{F_\mathfrak{q}/F}} H^1_s(F_\mathfrak{q}, T/I_{c^\sharp \mathfrak{q}} T) \otimes_{\mathbb{Z}} \mathfrak{G}_{c^\sharp \mathfrak{q}} \end{array} \quad (7.6.15)$$

The element κ_c belongs to $H^1_{\mathcal{F}(c^\sharp)}(F, T/I_{c^\sharp} T) \otimes_{\mathbb{Z}} \mathfrak{G}_{c^\sharp}$ at the top of this diagram, by Lemma 7.5.3. Its restriction at $\mathfrak{q}$ and taken modulo $I_{c^\sharp \mathfrak{q}}$ belongs to $H^1_f(F_\mathfrak{q}, T/I_{c^\sharp \mathfrak{q}} T) \otimes_{\mathbb{Z}} \mathfrak{G}_{c^\sharp}$ in the middle row of the diagram.

From (7.4.7) and (7.3.4), we have

$$T/I_{c^\sharp \mathfrak{q}} T \otimes_{\mathbb{Z}} \mathfrak{G}_{c^\sharp} \cong T/I_{c^\sharp \mathfrak{q}} T \otimes_{\mathbb{Z}} \mathfrak{G}_{c^\sharp \mathfrak{q}}$$

because $I_{c^\sharp \mathfrak{q}} \supseteq I_{c^\sharp}$ and also $\mathfrak{G}_{c^\sharp \mathfrak{q}}$ is a homomorphic image of $\mathfrak{G}_{c^\sharp}$.

7.6 Heegner Point and CM Point Kolyvagin Systems

Furthermore, as Fr_q acts trivially on $T/I_{c\sharp}T$, the finite singular comparison isomorphism ϕ_q^{fs} of the diagram is an isomorphism and factors as the composite of two isomorphisms in the next diagram provided by Corollary 4.5.3

$$\begin{array}{ccc} H_f^1(F_q, T/I_{c\sharp q}T) \otimes_{\mathbb{Z}} \mathfrak{G}_{c\sharp} & \stackrel{\text{ev}}{\cong} & T/I_{c\sharp q}T \otimes_{\mathbb{Z}} \mathfrak{G}_{c\sharp} \\ \downarrow \phi_q^{fs} & f \nearrow \cong & \\ H_s^1(F_q, T/I_{c\sharp q}T) \otimes_{\mathbb{Z}} \mathfrak{G}_{c\sharp q} & & \end{array} \qquad (7.6.16)$$

Here the map ev is evaluation of a cocycle at the Frobenius Fr_q; the map f is given as follows: the group $\mathfrak{G}_{c\sharp q}$ is a homomorphic image of the group $\kappa(q)^*$ and f is given by $f : c \otimes \alpha \mapsto c(\sigma_\alpha)$, for $\alpha \in \kappa(q)^*$, which is the evaluation of a cocycle c at σ_α where $\sigma_\alpha \in \text{Gal}(F_q^{ab}/F_q^{unr})$ is the image via the reciprocity isomorphism of any lift of α to F_q^* and where F_q^{ab} is the maximal separable abelian extension of F_q and F_q^{unr} is the maximal unramified abelian extension of F_q (Corollary 4.5.3). The finite singular comparison isomorphism is not uniquely determined in general, but we have made this choice of ϕ_q^{fs} in diagram (7.6.15).

The restriction $\text{res}_{F_q/F}(\kappa_c)$ of the element $\kappa_c \in H^1_{\mathcal{F}(c\sharp)}(F, T/I_{c\sharp}T) \otimes_{\mathbb{Z}} \mathfrak{G}_{c\sharp}$ transforms in (7.6.16) under the top horizontal isomorphism by evaluation of θ_c at the Frobenius Fr_q (Corollary 4.5.3), that is to say

$$\kappa_c = \chi_c^{-1}(\theta_c) \otimes (\bigotimes_{v|c} \phi_{v\sharp}) \mapsto \text{res}_{F_q/F}(\chi_c^{-1}(\theta_c)) \otimes (\bigotimes_{v|c} \phi_{v\sharp})$$

$$\mapsto \text{res}_{F_q/F}(\chi_c^{-1}(\theta_c))(\text{Fr}_q) \otimes (\bigotimes_{v|c} \phi_{v\sharp}) \in T/I_{c\sharp q}T \otimes_{\mathbb{Z}} \mathfrak{G}_{c\sharp}. \qquad (7.6.17)$$

On the other hand, κ_{cp} belongs to $H^1_{\mathcal{F}(c\sharp q)}(F, T/I_{c\sharp q}T) \otimes_{\mathbb{Z}} \mathfrak{G}_{c\sharp q}$ in the lower left of the diagram (7.6.15), by Lemma 7.5.3, and its image under the restriction at q is

$$\text{res}_{F_q/F}(\chi_{cp}^{-1}(\theta_{cp})) \otimes (\bigotimes_{v|cp} \phi_{v\sharp}) = \chi_c^{-1} \text{res}_{F_q/F}(\chi_p^{-1}\theta_{cp}) \otimes \phi_q \otimes (\bigotimes_{v|c} \phi_{v\sharp})$$

$$\in H_{tr}^1(F_q, T/I_{c\sharp q}T) \otimes_{\mathbb{Z}} \mathfrak{G}_{c\sharp q}$$

where this last group $H_{tr}^1(F_q, T/I_{c\sharp q}T) \otimes_{\mathbb{Z}} \mathfrak{G}_{c\sharp q}$ maps isomorphically to the singular quotient $H_s^1(F_q, T/I_{c\sharp q}T) \otimes_{\mathbb{Z}} \mathfrak{G}_{c\sharp q}$.

From Proposition 7.6.2, where the map Ω and the element Φ_p are defined in that proposition, we have

$$\theta_{cp}(\Phi_p) = \text{Fr}_p(\Omega(\theta_c(\text{Fr}_q)))$$

where this is a relation in the quotient $T/I_{c^\sharp q}T$ of $T/I_q T$. Hence we have $\Omega(\theta_c(\mathrm{Fr}_q)) = \chi_\mathfrak{p}^{-1}(\theta_{c\mathfrak{p}}(\Phi_q))$, and we obtain

$$\mathrm{res}_{F_q/F}(\chi_{c\mathfrak{p}}^{-1}(\theta_{c\mathfrak{p}}(\Phi_q))) \otimes \left(\bigotimes_{v|c\mathfrak{p}} \phi_{v^\sharp}\right) = \chi_c^{-1}\mathrm{res}_{F_q/F}(\Omega(\theta_c(\mathrm{Fr}_q))) \otimes \left(\bigotimes_{v|c} \phi_{v^\sharp}\right).$$

This last element equals the element $\mathrm{res}_{F_q/F}(\chi_c^{-1}(\theta_c))(\mathrm{Fr}_q) \otimes (\bigotimes_{v|c} \phi_{v^\sharp})$ of $T/I_{c^\sharp q}T \otimes_\mathbb{Z} \mathfrak{G}_{c^\sharp}$ given by (7.6.17). Therefore, the image of κ_c in $T/I_{c^\sharp q}T \otimes_\mathbb{Z} \mathfrak{G}_{c^\sharp}$ under the maps of (7.6.15) and (7.6.16) equals the image of $\kappa_{\mathfrak{p}c} \in H^1_{\mathcal{F}(c^\sharp q)}(F, T/I_{c^\sharp q}T) \otimes_\mathbb{Z} \mathfrak{G}_{c^\sharp q}$ in $T/I_{c^\sharp q}T \otimes_\mathbb{Z} \mathfrak{G}_{c^\sharp}$ under the maps of these same diagrams (7.6.15) and (7.6.16). This combined with Lemma 7.5.3, namely, that $\theta_c \in H^1_{\mathcal{F}(c^\sharp)}(F, T/I_{c^\sharp}T) \otimes \Gamma_{c^\sharp}$ for all c, shows that the classes κ_c form a Kolyvagin system.

The class κ_1 is the image in $H^1(F, T)$ via the Kummer map of $\mathrm{Tr}_{F_1/F} x_1$, the norm of the point $x_1 \in E(F_1)$, and this coincides with the class θ_1 (see (7.4.12) and (7.4.18)). □

8 Selmer Groups and Kolyvagin Systems

Contents

8.1	Hypotheses H.0–H.5 for Selmer Triples	218
8.2	The Cassels-Tate Pairing	230
8.3	Kolyvagin Systems over Artinian Principal Ideal Rings	237
8.4	Kolyvagin Systems over Discrete Valuation Rings	256
8.5	Selmer Groups over Artinian Principal Ideal Rings	261
8.6	Selmer Groups over Discrete Valuation Rings	284
8.7	Shafarevich-Tate Groups of Elliptic Curves	294

In this chapter, we demonstrate the main results on finiteness and determine the structure of Selmer groups and Shafarevich-Tate groups under suitable hypotheses using Heegner point and CM point Kolyvagin systems.

This chapter starts with a list of hypotheses that are imposed on a Galois representation T (Sect. 8.1). These hypotheses are shown to hold for the Galois representations that we consider (Proposition 8.1.18).

In Sect. 8.2, the Cassels-Tate pairing for Selmer modules is presented in a form essentially due to Flach.

In Sects. 8.3, 8.4, Kolyvagin systems are considered where the ground ring is either an artinian principal ideal ring or a discrete valuation ring. We have already constructed a Kolyvagin system starting from a Heegner system on an elliptic curve (Sect. 6.5 and Sects. 7.3–7.6). From this Kolyvagin system, we obtain results on Selmer modules in Sects. 8.5–8.6.

The principal results for elliptic curves are given in Sect. 8.7 and are obtained by specializing the previous results of Sects. 8.3–8.6. The elliptic curves considered are assumed to be parametrized by classical modular curves over $\mathbb{Q}$ or Drinfeld modular curves over global fields or Shimura curves over totally real number fields.

8.1 Hypotheses H.0–H.5 for Selmer Triples

(8.1.1) In this section, we state and give comments on some hypotheses imposed on the Galois representation which are required for later results of this chapter on Selmer groups. Let

- k be a global field;
- F be a finite Galois field extension of k;
- R be a coefficient ring where the residue characteristic l of R is different from the characteristic of k (see (4.1.2));
- $\mathfrak{m}$ be the maximal ideal of R;
- T be an R-representation of G_F;
- $(T, \mathcal{F}, \mathcal{L})$ be a Selmer triple on F (see (7.1.3)).

(8.1.2) We have the following hypotheses which may be imposed on the Selmer triple $(T, \mathcal{F}, \mathcal{L})$.

H.0 T is a free R-module of rank 2;
H.1 $T/\mathfrak{m}T$ is an absolutely irreducible representation of $(R/\mathfrak{m})[[\mathrm{Gal}(F^{\mathrm{sep}}/F)]]$;
H.2 We have

$$H^1(F(T, \mu_{l^\infty})/F, T/\mathfrak{m}T) = 0;$$

here $F(T, \mu_{l^\infty})$ is the smallest galois extension of k containing F where $G_{F(T,\mu_{l^\infty})}$ acts trivially on μ_{l^∞} and on T;

H.3 For every $v \in S(\mathcal{F})$, the local condition $\mathcal{F}$ at v is Cartesian on the category $\mathrm{Quot}(T)$ (see (4.1.7) and (4.3.3));

H.4 (a) The representation T extends to an R-representation of G_k. There is an alternating non-degenerate pairing of G_k-modules

$$e : T \times T \to R(1), \quad e(a^\sigma, b^\sigma) = e(a, b)^\sigma \quad (a, b \in T, \sigma \in G_k);$$

(b) There is an element $\tau \in G_k$ whose images in $\mathrm{Gal}(F(T)/k)$ and in $\mathrm{Gal}(F/k)$ have order 2 and which acts as multiplication by -1 on $R(1)$;

(c) The R-bilinear pairing of R-modules

$$(,) : T \times T \to R(1)$$

defined by $(,) = e(a, b^\tau)$ is non-degenerate and *symmetric* and satisfies

8.1 Hypotheses H.0–H.5 for Selmer Triples

$$(s^\sigma, t^{\tau\sigma\tau^{-1}}) = (s,t)^\sigma \qquad (s,t \in T, \sigma \in \mathrm{Gal}(k^{\mathrm{sep}}/k)).$$

Furthermore, the local condition $\mathcal{F}$ for every place v of F is its own orthogonal complement under the symmetric local pairing induced from $(,)$

$$<,>_v^{\mathrm{H.4}} \colon H^1(F_v, T) \times H^1(F_v, T) \to R.$$

We write this local symmetric pairing as $<,>_v^{\mathrm{H.4}}$ to avoid confusion with the Tate pairing, to which it is closely related.

H.5 (a) The action of $\mathrm{Gal}(F^{\mathrm{sep}}/F)$ on $T/\mathfrak{m}T$ extends to an action of $\mathrm{Gal}(k^{\mathrm{sep}}/k)$ and the Selmer structure $\mathcal{F}$ propagated to $T/\mathfrak{m}T$ is stable under the action of $\mathrm{Gal}(k^{\mathrm{sep}}/k)$;

(b) If H.4 is assumed to hold, then the residual pairing obtained from the symmetric pairing $(,)$

$$T/\mathfrak{m}T \times T/\mathfrak{m}T \longrightarrow (R/\mathfrak{m})(1)$$

satisfies $(s^\tau, t^\tau) = (s,t)^\tau$ for all $s,t \in T/\mathfrak{m}T$ and the action of τ splits $T/\mathfrak{m}T$ into one-dimensional eigenspaces over $R/\mathfrak{m}$

$$T/\mathfrak{m}T \cong (T/\mathfrak{m}T)^+ \oplus (T/\mathfrak{m}T)^-$$

Remarks 8.1.3

(i) The set of primes $\mathcal{L}$ of the Selmer triple $(T, \mathcal{F}, \mathcal{L})$ plays no rôle in the six hypotheses H.0–H.5.

Under the change of rings $R \to R/\mathfrak{m}^i$, the Selmer structure $\mathcal{F}$ on T is propagated to $T \otimes_R R/\mathfrak{m}^i$.

If the hypotheses H.0–H.5 hold for the R-representation T where R is the given coefficient ring, then for the representation $T \otimes_R R/\mathfrak{m}^i$ under the change of rings $R \to R/\mathfrak{m}^i$ with its propagated Selmer structure, the six hypotheses H.0–H.5 evidently hold.

(ii) The hypotheses H.0 to H.5 correspond exactly to the hypotheses H.0 to H.5 of [30], but H.4 here is altered from that of H.4 in [30].

The hypotheses H.0, H.1 correspond exactly with the hypotheses H.0, H.1 of [54, §3.5] except that T is assumed here to be of rank 2. H.2 above corresponds to H.3 of [54, §3.5]. H.3 above corresponds to H.6 of [54, §3.5]. Hypotheses H.4 and H.5 above do not correspond to any of those of [54, §3.5], where H.4 implies the self-duality of T (see Remark 8.1.11 et seq.). The hypotheses H.4 and H.5 of [54, §3.5] do not correspond to any of those above.

(iii) Let T be the Tate module T of an elliptic curve over $\mathbb{Q}$ and $F/\mathbb{Q}$ a corresponding imaginary quadratic extension. Then the action of G_F on $T \otimes \Lambda$ where Λ is the Iwasawa algebra for the anti-cyclotomic $\mathbb{Z}_l$-extension of F does not extend to an action of $G_\mathbb{Q}$ on $T \otimes \Lambda$ (see also Remark 7.1.19(iv)). For this reason, the hypothesis H.4 is stated differently and, inequivalently, in [30] in order to apply the corresponding Kolyvagin system to the anticyclotomic Iwasawa theory of the elliptic curve.

(iv) The hypotheses above are applied when T has rank 2 over R which is the case when T is the Tate module of an elliptic curve.

Hypothesis H.1

Proposition 8.1.4 *Suppose that E/F is an elliptic curve. Then for all except finitely many prime numbers l, the module T/lT, where T is the l-adic Tate module of E, is an absolutely irreducible $\mathbb{F}_l[[G_F]]$-representation and H.1 holds.*

Proof Suppose first either that F is a number field and E/F is an elliptic curve without potential complex multiplication or that F is a function field of positive characteristic and that E/F is not isotrivial. Then for all but finitely many prime numbers l, the group Gal $(F(E_{l^\infty})/F)$ acting on T, where a $\mathbb{Z}_l$ basis of T is fixed, contains the subgroup $SL_2(\mathbb{Z}_l)$ (by Theorems 2.7.3, 2.7.10). It follows that for all but finitely many l, the module T/lT is an absolutely irreducible $\mathbb{F}_l[[G_F]]$ representation.

Suppose now either that F is a number field and E/F is an elliptic curve with potential complex multiplication or that F is a function field of positive characteristic and that E/F is isotrivial. Then there is a finite extension field F'/F such that $E \times_F F'$ either has complex multiplication or that $E \times_F F'$ is definable over a finite subfield of F'. Furthermore, an elliptic curve defined over a finite field has complex multiplication by an order of an imaginary quadratic extension field of $\mathbb{Q}$ where the Frobenius provides the extra endomorphism.

If F is a number field, then $E \times_F F'$ has complex multiplication by an order R in some imaginary quadratic extension field of $\mathbb{Q}$. By Theorem 2.7.2, the image of $G_{F'}$ in $\text{Aut}(T)$ is isomorphic to $(R \otimes \mathbb{Z}_l)^*$ for all except finitely many l. It follows that in both cases where k is a number field or a function field that the image of $\mathbb{Z}_l[[G_{F'}]]$ in $\text{End}_{\mathbb{Z}_l}(T)$ is a quadratic algebra over $\mathbb{Z}_l$. Hence for all but finitely many l, we have that T/lT is an absolutely irreducible $\mathbb{F}_l[[G_F]]$-representation. □

Hypothesis H.2

(8.1.5) With the notation of (8.1.1), we have the homomorphisms, giving the action of G_F on T and on $T/\mathfrak{m}T$ where $\overline{\phi}$ is the reduction mod $\mathfrak{m}$ of ϕ, T has rank n, and where an R-basis of T is fixed,

$$\phi : G_F \to \text{GL}_n(R), \quad \overline{\phi} : G_F \to \text{GL}_n(R/\mathfrak{m}).$$

8.1 Hypotheses H.0–H.5 for Selmer Triples

Lemma 8.1.6 *Suppose that the field $R/\mathfrak{m}$ is finite and one of the following holds.*

(a) The group $\overline{\phi}(G_F)$ as a subgroup of $\mathrm{GL}_n(R/\mathfrak{m})$ contains a non-trivial homothety (i.e. not equal to 1);
(b) The group $\overline{\phi}(G_F)$ is abelian and $(T/\mathfrak{m}T)^{\overline{\phi}(G_F)} = 0$.

Then we have $H^1(\phi(G_F), T/\mathfrak{m}T) = 0$.

Proof Assume one of (a) or (b) holds. If T has rank 0 over R, then the lemma trivially holds, so we may assume that rank $T \geq 1$. Then in either case, there is a non-trivial normal subgroup N of $\overline{\phi}(G_F)$ which is finite cyclic. For this in case (a), one can take N to be generated by any non-trivial homothety of $\overline{\phi}(G_F)$ and in case (b) we may take N to be generated by any non-trivial element of $\overline{\phi}(G_F)$ where this last group is non-trivial by hypothesis (b) and that T has positive rank.

We have the Hochschild-Serre spectral sequence

$$H^i(\overline{\phi}(G_F)/N, H^j(N, T/\mathfrak{m}T)) \Longrightarrow H^{i+j}(\overline{\phi}(G_F), T/\mathfrak{m}T).$$

As N is a non-trivial finite cyclic group and $T/\mathfrak{m}T$ is a finite N-module, then $H^j(N, T/\mathfrak{m}T) = 0$ for all $j \geq 0$ by properties of the Herbrand quotient [57, Chap. 1, Prop. 4.3]. Therefore, the spectral sequence degenerates, and we obtain

$$H^i(\overline{\phi}(G_F), T/\mathfrak{m}T) = 0 \text{ for all } i \geq 0. \tag{8.1.7}$$

Let K be the kernel of the surjective homomorphism $\phi(G_F) \to \overline{\phi}(G_F)$. We have the inflation restriction sequence

$$0 \to H^1(\overline{\phi}(G_F), T/\mathfrak{m}T) \to H^1(\phi(G_F), T/\mathfrak{m}T) \to H^1(K, T/\mathfrak{m}T)^{\overline{\phi}(G_F)}$$

$$\to H^2(\overline{\phi}(G_F), T/\mathfrak{m}T).$$

By the isomorphisms (8.1.7), we have $H^i(\overline{\phi}(G_F), T/\mathfrak{m}T) = 0$ for all $i \geq 0$.; therefore the following restriction homomorphism is an isomorphism

$$H^1(\phi(G_F), T/\mathfrak{m}T) \cong \mathrm{Hom}(K, T/\mathfrak{m}T)^{\overline{\phi}(G_F)}$$

where K acts trivially on $T/\mathfrak{m}T$.

Any homomorphism $h : K \to T/\mathfrak{m}T$ factors through K^{ab}, the abelianization of K, and hence writing additively the composition on K^{ab}, we have that h factors through $K^{\mathrm{ab}}/lK^{\mathrm{ab}}$ where l is the characteristic of $R/\mathfrak{m}$. We then have

$$H^1(\phi(G_F), T/\mathfrak{m}T) \cong \mathrm{Hom}(K^{\mathrm{ab}}/lK^{\mathrm{ab}}, T/\mathfrak{m}T)^{\overline{\phi}(G_F)}.$$

Writing h' for the image of $h \in H^1(\phi(G_F), T/\mathfrak{m}T)$ in $\mathrm{Hom}(K^{\mathrm{ab}}/lK^{\mathrm{ab}}, T/\mathfrak{m}T)$, we have that h' is $\overline{\phi}(G_F)$-invariant and hence satisfies

$$gh'(g^{-1}ag) = h'(a)$$

for all $a \in K^{\mathrm{ab}}/lK^{\mathrm{ab}}$ and all $g \in \overline{\phi}(G_F) = \overline{\phi}(G_F)$.

In case (a), by hypothesis $\overline{\phi}(G_F)$ as a subgroup of $\mathrm{GL}_n(R/\mathfrak{m})$ contains a non-zero homothety which can be considered as an element $\lambda \in (R/\mathfrak{m})^*$, $\lambda \neq 1$. It follows that $\lambda h'(a) = h'(a)$ for all $a \in K^{\mathrm{ab}}/lK^{\mathrm{ab}}$ and therefore $h = 0$. In conclusion, we have $H^1(\phi(G_F), T/\mathfrak{m}T) = 0$.

In case (b), $\overline{\phi}(G_F)$ is abelian, so that h' satisfies $gh'(a) = h'(a)$ for all $a \in K^{\mathrm{ab}}/lK^{\mathrm{ab}}$ and all $g \in \overline{\phi}(G_F)$, and we then have $\mathrm{Hom}(K^{\mathrm{ab}}/lK^{\mathrm{ab}}, T/\mathfrak{m}T)^{\overline{\phi}(G_F)} \cong \mathrm{Hom}(K^{\mathrm{ab}}/lK^{\mathrm{ab}}, (T/\mathfrak{m}T)^{\overline{\phi}(G_F)}) \cong 0$. □

Theorem 8.1.8 *Suppose that E/F is an elliptic curve and T is its l-adic Tate module for any prime number l different from the characteristic of F. Then for all but finitely many prime numbers l, we have that H.2 holds for T.*

Proof The action of G_F on T is given by a homomorphism $\phi : G_F \longrightarrow \mathrm{GL}_2(\mathbb{Z}_l)$ and the reduction mod l of ϕ is $\overline{\phi} : G_F \longrightarrow \mathrm{GL}_2(\mathbb{F}_l)$. We have evidently $\phi(G_F) = \mathrm{Gal}(F(T)/F)$ and there is the non-degenerate $\mathbb{Z}_l$-bilinear Weil pairing

$$T \times T \to \mu_{l^\infty}$$

where μ_{l^∞} is the group of lth power roots of unity of F^{sep}. It follows that because $G_{F(T)}$ acts trivially on T, then it also acts trivially on μ_{l^∞}; hence $F(T)$ contains all elements of μ_{l^∞} so that $F(T) = F(T, \mu_{l^\infty})$ and $\phi(G_F) = \mathrm{Gal}(F(T, \mu_{l^\infty})/F)$.

Suppose first that either F is a number field and E/F is an elliptic curve without potential complex multiplication or that F is a global function field and E/F is a non-isotrivial elliptic curve. Then the hypothesis of Lemma 8.1.6 that $\overline{\phi}(G_F)$ as a subgroup of $\mathrm{GL}_n(R/\mathfrak{m})$ contains a non-trivial homothety holds for all but finitely many prime numbers l by Theorems 2.7.3 and 2.7.10.

Suppose now that either F is a number field and E/F is an elliptic curve with potential complex multiplication or that F is a global function field and E/F is an isotrivial elliptic curve. Then there is a finite galois extension field F' of F such that either $E \times_F F'$ has complex multiplication and all elements of $R = \mathrm{End}_{F^{\mathrm{sep}}}(E)$ are defined over F and R is an order in an imaginary quadratic extension field of $\mathbb{Q}$ or that $E \times_F F'$ is defined over a finite subfield of F'.

The inflation-restriction sequence gives, where $N = \phi(G_{F'})$

8.1 Hypotheses H.0–H.5 for Selmer Triples

$$0 \to H^1(\phi(G_F)/N, (T/lT)^N) \to H^1(\phi(G_F), T/lT) \to$$

$$H^1(\phi(G_{F'}), T/lT)^{\phi(G_F)/N}.$$

By Theorem 2.7.2 in the complex multiplication case and evidently in the isotrivial function field case, we have $(T/lT)^N = 0$ for all but finitely many l. Therefore, the inflation-restriction sequence shows that in order to demonstrate that $H^1(\phi(G_F), T/lT) = 0$ for all but finitely many l, it is enough to demonstrate

$$H^1(\phi(G_{F'}), T/lT) = 0$$

for all except finitely many prime numbers l. So we are reduced to the one of the following two cases:

(i) F is an algebraic number field, and E/F has complex multiplication where all endomorphisms of $E \otimes_F F^{\text{sep}}$ are definable over F;
(ii) F is a global function field, and E/F is definable over a finite subfield of F.

Assume case (i) holds. We have that all elements of $R = \text{End}_{F^{\text{sep}}}(E)$ are defined over F and that R is an order in an imaginary quadratic extension field of $\mathbb{Q}$. By Theorem 2.7.2, we have that $\phi(G_F) \subseteq \text{GL}_2(\mathbb{Z}_l)$ is contained in $R_l = R \otimes_\mathbb{Z} \mathbb{Z}_l$ for all l and where $\phi(G_F) = R_l$ for all but finitely many l. It follows that $\overline{\phi}(G_F)$ contains a non-trivial homothety of $\text{GL}_n(\mathbb{F}_l)$ for all except finitely many l. Therefore, the hypothesis (a) of Lemma 8.1.6, holds for all but finitely many prime numbers l. Therefore $H^1(\phi(G_F), T/lT) = 0$ for all but finitely many l from this lemma and so T satisfies H.2 for all but finitely many l.

Assume case (ii) holds. Then $E \cong E_0 \otimes_k F$ where E_0/k is some elliptic curve over a finite subfield k of F. Then we have that $\overline{\phi}(G_F) \subseteq \text{GL}_2(\mathbb{F}_l)$ and $\overline{\phi}(G_F)$ is a finite abelian group. It follows from the Weil conjectures for elliptic curves over finite fields that $(T/lT)^{\overline{\phi}(G_F)} = 0$ for all except finitely many prime numbers l. Therefore, hypothesis (b) of Lemma 8.1.6 holds and that lemma gives $H^1(\phi(G_F), T/lT) = 0$ for all but finitely many l and so T satisfies H.2 for all but finitely many l. □

Hypothesis H.4

Proposition 8.1.9 *Suppose that T satisfies H.4(a) and H.4(b) and $e : T \times T \to R(1)$ is the alternating non-degenerate pairing of G_k-modules provided by H.4(a) and $\tau \in G_k$ is the element provided by H.4(b).*

Then the pairing of R-modules

$$(s, t) = e(s, t^\tau), T \times T \to R(1) \tag{8.1.10}$$

is non-degenerate and symmetric.

Furthermore, if the local condition $\mathcal{F}$ for every place v of F is its own orthogonal complement under the induced local symmetric pairing obtained from $(,)$

$$(,)_v : H^1(F_v, T) \times H^1(F_v, T) \to R$$

then $(T, \mathcal{F}, \mathcal{L})$, the pairing $(,)$ and τ satisfy H.4(c) and therefore satisfy H.4.

Proof We have $e(a^\sigma, b^\sigma) = e(a, b)^\sigma$ for all $a, b \in T, \sigma \in G_k$. It is straightforward to check that the pairing of R-modules $(s, t) = e(s, t^\tau)$ is non-degenerate and satisfies

$$(s^\sigma, t^{\tau\sigma\tau^{-1}}) = (s, t)^\sigma \qquad (s, t \in T, \sigma \in \mathrm{Gal}(k^{\mathrm{sep}}/k)).$$

We have

$$(s, t) = e(s, t^\tau) = -e(t^\tau, s) = -e(t, s^{\tau^{-1}})^\tau = -e(t, s^\tau)^\tau = -(t, s)^\tau$$

where $s^{\tau^2} = s$ for all $s \in T$. As the action of τ on $R(1) = R[[\mu_{l^\infty}]]$ is by multiplication by -1, we then have $(s, t) = (t, s)$, and the pairing is symmetric. This last part follows. □

Remark 8.1.11 The existence in H.4 of the R-bilinear alternating pairing $e(,)$ shows that there is an isomorphism $T \cong T^*$ of R-representations of G_k, so that H.4 implies that T is self-dual.

(8.1.12) (The twisted module $\mathrm{Tw}(T)$.) Suppose that H.4 holds for T with the element $\tau \in G_k$ and pairing $e(,) : T \times T \to R(1)$. Let $\mathrm{Tw}(T)$ be the R-representation T of G_k with underlying group equal to T and where the G_k-action is through conjugation by the element $\tau \in G_k$; that is to say for $t \in T, g \in G_F$ then $g * t = \tau g \tau^{-1} t$. Then the pairing provided by H.4 $(s, t) = e(s, t^\tau)$ can be considered as a pairing of G_F-modules

$$T \times \mathrm{Tw}(T) \to R(1).$$

For a place q of F, we have natural isomorphisms

$$H^1(F_{\mathfrak{s}}, T) \cong H^1(F_{\mathfrak{q}}, \mathrm{Tw}(T))$$

$$H^1(F, T) \cong H^1(F, \mathrm{Tw}(T)).$$

The local pairing induced by the symmetric pairing (s, t) on T of H.4

$$<,>_{\mathfrak{q}}^{\mathrm{H.4}} : H^1(F_{\mathfrak{q}}, T) \times H^1(F_{\mathfrak{q}}, T) \to R$$

8.1 Hypotheses H.0–H.5 for Selmer Triples

is symmetric. For the Tate module T of an elliptic curve over $\mathbb{Q}$, the symmetric pairing $<,>_q^{H.4}$ is roughly speaking a τ-twisted form of the local Tate pairing.

Construction of the Element τ in G_k

(8.1.13) The element τ is given as follows in these two cases (a) and (b) which are the only ones that we require.

(a) Let k be a totally real number field and F be a totally imaginary quadratic extension of k. Let T be the l-adic Tate module of an elliptic curve E/k defined over k, $\tau \in G_k$ be complex conjugation, and $e(,) : T \times T \to \mathbb{Z}_l(1)$ be the Weil pairing. Then T is evidently a G_F-representation which extends in the natural way to a G_k-representation, and we have that $(T, \mathcal{F}, \mathcal{L})$ where $\mathcal{F}$ is the classical Selmer structure on T, together with $e(,)$ and τ, satisfies H.4;

[This follows from Proposition 8.1.9 and evident properties of the Weil pairing.]

(b) Suppose that k is a global of positive characteristic equipped with a special set of places, K/k is an imaginary quadratic extension with respect to ∞, and c is a cycle on k supported only at non-Archimedean primes outside ∞. Let $F = K[c]$ be the ring class field with conductor c (Sect. 1.5).

The group $\mathrm{Gal}(F/k)$ is generalized dihedral and hence contains an element τ_F of order 2 whose restriction to K is the non-trivial element of $\mathrm{Gal}(K/k)$.

Let E/k be an elliptic curve. Then there is an infinite set $\mathcal{E}$ of prime numbers l of positive Dirichlet density such that these two conditions hold (see Proposition 2.7.13 and [4, Proposition 7.3.10]):

- $\mathrm{Gal}\,k(E_{l^\infty}/k)$ contains the element, relative to some fixed basis of E_{l^∞} over $\mathbb{Q}_l/\mathbb{Z}_l$,

$$\tau_{k(E_{l^\infty})} = \begin{pmatrix} 1 & 0 \\ 0 & -1 \end{pmatrix};$$

- The fields F and $k(E_{l^\infty})$ are linearly disjoint over k.

There is then an unique element σ of order 2 in $\mathrm{Gal}(F(E_{l^\infty})/k)$ satisfying the two conditions

$$\sigma|_{k(E_{l^\infty})} = \tau_{k(E_{l^\infty})}, \quad \sigma|_F = \tau_F.$$

We may then take $\tau \in G_k$ to be any element lifting the element $\sigma \in \mathrm{Gal}(F(E_{l^\infty})/k)$. This element τ satisfies H.4(b) and H.5(b) in this function field case. Let T be the l-adic Tate module of E/k, and $e(,) : T \times T \to \mathbb{Z}_l(1)$ be the Weil pairing. Then τ acts on $R(1)$ as -1 because of the form of $\tau_{k(E_{l^\infty})}$ specified above and because of the Weil pairing.

It follows that T is a G_F-representation which extends in the natural way to a G_k-representation and the Selmer triple $(T, \mathcal{F}, \mathcal{L})$ where $\mathcal{F}$ is the classical Selmer structure on T, together with $e(,)$ and τ, satisfy H.4, provided that l lies in $\mathcal{E}$.

(8.1.14) (Action of $\tau \in G_k$ on cohomology groups.) Suppose that H.4 holds and that $\tau \in G_k$ is the element given by H.4. Then $H^1(F, T)$ is acted naturally upon by $\text{Gal}(F/k)$. Assuming that $l \geq 3$, where l is the residue characteristic of R, then because the image of τ in $\text{Gal}(F/k)$ has order 2 we have that $H^1(F, T)$ decomposes into eigenspaces under the action of τ, namely, (see Sect. 1.6)

$$H^1(F, T) \cong H^1(F, T)^+ \oplus H^1(F, T)^-.$$

Basic Consequences of the Hypotheses

Lemma 8.1.15 *Suppose that H.2 holds. Then for all $i \in \mathbb{N}$, we have $(T/\mathfrak{m}^i T)^{G_F} = T^{G_F} = 0$.*

Proof From H.2, we have $H^1(F(T, \mu_{l^\infty})/F, T/\mathfrak{m}T) = 0$. A non-zero submodule of $(T/\mathfrak{m}T)^{G_F}$ would then provide a non-zero submodule of $H^1(F(T, \mu_{l^\infty})/F, T/\mathfrak{m}T)$ which is a contradiction; therefore we have $(T/\mathfrak{m}T)^{G_F} = 0$. Furthermore, there is an isomorphism of $R/\mathfrak{m}[[G_F]]$-modules $\mathfrak{m}^i T/\mathfrak{m}^{i+1} T \cong T/\mathfrak{m}T$ for $i \in \mathbb{N}$, so that an immediate induction shows that

$$(T/\mathfrak{m}^i T)^{G_F} = 0 \text{ for all } i.$$

As $T \cong \varprojlim T/\mathfrak{m}^i T$, we have that $T^{G_F} = 0$. □

Lemma 8.1.16 *Propagate the Selmer structure $\mathcal{F}$ to all objects of $\text{Quot}(T)$ (see (4.1.7), (4.3.2)) and to all submodules of T. Suppose also that the coefficient ring R is an artinian principal ideal ring of length h and that H.0, H.2, and H.3 hold. If $0 \leq i \leq h$ and π are a generator of the maximal ideal $\mathfrak{m}$ of R, then the maps*

$$T/\mathfrak{m}^i T \xrightarrow{\pi^{h-i}} T[\mathfrak{m}^i] \to T$$

induce isomorphisms

$$H^1_{\mathcal{F}}(F, T/\mathfrak{m}^i T) \longrightarrow H^1_{\mathcal{F}}(F, T[\mathfrak{m}^i]) \longrightarrow H^1_{\mathcal{F}}(F, T)[\mathfrak{m}^i]$$

$$H^1(F, T/\mathfrak{m}^i T) \longrightarrow H^1(F, T)[\mathfrak{m}^i].$$

8.1 Hypotheses H.0–H.5 for Selmer Triples

Proof A local condition on the quotient category Quot(T) of T (see (4.1.7)) that is Cartesian (see (4.3.3)) on this category has the property that for $i \leq h$, the local condition on $T[\mathfrak{m}^i]$, propagated from T as a submodule of T, agrees with the local condition on $T/\mathfrak{m}^i T$, propagated from T as a quotient of T in Quot(T), when these two representations are identified by the isomorphism

$$T/\mathfrak{m}^i T \xrightarrow{\pi^{h-i}} T[\mathfrak{m}^i].$$

This last map is an isomorphism as T is a free R-module of finite rank by H.0. We have then shown via H.3 that this isomorphism induces an an isomorphism of local conditions for all places $v \in S(\mathcal{F})$ of F, that is to say, we have

$$H^1_{\mathcal{F}}(F_v, T/\mathfrak{m}^i T) \cong H^1_{\mathcal{F}}(F_v, T[\mathfrak{m}^i]) \qquad (v \in S(\mathcal{F})).$$

Suppose that v is a place of F where $v \notin S(\mathcal{F})$. Then the local conditions on $T, T/\mathfrak{m}^i T, T[\mathfrak{m}^i]$ at v are the finite conditions (see (5.1.1)); furthermore, T and the objects of Quot(T) as well as the sub-representations of T are unramified at v. From lemma 4.3.8, the finite condition is Cartesian on any category of unramified representations with coefficients in R, so we then have the isomorphisms

$$H^1_{\mathcal{F}}(F_v, T/\mathfrak{m}^i T) \cong H^1_{\mathcal{F}}(F_v, T[\mathfrak{m}^i]) \qquad (v \notin S(\mathcal{F})).$$

This provides the isomorphism of Selmer groups

$$H^1_{\mathcal{F}}(F, T/\mathfrak{m}^i T) \cong H^1_{\mathcal{F}}(F, T[\mathfrak{m}^i]).$$

We have the exact sequence of G_F-modules, for any $0 \leq j \leq h$,

$$0 \longrightarrow T/\mathfrak{m}^j T \xrightarrow{\pi^{h-j}} T \longrightarrow T/\mathfrak{m}^{h-j} T \longrightarrow 0$$

where the map labelled π^{h-j} is multiplication by π^{h-j}. Taking $j = i$ and $j = h - i$, the cohomology of this exact sequence provides the exact sequences, where the left-hand zeros arise because $(T/\mathfrak{m}^j T)^{G_F} = 0$ for all j by Lemma 8.1.15,

$$0 \longrightarrow H^1(F, T/\mathfrak{m}^i T) \xrightarrow{\pi^{h-i}} H^1(F, T) \longrightarrow H^1(F, T/\mathfrak{m}^{h-i} T)$$

$$0 \longrightarrow H^1(F, T/\mathfrak{m}^{h-i} T) \xrightarrow{\pi^i} H^1(F, T) \longrightarrow H^1(F, T/\mathfrak{m}^i T)$$

It follows from this last pair of exact sequences that the map induced by multiplication by π^{h-i}

$$[\pi^{h-i}] : H^1(F, T/\mathfrak{m}^i T) \longrightarrow H^1(F, T)[\mathfrak{m}^i]$$

is an isomorphism.

The map of multiplication by π^{h-i} takes $H^1_{\mathcal{F}}(F, T/\mathfrak{m}^i T)$ to a subgroup of $H^1_{\mathcal{F}}(F, T)[\mathfrak{m}^i]$, because both Selmer groups are obtained by propagation of the local conditions $H^1_{\mathcal{F}}(F_v, T)$. It is enough to show that $[\pi^{h-i}]^{-1} H^1_{\mathcal{F}}(F, T) \subseteq H^1_{\mathcal{F}}(F, T/\mathfrak{m}^i T)$, and hence it is enough to show this inclusion for the corresponding local conditions. For the places v in $S(\mathcal{F})$, this inclusion holds by the Cartesian property H.3.

Suppose then that v is a place of F where $v \notin S(\mathcal{F})$. Let I_v be the inertia group of G_{F_v}, then we have the commutative diagram with exact rows

$$\begin{array}{ccccccc}
0 & \longrightarrow & H^1_f(F_v, T/\mathfrak{m}^i T) & \longrightarrow & H^1(F_v, T/\mathfrak{m}^i T) & \longrightarrow & \operatorname{Hom}(I_v, T/\mathfrak{m}^i T) \\
 & & {\scriptstyle [\pi^{h-i}]}\Big\downarrow & & {\scriptstyle [\pi^{h-i}]}\Big\downarrow & & {\scriptstyle [\pi^{h-i}]}\Big\downarrow \\
0 & \longrightarrow & H^1_f(F_v, T) & \longrightarrow & H^1(F_v, T) & \longrightarrow & \operatorname{Hom}(I_v, T)
\end{array}$$

The right-hand vertical arrow is injective so if $c \in H^1(F_v, T/\mathfrak{m}^i T)$ and $[\pi^{h-i}]c$ is unramified then c is unramified and so $c \in H^1_f(F_v, T/\mathfrak{m}^i T)$. We have then shown that $[\pi^{h-i}]^{-1} H^1_f(F_v, T) \subseteq H^1_f(F_v, T/\mathfrak{m}^i T)$, and therefore we have the isomorphism

$$H^1_{\mathcal{F}}(F, T/\mathfrak{m}^i T) \cong H^1_{\mathcal{F}}(F, T)[\mathfrak{m}^i].$$

$\square$

The Hypotheses H.0-H.5 and the Classical Selmer Structure on Elliptic Curves

(8.1.17) The application we make to elliptic curves of the hypotheses H.0-H.5 depends on the following result.

Proposition 8.1.18 *Suppose that k is a global field and E/k is an elliptic curve. Assume one of the following:*

(a) *Suppose k is a totally real number field, and F is a totally imaginary quadratic extension field of k.*
(b) *Suppose that k is a global of positive characteristic equipped with a special set ∞ of places and F/k is an imaginary quadratic extension with respect to ∞.*

For any prime number l different from the characteristic of k, let T be the l-adic Tate module of E and $(,)^{\text{Weil}} : T \times T \to \mathbb{Z}_l(1)$ be the Weil pairing. We have the Selmer triple $(T, \mathcal{F}, \mathcal{L})$, where $\mathcal{F}$ is the classical Selmer structure on T as a G_F-module and $\mathcal{L}$ is any set of prime places of F.

8.1 Hypotheses H.0–H.5 for Selmer Triples

There is a set $\mathcal{E}$ of prime numbers $l > 2$, and an element $\tau \in G_k$ such that for the corresponding pairing $(s, t) = (s, t^\tau)^{\text{Weil}}$ on T then $(T, \mathcal{F}, \mathcal{L})$ together with l, $(,)$ and τ satisfies satisfies the hypotheses H.0 to H.5. The set $\mathcal{E}$ contains all but finitely many prime numbers, if k is a number field, and has positive Dirichlet density, if k is a function field.

Proof Suppose that T is the l-adic Tate module of an elliptic curve where l is different from the characteristic of k and $F \supseteq k$ is the given quadratic field extension. Let $\mathcal{F}$ be the classical Selmer structure on $E \times_k F/F$. As the hypotheses H.0-H.5 do not involve $\mathcal{L}$, we may take $\mathcal{L}$ to be any set of prime places of F.

Then we have for the Selmer triple $(T, \mathcal{F}, \mathcal{L})$ given by $E \times_k F/F$:

- H.0 evidently holds for $(T, \mathcal{F}, \mathcal{L})$ for all prime numbers l different from the characteristic of k;
- H.1 holds for all but finitely many prime numbers l by Proposition 8.1.4;
- H.2. By Theorem 8.1.8, for all but finitely many prime numbers l, H.2 holds;
- H.3. It is easily checked that this Cartesian property holds for the classical Selmer structure on the Tate module of an abelian variety over any global field (see (4.1.7), (4.3.3), (5.5.7), (5.5.10));
- H.4. If (a) is satisfied so that k is a number field, then take $\tau \in G_k$ to be complex conjugation. If (b) is satisfied, then the paragraph (8.1.13)(b) provides a set $\mathcal{E}$ of prime numbers l different from 2 where $\mathcal{E}$ has positive Dirichlet density and an element $\tau \in G_k$ which may depend on l.

  In both cases (a) and (b), the pairing $(,)$ is non-degenerate and symmetric follows from paragraph (8.1.13)(b) and Proposition 8.1.9. As the pairing $(,)$ is a τ-twisted form of the Weil pairing, the classical Selmer structure $\mathcal{F}$ satisfies the property that the local condition $\mathcal{F}$ for every place v of F is its own orthogonal complement under the symmetric local pairing induced from $(,)$

$$<,>_v \colon H^1(F_v, T) \times H^1(F_v, T) \to R.$$

  Therefore H.4 holds for the Selmer triple $(T, \mathcal{F}, \mathcal{L})$;
- H.5. The hypothesis H.5(a) evidently holds. The hypothesis H.5(b) holds for $(T, \mathcal{F}, \mathcal{L})$ for the prime number $l > 2$ and element $\tau \in G_k$ as chosen above for H.4 either where τ is complex conjugation in the number field case or where τ is given by paragraph (8.1.13)(b) in the function field case.

In summary, the prime number $l > 2$ and $\tau \in G_k$ can be selected such that H.0 to H.5 hold for $(T, \mathcal{F}, \mathcal{L})$, and in fact, the set of such prime numbers l contains all but finitely many prime numbers, if k is a number field, and has positive Dirichlet density, if k is a function field. □

8.2 The Cassels-Tate Pairing

(8.2.1) The Cassels-Tate pairing is presented in this section in a version for Selmer modules.

(8.2.2) Let

- k be a global field;
- F be a finite Galois field extension of k;
- R be a coefficient ring which is an artin principal ideal ring and where the residue characteristic l of R is different from the characteristic of k;
- $\mathfrak{m}$ be the maximal ideal of R with generator π;
- T be an R-representation of G_F which is a finite free R-module with dual $T^* = \mathrm{Hom}(T, \mu_{l^\infty})$;
- $\mathcal{F}$ be a Selmer structure on T and $\mathcal{F}^*$ be the dual Selmer structure on T^*.

Proposition 8.2.3 *Let s, t be positive integers such that $s + t \leq \mathrm{length}(R)$. Suppose that $(T, \mathcal{F})$ and $(T^*, \mathcal{F}^*)$ satisfy hypothesis H.3. Then there is a functorial pairing*

$$(,) : H^1_{\mathcal{F}}(F, T/\mathfrak{m}^s T) \times H^1_{\mathcal{F}^*}(F, T^*[\mathfrak{m}^t]) \longrightarrow R$$

whose kernels on the left and right are, respectively, the images of the maps

$$H^1_{\mathcal{F}}(F, T/\mathfrak{m}^{s+t} T) \longrightarrow H^1_{\mathcal{F}}(F, T/\mathfrak{m}^s T)$$

$$H^1_{\mathcal{F}^*}(F, T^*[\mathfrak{m}^{s+t}]) \longrightarrow H^1_{\mathcal{F}^*}(F, T^*[\mathfrak{m}^t]).$$

The group $\mathrm{Gal}(F/k)$ acts on the terms of this pairing where $\mathrm{Gal}(F/k)$ acts trivially on the target R and the pairing satisfies $(ga, gb) = (a, b)$ for $g \in G, a \in H^1_{\mathcal{F}}(F, T/\mathfrak{m}^s T), b \in H^1_{\mathcal{F}^*}(F, T^*[\mathfrak{m}^t])$.

Proof The proof follows largely the original construction of Cassels and Tate for the case of the Shafarevich-Tate group of elliptic curves. Let

$$a \in H^1_{\mathcal{F}}(F, T/\mathfrak{m}^s T), \quad b \in H^1_{\mathcal{F}^*}(F, T^*[\mathfrak{m}^t])$$

be given cohomology classes. Then a, b are represented by continuous cocycles $a^{(1)}, b^{(1)}$, respectively, where

$$a^{(1)} \in Z^1(F, T/\mathfrak{m}^s T), \quad b^{(1)} \in Z^1(F, T^*[\mathfrak{m}^t]).$$

Because the continuous cochain functor $C^1(F, -)$ from the category of R-modules to itself is exact, the following maps are surjective

8.2 The Cassels-Tate Pairing

$$C^1(F, T/\mathfrak{m}^{s+t}T) \longrightarrow C^1(F, T/\mathfrak{m}^s T)$$

$$C^1(F, T^*[\mathfrak{m}^{s+t}]) \xrightarrow{\pi^s} C^1(F, T^*[\mathfrak{m}^t]).$$

Therefore, we may select continuous cochains

$$\alpha \in C^1(F, T/\mathfrak{m}^{s+t}T), \quad \beta \in C^1(F, T^*[\mathfrak{m}^{s+t}])$$

which map to $a^{(1)}, b^{(1)}$ under these two maps, respectively. Letting d be the coboundary operator on cochains, we have $\pi^s d\beta = db^{(1)} = 0$. Furthermore, as $d\alpha$ mod $\mathfrak{m}^s$ is zero in $C^2(F, T/\mathfrak{m}^s T)$, we have that $d\alpha$ is divisible by $\mathfrak{m}^s$ in $C^2(F, T/\mathfrak{m}^{s+t}T)$. We then obtain that the cup-product $d\alpha \cup d\beta$ is zero in $C^4(F, R(1))$. Therefore, we have in $C^4(F, R(1))$ the relation

$$d(d\alpha \cup \beta) = d^2\alpha \cup \beta + d\alpha \cup d\beta = 0.$$

Therefore, $d\alpha \cup \beta$ is a cocycle lying in $Z^3(F, R(1))$. But we have $H^3(F, R(1)) = 0$ by [48, Corollary I.4.18 or Corollary I.4.21]. It follows that there is a cochain $\gamma \in C^2(F, R(1))$ such that

$$d\gamma = d\alpha \cup \beta.$$

We have the exact sequence, where the map labelled π^s is induced by multiplication by π^s on T

$$0 \longrightarrow T^*[\mathfrak{m}^s] \longrightarrow T^*[\mathfrak{m}^{s+t}] \xrightarrow{\pi^s} T^*[\mathfrak{m}^t] \longrightarrow 0.$$

This has the long exact sequence of cohomology

$$\cdots \longrightarrow H^1(F_v, T^*[\mathfrak{m}^s]) \longrightarrow H^1(F_v, T^*[\mathfrak{m}^{s+t}]) \longrightarrow H^1(F_v, T^*[\mathfrak{m}^t]) \longrightarrow \cdots$$

Hypothesis H3 implies that for any place v of F, this long exact sequence restricts to an exact sequence, and where the map f exists because of hypothesis H.3 and that g is surjective results from the definition of the propagation of $\mathcal{F}^*$ to $T^*[\mathfrak{m}^t]$

$$H^1_{\mathcal{F}^*}(F_v, T^*[\mathfrak{m}^s]) \xrightarrow{f} H^1_{\mathcal{F}^*}(F_v, T^*[\mathfrak{m}^{s+t}]) \xrightarrow{g} H^1_{\mathcal{F}^*}(F_v, T^*[\mathfrak{m}^t]) \longrightarrow 0.$$

Therefore, there is a cocyle $\beta'_v \in Z^1_{\mathcal{F}^*}(F_v, T^*[\mathfrak{m}^{s+t}])$ such that $\pi^s \beta'_v = b_v^{(1)}$, where $b_v^{(1)}$ is the localization of $b^{(1)}$ at v and where $Z^1_{\mathcal{F}^*}(F_v, T^*[\mathfrak{m}^{s+t}]) \subseteq Z^1(F_v, T^*[\mathfrak{m}^{s+t}])$ is the preimage of $H^1_{\mathcal{F}^*}(F_v, T^*[\mathfrak{m}^t])$ under multiplication by π^s.

Write $\alpha_v \in C^1(F_v, T/\mathfrak{m}^{s+t}T)$ for the continuous cochain induced by localization of $\alpha \in C^1(F, T/\mathfrak{m}^{s+t}T)$. We then have that the cochain $\alpha_v \cup \beta'_v - \gamma_v \in C^2(F_v, R(1))$ satisfies

$$d(\alpha_v \cup \beta'_v - \gamma_v) = d\alpha_v \cup \beta'_v + \alpha_v \cup d\beta'_v - d\gamma_v$$

$$= d\alpha_v \cup \beta'_v - d\gamma_v = d\alpha_v \cup \beta_v - d\gamma_v + d\alpha_v \cup (\beta'_v - \beta_v) = d\alpha_v \cup (\beta'_v - \beta_v)$$

We have already seen that $d\alpha$ is divisible by $\mathfrak{m}^s$ in $C^2(F, T/\mathfrak{m}^{s+t}T)$, and we have $\pi^s \beta'_v = b_v^{(1)}$ by the choice of β'_v; we also have $\pi^s \beta = b^{(1)} \in C^1(F, T^*[\mathfrak{m}^t])$ by construction of β; therefore, we have

$$d(\alpha_v \cup \beta'_v - \gamma_v) = d\alpha_v \cup (\beta'_v - \beta_v) = 0,$$

' and therefore $\alpha_v \cup \beta'_v - \gamma_v$ is a cocyle in $Z^2(F_v, R(1))$.

Denote by

$$\mathrm{inv}_v : \mathrm{Br}(k_v) = H^2(F_v, \mathbb{G}_m) = H^2(F_v^{\mathrm{sep}}/k_v, (F_v^{\mathrm{sep}})^*) \to \mathbb{Q}/\mathbb{Z}$$

the canonical isomorphism (the invariant map) given by local class field theory. The group G_{k_v} acts trivially on R, so we have canonical isomorphisms

$$H^2(F_v, R(1)) \cong R \otimes_{\mathbb{Z}} H^2(F_v, \mathbb{G}_m) \cong R \otimes_{\mathbb{Z}} \mathbb{Q}/\mathbb{Z}.$$

The artin ring R is a quotient of a discrete valuation ring and $\mathbb{Q}/\mathbb{Z}$ can be written as a direct limit $\varinjlim (\frac{1}{n}\mathbb{Z})/\mathbb{Z}$ where n runs over all positive integers. It follows that there are canonical isomorphisms of R-modules, where inv_v is induced by the invariant map of local class field theory

$$R \otimes_{\mathbb{Z}} \mathbb{Q}/\mathbb{Z} \cong R, \qquad \mathrm{inv}_v : H^2(F_v, R(1)) \cong R.$$

We then define the pairing (a, b) to be

$$(a, b) = \sum_v \mathrm{inv}_v [\alpha_v \cup \beta'_v - \gamma_v]$$

where $[\epsilon]$ denotes the cohomology class in $H^2(F_v, R(1))$ represented by a cocycle ϵ and where the sum runs over all places v of F.

It can be checked the sum for $(,)$ runs over only finitely many places v for given a, b and that the pairing is independent of the choices made. That the kernels are as stated follows as in [19].

Note that if K/L is a finite Galois extension of local non-Archimedean fields of characteristic different from l, then $\mathrm{Gal}(K/L)$ acts trivially on the invariant map of

8.2 The Cassels-Tate Pairing

$H^2(L, L^{\text{sep},*})$ of local class field theory for all places v. Hence, the galois group G acts trivially on the target R of the pairing which then satisfies $(ga, gb) = g(a, b) = (a, b)$ for $g \in G$. □

Proposition 8.2.4 *Suppose that $(T, \mathcal{F})$ and $(T^*, \mathcal{F}^*)$ satisfy the hypotheses H.0, H.1, H.2, H.3, and H.4 of Sect. 8.1. Write $\mathcal{H}$ for $H^1_{\mathcal{F}}(F, T)$. For an integer s with $1 \leq s < \text{length}(R)$ put*

$$V_s = \mathcal{H}[\mathfrak{m}^s]/\mathfrak{m}(\mathcal{H}[\mathfrak{m}^{s+1}]).$$

Then there is a functorial alternating non-degenerate pairing of R-modules, where G acts trivially on the target $R[\mathfrak{m}]$ and $\{gv, gw\} = \{v, w\}$ for $g \in G$, $v, w \in V_s/V_{s-1}$,

$$\{,\} : V_s/V_{s-1} \times V_s/V_{s-1} \to R[\mathfrak{m}].$$

Proof For an integer s with $1 \leq s < \text{length}(R)$ put

$$V_s = \mathcal{H}[\mathfrak{m}^s]/\mathfrak{m}(\mathcal{H}[\mathfrak{m}^{s+1}]), \qquad W_s = \mathcal{H}[\mathfrak{m}]/\mathfrak{m}^s(\mathcal{H}[\mathfrak{m}^{s+1}]).$$

Here the notation means that $\mathfrak{m}(\mathcal{H}[\mathfrak{m}^{s+1}])$ is a submodule of $\mathcal{H}[\mathfrak{m}^s]$ where the latter is the submodule of elements of $\mathcal{H}$ annihilated by $\mathfrak{m}^s$ and similarly for $\mathfrak{m}^s(\mathcal{H}[\mathfrak{m}^{s+1}]) \subseteq \mathcal{H}[\mathfrak{m}]$.

By hypothesis H.4 and Remark 8.1.11, there is an isomorphism $T \cong T^*$ of R-representations of G_k. By Lemma 8.1.16 and because hypotheses H.0, H.1, H.2, and H.3 hold, we have the isomorphisms

$$H^1_{\mathcal{F}^*}(F, T^*[\mathfrak{m}]) \cong H^1_{\mathcal{F}}(F, T[\mathfrak{m}]) \cong \mathcal{H}[\mathfrak{m}]$$

$$H^1_{\mathcal{F}}(F, T/\mathfrak{m}^s T) \cong \mathcal{H}[\mathfrak{m}^s].$$

By the previous Proposition 8.2.3 as $s + 1 \leq \text{length}(R)$, we obtain a pairing

$$H^1_{\mathcal{F}}(F, T/\mathfrak{m}^s T) \times H^1_{\mathcal{F}^*}(F, T^*[\mathfrak{m}]) \to R[\mathfrak{m}]$$

which, by reason of the kernels of this pairing given in Proposition 8.2.3, then induces a non-degenerate pairing of of $R/\mathfrak{m}$-vector spaces spaces

$$(,) : V_s \times W_s \longrightarrow R[\mathfrak{m}].$$

We have the exact sequence of $R/\mathfrak{m}$-vector spaces

$$0 \longrightarrow V_{s-1} \longrightarrow V_s \xrightarrow{\pi^{s-1}} W_s$$

where the map labelled π^{s-1} is obtained by multiplication by π^{s-1} on $\mathcal{H}[\mathfrak{m}^s]$.

We may then define a pairing

$$<,>: V_s \times V_s \to R[\mathfrak{m}]$$

by

$$<a, b> = (a, \pi^{s-1}b)$$

where $(,)$ is the pairing above and where the kernel on the right of $<,>$ is V_{s-1} by the above exact sequence. That $<,>$ is an alternating pairing can be shown exactly as in the end of the proof of [30, Theorem 1.4.2] where τ becomes the element given by H.4 above and the symmetry of the pairing $<,>_v^{H.4}$ is required (the hypothesis H.5 stated in [30] is superfluous for [30, Theorem 1.4.2]); it follows that the kernel on the left of $<,>$ is also V_{s-1} and hence $<,>$ induces an alternating non-degenerate pairing on $V_s/V_{s-1} \times V_s/V_{s-1}$ as stated where the galois property follows from Proposition 8.2.3. □

Lemma 8.2.5 *Let G be a group of order 2, and let $R[G]$ be the corresponding group algebra. Assume that the residue characteristic of R is different from 2. Suppose that M is noetherian $R[G]$-module and there is a non-degenerate alternating pairing of R-modules*

$$[,]: M \times M \longrightarrow R$$

which satisfies the relation

$$[gx, gy] = [x, y] \quad (g \in G, x, y \in M).$$

Then there is a decomposition of $R[G]$-modules

$$M \cong N \oplus N.$$

Proof Let $\sigma \in G$ be a the generator of G. First note that an indecomposable character $\chi: G \to R^*$ is either trivial or non-trivial denoted by $+, -$, respectively, and that a noetherian $R[G]$-module P decomposes into isotypical components $P \cong P^+ \oplus P^-$ where σ acts trivially on P^+ and acts as -1 on P^- (see Sect. 1.6). Evidently if $v, u \in M$ are in opposite isotypical components under G, then we have $[v, u] = 0$ and hence M^+ and M^- are orthogonal submodules.

If the required decomposition exists for $M \otimes_R R/\mathfrak{m}$ over the residue field $R/\mathfrak{m}$, namely, $M \otimes_R R/\mathfrak{m} \cong N_1^+ \oplus N_1^- \oplus N_2^+ \oplus N_2^-$ where for $\delta = \pm$, we have $N_1^\delta \cong N_2^\delta$, then one takes bases of the two spaces N_i, $i = 1, 2$ over $R/\mathfrak{m}$ and one lifts these basis elements to elements of M with the required properties. We may then reduce to the case where R is a field.

8.2 The Cassels-Tate Pairing

The space M decomposes as a sum of eigenspaces $M^+ \oplus M^-$ under the action of G. We have shown that M^+ and M^- are orthogonal spaces under the pairing $<,>$. It follows that $<,>$ restricts to non-degenerate alternating pairings $<,>^+$ and $<,>^-$ on the subspaces M^+ and M^-, respectively,

$$<,>^+ : M^+ \times M^+ \to R, \quad <,>^- : M^- \times M^- \to R.$$

Hence M^+ is even dimensional over R and decomposes as $N_1 \oplus N_2$ where N_1, N_2 are isomorphic and inherit the trivial action of the group G. Similarly, M^- is even dimensional over R, and there is a decomposition $M^- \cong N_3 \oplus N_4$ where N_3 and N_4 are isomorphic $R[G]$-modules on which σ acts as -1. The lemma then holds by taking $N = N_1 \oplus N_3$. $\square$

Theorem 8.2.6 *Under the hypotheses of Proposition 8.2.4, suppose that $G = \mathrm{Gal}(F/k)$ has order 2 and the residue characteristic of R is different from 2. For ψ an indecomposable character $G \to R^*$, let $R(\psi)$ denote a $R[G]$-module which is a free R-module of rank 1 on which G acts via ψ. Then there is an $R[G]$-module N, annihilated by $\mathfrak{m}^{L-1}$ where $L = \mathrm{length}(R)$, a pair of integers $\epsilon(\phi) \in \mathbb{N}$ for $\phi = \pm$, such that there is an isomorphism of $R[G]$-modules*

$$H^1_{\mathcal{F}}(F, T) \cong \bigoplus_{\phi} \left(R(\phi)^{\epsilon(\phi)} \oplus (N \oplus N)^{\phi} \right)$$

where the sum runs over $\phi = +, -$.

Proof Write $\mathcal{H}$ for $H^1_{\mathcal{F}}(F, T)$. For an integer s with $1 \leq s < L$ where L is the length of R, we put as in Proposition 8.2.4

$$V_s = \mathcal{H}[\mathfrak{m}^s]/\mathfrak{m}(\mathcal{H}[\mathfrak{m}^{s+1}]).$$

From the Proposition 8.2.4, for any integer s with $1 \leq s < L$, there is a functorial alternating non-degenerate pairing of R-modules

$$\{,\} : V_s/V_{s-1} \times V_s/V_{s-1} \to R[\mathfrak{m}].$$

where this satisfies $\{gv, gw\} = \{v, w\}$ for $g \in G$, $v, w \in V_s/V_{s-1}$ where G acts trivially on the term $R[\mathfrak{m}]$ of the pairing.

The group G acts on $\mathcal{H}$ and provides a decomposition $\mathcal{H} = \mathcal{H}^+ \oplus \mathcal{H}^-$, where $+$ denotes the isotypical component on which G acts trivially and $-$ denotes the isotypical component on which the generator of G acts as -1; there is a similar decomposition of V_s/V_{s-1}.

For a character $\phi : G \to (R/\mathfrak{m})^*$, take a lifting of ϕ to $\phi' : G \to (R/\mathfrak{m}^d)^*$ by taking Teichmüller lifts of the image of ϕ. Let $(R/\mathfrak{m}^d)(\phi)$ be the free $R/\mathfrak{m}^d$-module of rank 1 on which G acts like ϕ'. Then $(R/\mathfrak{m}^d)(\phi)$ is an indecomposable $R[G]$-module for all d, ϕ and every indecomposable $R[G]$-module is of this type for some d, ϕ (see Sect. 1.6).

We may then decompose $\mathcal{H}$ as a sum of indecomposable $R[G]$-modules as follows

$$\mathcal{H} \cong \bigoplus_d \bigoplus_\phi U(d, \phi)$$

where $U(d, \phi)$ is the direct sum of the components all of which are $R[G]$-isomorphic to $(R/\mathfrak{m}^d)(\phi)$ so that $U(d, \phi)$ is a finite free $R/\mathfrak{m}^d$ module on which G acts like ϕ' where ϕ' is a Teichmüller lifting of $\phi : G \to (R/\mathfrak{m})^*$.

It follows from the decomposition of $\mathcal{H}$ that V_s/V_{s-1} can be written as a sum of indecomposable $R[G]$-modules as follows

$$V_s/V_{s-1} \cong \bigoplus_\phi U(s, \phi) \otimes_R R/\mathfrak{m}.$$

As V_s/V_{s-1} is equipped with the alternating pairing $\{,\}$, it follows from Lemma 8.2.5 that there is an $(R/\mathfrak{m})[G]$-submodule N of V_s/V_{s-1} such that there is an isomorphism of $(R/\mathfrak{m})[G]$-modules

$$V_s/V_{s-1} \cong N \oplus N.$$

It follows that

$$\bigoplus_\phi U(s, \phi) \otimes_R R/\mathfrak{m} \cong N \oplus N.$$

Comparing isotypical components, we then have isomorphisms of $(R/\mathfrak{m})[G]$-modules, where we write $\phi = +$ or $\phi = -$,

$$U(s, \phi) \otimes_R R/\mathfrak{m} \cong N^\phi \oplus N^\phi.$$

It then follows from the structure of $R[G]$-modules and as 2 is a unit of R (see (1.6)) that there is an $R[G]$-module N_s which is a free $R/\mathfrak{m}^s$-module of finite rank and an isomorphism of $R[G]$-modules where $\phi = \pm$

$$U(s, \phi) \cong N_s^\phi \oplus N_s^\phi.$$

From the decomposition $\mathcal{H} \cong \bigoplus_d \bigoplus_\phi U(d, \phi)$, we then derive the $R[G]$-module decomposition

$$\mathcal{H} \cong \left(\bigoplus_\phi U(L, \phi) \right) \oplus \bigoplus_{s<L} \left(N_s^+ \oplus N_s^+ \oplus N_s^- \oplus N_s^- \right).$$

The decomposition of $\mathcal{H}$ stated in the theorem now follows. □

8.2.7. Bibliographical Remarks. This section partly follows [30, §1.4] and the generalization of the Cassels-Tate pairing due to Flach [19], although both these references consider only number fields.

Proposition 8.2.3 is a pairing on Selmer groups of galois representations. It is straightforward to derive from this the classical Cassels-Tate pairing (for the parts coprime to the characteristic of F) on the Shafarevich-Tate groups of abelian varieties over F.

Many different constructions of the pairing are known such as [19] or [55]. For abelian varieties, a number of different constructions of the Cassels-Tate pairing are given in [48, Chap. I, Remark 6.11, p. 98; Chap. I, Remark 6.12, p. 100; Chap. II, Theorem 5.6, pp. 247–248].

The construction and properties of the Cassels-Tate pairing are well documented, so that the proofs of Propositions 8.2.3, 8.2.4 are abbreviated.

8.3 Kolyvagin Systems over Artinian Principal Ideal Rings

(8.3.1) In this section Sect. 8.3, we assume that

- k is a global field equipped with the special set of places ∞;
- F is an imaginary quadratic field extension of k with respect to ∞;
- R is a coefficient ring which is an artinian principal ideal ring with maximal ideal $\mathfrak{m}$ and with residue field of characteristic l different from 2 and the characteristic of k.

Let $(T, \mathcal{F}, \mathcal{L})$ be a Selmer triple on F of the Heegner point and CM point type given by Definition 7.3.3, (7.3.4), (7.3.2)) and for the above imaginary quadratic extension F/k and the integers $d_\mathfrak{q}$ for all $\mathfrak{q} \in \mathcal{L}$, defined in (7.3.4).

We assume further that this Selmer triple satisfies:

- $(T, \mathcal{F}, \mathcal{L})$ is a Selmer triple on F which satisfies hypotheses H.0-H.5 and that $\tau \in G_k$ is the element given by hypothesis H.4(b);
- $\mathcal{L}$ is so small that $\mathcal{L} = \mathcal{L}_L(T)$, where $L = \text{length}(R)$.

Remarks 8.3.2

(i) This Selmer triple $(T, \mathcal{F}, \mathcal{L})$ on F of Heegner point and CM point type Sect. 7.3 with imaginary quadratic extension F/k then has the integers $d_\mathfrak{q}$ for all $\mathfrak{q} \in \mathcal{L}$ defined in (7.3.4).

Associated with these $(T, \mathcal{F}, \mathcal{L})$, d_q, there is an abelian variety J/k whose l-adic Tate module is T_l where $T = T_l \otimes_{\mathbb{Z}_l} R$, and the uniquely determined objects $\mathcal{L}_i$, Γ_n, $\mathcal{M}_{i,F}$, I_n which are constructed exactly as in Sect. 7.1 or (7.3.4) from these $(T, \mathcal{F}, \mathcal{L})$, F/k and d_q.

We recall in particular that $\mathcal{L}_i$ are the subsets of $\mathcal{L}$ defined in (7.1.4), for all $i \in \mathbb{N}$ where $\mathcal{L}_0 = \mathcal{L}$; $\mathcal{M}_{i,F}$, for $i \in \mathbb{N}$, is the set of squarefree cycles which are products of distinct primes q of $\mathcal{L}_i$ lying over primes p of k which are inert and and unramified in F/k.

(ii) As T is obtained from the Tate module of an abelian variety J/k, the hypothesis H.0 shows that J/k is an elliptic curve.

(iii) The hypothesis that $\mathcal{L} = \mathcal{L}_L(T)$, where $L = \text{length}(R)$, implies that $I_n = 0$ for every $n \in \mathcal{M}_{0,F} = \mathcal{M}_{0,F}(\mathcal{L})$ (the sets $\mathcal{L}_i(T)$ are defined in Definition 7.1.4). The last assumption $\mathcal{L} = \mathcal{L}_L(T)$ can always be satisfied by replacing $\mathcal{L}$ by a subset and furthermore that $\mathcal{L} = \mathcal{L}_L(T)$ implies $\mathcal{L} = \mathcal{L}_s$ for all $s \in \mathbb{N}$.

(8.3.3) If M is an R-module with an action by the element $\tau \in G_k$ such that τ restricted to M has order 2, then denote by M^+, M^- the isotypical components of M on which τ acts like $+1$, -1, respectively; we write $M^\pm$ to denote either one of these components. Write

$$\mathcal{H}_b^a(c) = H^1_{\mathcal{F}_b^a(c)}(F, T), \overline{\mathcal{H}}_b^a(c) = H^1_{\mathcal{F}_b^a(c)}(F, T/\mathfrak{m}T).$$

If anyone of a, b, c is equal to 1, then it will be omitted from the notation.

Lemma 8.3.4 *If $q \in \mathcal{L}$, then there are isomorphisms of $R[\text{Gal}(F/k)]$-modules which are free R-modules of rank 2*

$$H^1_f(F_q, T) \cong T \cong H^1_{tr}(F_q, T).$$

In particular $H^1_f(F_q, T)^\pm$, $H^1_{tr}(F_q, T)^\pm$ are free R-modules of rank 1.

Proof By H.0, T is a free R-module of rank 2. By Definition 7.1.4, the ideal I_q for $q \in \mathcal{L}$ contains the positive integer d_q where d_q divides $|\kappa(q)^*|$. As $\mathcal{L} = \mathcal{L}_L(T)$, where $L = \text{length}(R)$, we have $I_q = 0$ for $q \in \mathcal{L}$. All elements of $\mathcal{L}$ are by definition places where T is unramified (see (5.1.1)). By definition of the ideals I_q for $q \in \mathcal{L}$, it follows that Fr_q acts trivially on T (see (7.3.4); the case where $k = \mathbb{Q}$ is explained in detail in Remark 7.3.6(ii)). It then follows from Corollary 4.5.3 that for all $q \in \mathcal{L}$, the groups $H^1_f(F_q, T)$ and $H^1_{tr}(F_q, T)$ are free R-modules of rank 2, and furthermore they are canonically isomorphic as $R[\text{Gal}(F/k)]$-modules to T. The last part follows from H.5(b). □

Lemma 8.3.5 *For any $a, b, c \in \mathcal{M}_{0,F}$ which are mutually coprime, the Selmer triple $(T, \mathcal{F}_a^b(c), \mathcal{L}(abc))$ (see (7.1.5)) satisfies the hypotheses H.0-H.5.*

Proof For the hypotheses H.0, H.1, H.2, and H.5, the result is obvious. For H.4, this follows from Proposition 4.5.4.

8.3 Kolyvagin Systems over Artinian Principal Ideal Rings

Recall that the exceptional set of the $\mathcal{F}_a^b(c)$ is the finite set

$$S(\mathcal{F}_a^b(c)) = S(\mathcal{F}) \cup \text{Supp}(abc).$$

Furthermore, $\mathcal{L}(abc)$ is the set $\mathcal{L}$ with the primes dividing abc removed.

It remains to verify H.3. If $v|b$, that is to say v is a place of the relaxed local condition, then the result is obvious. If $v|a$ that is to say v is a place of the strict local condition, then that H.3 holds follows from Corollary 4.5.3.

Let q be a prime place of F dividing c. Let $L = \text{length}(R)$ and π be a generator of $\mathfrak{m}$. The objects of the category $\text{Quot}(T)$ (see (4.1.7)) are simply the quotients $T/\mathfrak{m}^i T$ for $i \in \mathbb{N}$, and the morphisms of this category are multiplication by suitable elements of R. To check the Cartesian property of $\mathcal{F}_a^b(c)$, we may then reduce to the case of the injection $T/\mathfrak{m}^i T \to T$ given by multiplication by π^{L-i} for i a positive integer $\leq L$. There is a commutative diagram, where the left-hand horizontal maps come from the splitting into finite and transverse components of Proposition 4.5.4, and the isomorphisms on the right are the descriptions of the finite components from Corollary 4.5.3, and the three vertical maps all arise by multiplication by π^{L-i}

$$\begin{array}{ccccc} H^1(F_\mathfrak{q}, T/\mathfrak{m}^i T) & \longrightarrow & H^1_f(F_\mathfrak{q}, T/\mathfrak{m}^i T) & \xrightarrow{\cong} & T/(\mathfrak{m}^i, \text{Fr}_\mathfrak{q} - 1)T \\ h \downarrow & & \downarrow & & \downarrow \\ H^1(F_\mathfrak{q}, T) & \longrightarrow & H^1_f(F_\mathfrak{q}, T) & \xrightarrow{\cong} & T/(\text{Fr}_\mathfrak{q} - 1)T \end{array}$$

By definition of the ideals $I_\mathfrak{q}$ for $\mathfrak{q} \in \mathcal{L}_L(T)$, we have that $I_\mathfrak{q} = 0$ and that $\text{Fr}_\mathfrak{q}$ acts trivially on T (see (7.3.4) or the proof of Lemma 8.3.4 or Remark 7.3.6(ii)). It then follows from Corollary 4.5.3. Therefore, this diagram shows that $H^1_f(F_\mathfrak{q}, T)$ and $H^1_f(F_\mathfrak{q}, T/\mathfrak{m}^i T)$ are modules that are free of rank 2 over R and $R/\mathfrak{m}^i$, respectively. Hence, the right-hand and centre vertical maps are injective. Therefore, if $x \in H^1(F_\mathfrak{q}, T/\mathfrak{m}^i T)$ and $h(x)$ projects to zero in $H^1_f(F_\mathfrak{q}, T)$, then x projects to zero in $H^1_f(F_\mathfrak{q}, T/\mathfrak{m}^i T)$.

It follows that $\mathcal{F}_a^b(c)$ is Cartesian on $\text{Quot}(T)$. □

Proposition 8.3.6 *For $mn \in \mathcal{M}_{0,F}$, the image of $\mathcal{H}^m(n)$ in $\bigoplus_{\mathfrak{q}|m} H^1(F_\mathfrak{q}, T)$ is maximal isotropic for the sum of the local pairings $\sum_{\mathfrak{q}|m} <,>_\mathfrak{q}^{H.4}$.*

Proof Define an inner product $<,>$ on $\mathcal{H}^m(n)$ by taking

$$<c, d> = \sum_v <c_v, d_v>_v^{H.4}$$

where v runs over all places of F and $<,>_v^{H.4}$ is the local pairing provided by H.4(c).

Let A be the image of $\mathcal{H}^m(n)$ in $\bigoplus_{\mathfrak{q}|m} H^1(F_\mathfrak{q}, T)$. The local condition $\mathcal{F}^m(n)$ is maximal isotropic for the local pairing $<,>_v^{H.4}$ at any prime v not dividing m. Then class field theory gives that for the global classes $c, d \in \mathcal{H}^m(n)$, we have

$$\sum_{q|m} <c_q, d_q>_q^{H.4} = \sum_{\text{all places } q \text{ of } F} <c_q, d_q>_q^{H.4} = 0$$

and this shows that $A \subseteq A^\perp$ where $A^\perp$ is the orthogonal complement of A for the inner product $<,>$.

From Proposition 5.3.2, we obtain the pair of exact sequences

$$0 \longrightarrow H^1_{\mathcal{F}(n)}(F,T) \longrightarrow H^1_{\mathcal{F}^m(n)}(F,T) \longrightarrow \bigoplus_{v|m} \frac{H^1_{\mathcal{F}^m(n)}(F_v,T)}{H^1_{\mathcal{F}(n)}(F_v,T)}$$

$$0 \longrightarrow H^1_{\mathcal{F}_m(n)}(F,T^*) \longrightarrow H^1_{\mathcal{F}(n)}(F,T^*) \longrightarrow \bigoplus_{v|m} \frac{H^1_{\mathcal{F}(n)}(F_v,T^*)}{H^1_{\mathcal{F}_m(n)}(F_v,T^*)}$$

where $T \cong T^*$ and the images of the right-hand localization arrows are orthogonal complements under the sum of local pairings $\sum_{v|m} <,>_v^{H.4}$.

On the one hand, it follows from the definition of $\mathcal{H}^m(n)$, and these exact sequences that length(A) is given by the formula

$$\text{length}(A) = \text{length}(\mathcal{H}^m(n)/\mathcal{H}(n)) + \text{length}(\mathcal{H}(n)/\mathcal{H}_m(n)).$$

On the other hand, let $\nu(n)$ be the number of prime divisors of F dividing n and L be the length of R. Then the right-hand groups of the exact sequences $\bigoplus_{v|m} \frac{H^1_{\mathcal{F}^m(n)}(F_v,T)}{H^1_{\mathcal{F}(n)}(F_v,T)}$ and $\bigoplus_{v|m} \frac{H^1_{\mathcal{F}(n)}(F_v,T^*)}{H^1_{\mathcal{F}_m(n)}(F_v,T^*)}$ both have length $2L\nu(m)$ by Lemma 8.3.4. The duality under $<,>$ of the images of the right-hand arrows of these exact sequences implies that

$$\text{length}(A) = 2L\nu(m).$$

The sum of the lengths of A and $A^\perp$ equals

$$\text{length}(A) + \text{length}(A^\perp) = \sum_{q|m} \text{length}(H^1(F_q,T)) = 4L\nu(m)$$

because length $H^1(F_q, T) = 4L$ by Lemma 8.3.4. Therefore, as $A \subseteq A^\perp$, we must have $A = A^\perp$. □

Definition 8.3.7 For mutually coprime cycles $a, b, n \in \mathcal{M}_{0,F}$ and $\delta = +$ or $-$, we denote, where $\pm$ means the isotypical component under the action of the element $\tau \in G_k$ of H.4(b),

$$\rho_a^{b,\delta}(n) = \dim_{R/\mathfrak{m}} \overline{\mathcal{H}}_a^b(n)^\delta.$$

8.3 Kolyvagin Systems over Artinian Principal Ideal Rings

Here if one of a, b, n equals 1, then it is omitted. We also put

$$\rho_a^b(n) = \dim_{R/\mathfrak{m}} \overline{\mathcal{H}}_a^b(n) = \rho_a^{b,+}(n) + \rho_a^{b,-}(n).$$

Proposition 8.3.8 *For any prime $\mathfrak{q}$ of F, let $\mathrm{res}_{F_\mathfrak{q}/F} : \overline{\mathcal{H}}(n) \to H^1_{\mathcal{F}(n)}(F_\mathfrak{q}, T/\mathfrak{m}T)$ be the restriction homomorphism. For any $n\mathfrak{q} \in \mathcal{M}_{0,F}$ where $\mathfrak{q}$ is a prime divisor of F, coprime to n, and for $\delta = +$ or $-$, we have*

(i) *if $\mathrm{res}_{F_\mathfrak{q}/F}(\overline{\mathcal{H}}(n)^\delta) \neq 0$ then $\rho^\delta(n\mathfrak{q}) = \rho^\delta(n) - 1$ and $\mathrm{res}_{F_\mathfrak{q}/F}(\overline{\mathcal{H}}(n\mathfrak{q})^\delta) = 0$;*
(ii) *if $\mathrm{res}_{F_\mathfrak{q}/F}(\overline{\mathcal{H}}(n)^\delta) = 0$ then $\rho^\delta(n\mathfrak{q}) = \rho^\delta(n) + 1$.*
(iii) *$\rho(n)$ mod 2 is independent of $n \in \mathcal{M}_{0,F}$.*

Proof Let δ be either $+$ or $-$. We have the pair of exact sequences, from Proposition 5.3.2(i),

$$0 \longrightarrow \overline{\mathcal{H}}_\mathfrak{q}(n) \longrightarrow \overline{\mathcal{H}}(n) \longrightarrow H^1_f(F_\mathfrak{q}, T/\mathfrak{m}T)$$

$$0 \longrightarrow \overline{\mathcal{H}}(n) \longrightarrow \overline{\mathcal{H}}^\mathfrak{q}(n) \longrightarrow H^1_s(F_\mathfrak{q}, T/\mathfrak{m}T) \tag{8.3.9}$$

By the isomorphisms of Corollary 4.5.3, we have

$$H^1_f(F_\mathfrak{q}, T/\mathfrak{m}T) \cong T/\mathfrak{m}T \cong H^1_s(F_\mathfrak{q}, T/\mathfrak{m}T) \otimes \kappa(\mathfrak{q})^*.$$

It follows using the hypothesis H.5 and Lemma 8.3.4 that the action of the element τ splits each of $H^1_f(F_\mathfrak{q}, T/\mathfrak{m}T)$ and $H^1_s(F_\mathfrak{q}, T/\mathfrak{m}T)$ into two one-dimensional eigenspaces over $R/\mathfrak{m}$.

By Theorem 5.3.2, a consequence of global duality, we have that the images of the far right arrows of the above two exact sequences are orthogonal complements under the local pairing $<,>_\mathfrak{q}^{H.4}$ provided by H.4(c). This global duality then gives

$$\rho^{\mathfrak{q},\delta}(n) - \rho_\mathfrak{q}^\delta(n) = 1. \tag{8.3.10}$$

By definition, we have

$$\overline{\mathcal{H}}(n) \cap \overline{\mathcal{H}}(n\mathfrak{q}) = \overline{\mathcal{H}}_\mathfrak{q}(n)$$

and it follows that the following commutative square, where all arrows are the natural inclusions of subspaces, is a pullback diagram

$$\begin{array}{ccc} \overline{\mathcal{H}}_\mathfrak{q}(n)^\delta & \longrightarrow & \overline{\mathcal{H}}(n\mathfrak{q})^\delta \\ \downarrow & & \downarrow \\ \overline{\mathcal{H}}(n)^\delta & \longrightarrow & \overline{\mathcal{H}}^\mathfrak{q}(n)^\delta \end{array} \tag{8.3.11}$$

The above relation $\rho^{q,\delta}(n) - \rho_q^\delta(n) = 1$ of (8.3.10) means that the NW and SE diagonally opposite spaces in this diagram have a dimension difference equal to 1.
(i) Suppose that $\mathrm{res}_{F_q/F}(\overline{\mathcal{H}}(n)^\delta) \neq 0$. Then the image of $\overline{\mathcal{H}}(n)^\delta$ in $H_f^1(F_q, T/\mathfrak{m}T)^\delta$ is non-zero, and it follows from the first exact sequence of (8.3.9) that

$$\overline{\mathcal{H}}_q(n)^\delta \neq \overline{\mathcal{H}}(n)^\delta.$$

It then follows from the pullback diagram (8.3.11) that we have the equalities of subspaces

$$\overline{\mathcal{H}}_q(n)^\delta = \overline{\mathcal{H}}(n\mathfrak{q})^\delta$$

$$\overline{\mathcal{H}}(n)^\delta = \overline{\mathcal{H}}^q(n)^\delta.$$

Therefore from the global duality relation (8.3.10), we have

$$\rho^\delta(n\mathfrak{q}) = \rho^\delta(n) - 1$$

and also from the exact sequences (8.3.9) $\mathrm{res}_{F_q/F}(\overline{\mathcal{H}}(n\mathfrak{q})^\delta) = 0$ which proves (i).
(ii) By H.4(c), there is a perfect symmetric R-bilinear pairing of G_F-modules

$$(,); T \times T \to R(1)$$

which satisfies $(s^\sigma, t^{\tau\sigma\tau^{-1}}) = (s,t)^\sigma$ for every $s, t \in T$ and $\sigma \in \mathrm{Gal}(k^{\mathrm{sep}}/k)$ and the local condition $\mathcal{F}$ for every place v of F is its own orthogonal complement under the induced symmetric local pairing $<,>_v^{\mathrm{H}4}$ from $(,)$

$$<,>_v^{\mathrm{H}4}: H^1(F_v, T) \times H^1(F_v, T) \to R.$$

It follows that $\mathcal{F}^q(n)$ is the dual Selmer structure of $\mathcal{F}_q(n)$ on $T/\mathfrak{m}T$.
Suppose now that $\mathrm{res}_{F_q/F}(\overline{\mathcal{H}}(n)^\delta) = 0$. By the first exact sequence of (8.3.9), we then have

$$\overline{\mathcal{H}}_q(n)^\delta = \overline{\mathcal{H}}(n)^\delta. \tag{8.3.12}$$

From the pullback diagram (8.3.11) with the relation (8.3.10), we then obtain

$$\dim_{R/\mathfrak{m}} \overline{\mathcal{H}}^q(n)^\delta = \rho^\delta(n) + 1. \tag{8.3.13}$$

In consequence, to show part (ii), it is enough to show that there is an equality of subspaces of $\overline{\mathcal{H}}^q(n)^\delta$

$$\overline{\mathcal{H}}^q(n)^\delta = \overline{\mathcal{H}}(n\mathfrak{q})^\delta.$$

8.3 Kolyvagin Systems over Artinian Principal Ideal Rings

From Proposition 8.3.6 applied to $T/\mathfrak{m}T$, the restriction at $\mathfrak{q}$ of $\overline{\mathcal{H}}^\mathfrak{q}(n)$ is a maximal isotropic subspace of $H^1(F_\mathfrak{q}, T/\mathfrak{m}T)$ for the local pairing $<,>_\mathfrak{q}^{H.4}$.

The space $H^1(F_\mathfrak{q}, T/\mathfrak{m}T)^\delta$ is a 2-dimensional $R/\mathfrak{m}$-space by Lemma 8.3.4. Therefore, $\mathrm{res}_{F_\mathfrak{q}/F}(\overline{\mathcal{H}}^\mathfrak{q}(n)^\delta)$ is an $R/\mathfrak{m}$-subspace of $H^1(F_\mathfrak{q}, T/\mathfrak{m}T)^\delta$ which maps surjectively to $H^1_s(F_\mathfrak{q}, T/\mathfrak{m}T)^\delta$ and in particular we have

$$\mathrm{res}_{F_\mathfrak{q}/F}(\overline{\mathcal{H}}^\mathfrak{q}(n)^\delta) \neq 0.$$

The 2-dimensional $R/\mathfrak{m}$-space $H^1(F_\mathfrak{q}, T/\mathfrak{m}T)^\delta$, equipped with the symmetric non-degenerate pairing $<,>_\mathfrak{q}^{H.4}$ (hypothesis H.4), has two 1-dimensional isotropic subspaces for this pairing, namely, $H^1_f(F_\mathfrak{q}, T/\mathfrak{m}T)^\delta$ and $H^1_{\mathrm{tr}}(F_\mathfrak{q}, T/\mathfrak{m}T)^\delta$ by H.4; furthermore, $H^1(F_\mathfrak{q}, T/\mathfrak{m}T)^\delta$ is the direct sum of these two subspaces. As the pairing $<,>_\mathfrak{q}^{H.4}$ is non-degenerate and symmetric, there can be no other non-zero isotropic subspace. Therefore, as $\mathrm{res}_{F_\mathfrak{q}/F}(\overline{\mathcal{H}}^\mathfrak{q}(n)^\delta)$ is a non-zero isotropic subspace of $H^1(F_\mathfrak{q}, T/\mathfrak{m}T)^\delta$, it is equal to one of $H^1_f(F_\mathfrak{q}, T/\mathfrak{m}T)^\delta$ and $H^1_{\mathrm{tr}}(F_\mathfrak{q}, T/\mathfrak{m}T)^\delta$.

It follows that $\overline{\mathcal{H}}^\mathfrak{q}(n)^\delta$ is equal either to $\overline{\mathcal{H}}(n)^\delta$ or to $\overline{\mathcal{H}}(n\mathfrak{q})^\delta$ (see the diagram (8.3.11)). The first possibility does not occur by (8.3.13); therefore, we have $\overline{\mathcal{H}}^\mathfrak{q}(n)^\delta = \overline{\mathcal{H}}(\mathfrak{q}n)^\delta$. This proves (ii).

(iii) This follows immediately from parts (i) and (ii). □

The Stub Selmer Module

(8.3.14) Suppose for the rest of this section Sect. 8.3 that $(T^*, \mathcal{F}^*, \mathcal{L})$ as well as $(T, \mathcal{F}, \mathcal{L})$ satisfy the hypotheses H.0-H.5 of Sect. 8.1 where $T \cong T^*$ by Remark 8.1.11.

Proposition 8.3.15 *For ψ, an irreducible character of $G = \mathrm{Gal}(F/k)$ over $R/\mathfrak{m}$, let $R(\psi)$ denote the $R[G]$-module which is a free R-module of rank 1 on which G acts via the lifting of ψ from $(R/\mathfrak{m})^*$ to R^* and where ψ may be identified with $+$ or $-$ as in (8.3.1). Let $a, b, n \in \mathcal{M}_{0,F}$ be mutually coprime cycles.*

(i) *The Selmer triples $(T, \mathcal{F}_b^a(n), \mathcal{L}(abn))$ and $(T^*, \mathcal{F}_a^{*b}(n), \mathcal{L}(abn))$ (see (7.1.5)) satisfy the hypotheses H.0-H.5;*
(ii) *There are an $R[G]$-module N, annihilated by $\mathfrak{m}^{L-1}$ where $L = \mathrm{length}(R)$, and a pair of integers $\epsilon(\phi) \in \mathbb{N}$ for $\phi = \pm$, such that there is an isomorphism of $R[G]$-modules*

$$\mathcal{H}_b^a(n) \cong \bigoplus_\phi \left(R(\phi)^{\epsilon(\phi)} \oplus (N \oplus N)^\phi \right) \tag{8.3.16}$$

where the sum runs over $\phi = +, -$.

We have $\epsilon(+) + \epsilon(-) \equiv \rho_b^a(n) \mod 2$ and the parity of the integer $\epsilon(+) + \epsilon(-)$ is independent of n.

(iii) *There is an R[G]-module $P(n)$ and a pair of integers $0 \leq \eta(\phi) \leq 1$ for $\phi = \pm$, where $\eta(+) + \eta(-) \equiv \rho_b^a(n)$ mod 2, and where the parity of $\eta(+) + \eta(-)$ is independent of n, such that there is the decomposition into R[G]-modules*

$$\mathcal{H}_b^a(n) \cong R(+)^{\eta(+)} \oplus R(-)^{\eta(-)} \oplus P(n) \oplus P(n). \tag{8.3.17}$$

Proof

(i) This follows from Lemma 8.3.5.
(ii) From part (i) and Theorem 8.2.6, there is an $R[G]$-module N, annihilated by $\mathfrak{m}^{L-1}$ where $L = \text{length}(R)$, and a pair of integers $\epsilon(\phi) \in \mathbb{N}$ for $\phi = \pm$ such that there is an isomorphism of $R[G]$-modules

$$\mathcal{H}_b^a(n) \cong \bigoplus_\phi \left(R(\phi)^{\epsilon(\phi)} \oplus (N \oplus N)^\phi \right)$$

where the sum runs over $\phi = +, -$.
This gives the equation

$$\epsilon(+) + \epsilon(-) + 2\dim_{R/\mathfrak{m}} N[\mathfrak{m}] = \dim_{R/\mathfrak{m}} (\mathcal{H}_b^a(n)[\mathfrak{m}]) = \rho_b^a(n).$$

The first equality here follows from decomposition and the second from Lemma 8.1.16.

The congruence $\epsilon(+) + \epsilon(-) \equiv \rho_b^a(n)$ mod 2 follows immediately. The independence of n of the parity of $\epsilon(+) + \epsilon(-)$ now follows from Proposition 8.3.8, although the parity may well depend on a, b.

(iii) This follows from part (ii) where we define the $R[G]$-module $P(n)$ by

$$P(n) = R(+)^{m(+)/2} \oplus R(-)^{m(-)/2} \oplus N$$

where $m(\phi)$ is the greatest even integer less than or equal to $\epsilon(\phi)$, for $\phi = \pm$. We then put $\eta(\phi) = \epsilon(\phi) - 2m(\phi)$ for $\phi = \pm$ whence the result. □

Corollary 8.3.18 *Under the hypotheses of Proposition 8.3.15, there is an R-module M and an integer $\epsilon \in \{0, 1\}$ which is independent of n, such that there is an isomorphism of R-modules*

$$\mathcal{H}_b^a(n) \cong R^\epsilon \oplus M \oplus M.$$

Proof This follows immediately from part (iii) of the proposition. □

8.3 Kolyvagin Systems over Artinian Principal Ideal Rings

Definition 8.3.19 For $n \in \mathcal{M}$ and assuming that $(T^*, \mathcal{F}^*, \mathcal{L})$ as well as $(T, \mathcal{F}, \mathcal{L})$ satisfy the hypotheses H.0-H.5 of Sect. 8.1 (as in (8.3.14)), we have by Corollary 8.3.18 the decomposition of R-modules $\mathcal{H}(n) \cong R^\epsilon \oplus M(n) \oplus M(n)$ where $\epsilon \in \{0, 1\}$ is independent of n. We define

(a) $\lambda(n) = \text{length } M(n)$;
(b) the *stub Selmer module* to be $\text{St}(n) = \mathfrak{m}^{\lambda(n)} \mathcal{H}(n)$.

Note that the stub Selmer module $\text{St}(n)$ is a cyclic $R[\text{Gal}(F/k)]$-module and it forms a single isotypical component under the action of τ.

This holds because there is an $R[G]$-module decomposition (Proposition 8.3.15) $\mathcal{H}(n) \cong R(+)^{\eta(+)} \oplus R(-)^{\eta(-)} \oplus P(n) \oplus P(n)$, where $\eta(+), \eta(-) \in \{0, 1\}$, and then $\text{St}(n) \cong 0$ if $\eta(+)$ and $\eta(-)$ are equal and $\text{St}(n) \cong \mathfrak{m}^{\lambda(n)} R(+)$ or $\mathfrak{m}^{\lambda(n)} R(-)$ if $\eta(+)$ and $\eta(-)$ are not equal.

Definition 8.3.20 Suppose that $(T^*, \mathcal{F}^*, \mathcal{L})$ as well as $(T, \mathcal{F}, \mathcal{L})$ satisfy the hypotheses H.0-H.5 of Sect. 8.1. Then $(T, \mathcal{F}, \mathcal{L})$ is said to be *odd* (resp., *even*) if $\rho(n) \equiv 1 \mod 2$ (resp., $\rho(n) \equiv 0 \mod 2$) for all $n \in \mathcal{M}_{0,F}$, where the parity of $\rho(n)$ is independent of n (Proposition 8.3.15(iii)).

Remark 8.3.21 The above definition of the stub Selmer module applies when T is isomorphic to its dual T^* and is specific to the Selmer triples of this section. If T is not necessarily isomorphic to T^*, then the stub Selmer modules for T and T^* are defined in [54, Definition 4.3.1]. This latter definition of the stub Selmer module does not reduce to that of Definition 8.3.19 when T and T^* are isomorphic.

Proposition 8.3.22 *Put $G = \text{Gal}(F/k)$. For a character $\chi : G \to (R/\mathfrak{m})^*$ and any $r \in \mathbb{N}$, take a lifting of χ to $\chi' : G \to (R/\mathfrak{m}^r)^*$ by taking Teichmüller lifts of the image of χ. Let $(R/\mathfrak{m}^r)(\chi)$ be the free $R/\mathfrak{m}^r$-module of rank 1 on which G acts like χ'. Let $n\mathfrak{q} \in \mathcal{M}_{0,F}$ where $\mathfrak{q}$ is a prime divisor of F, coprime to n. Then for some $d \in \mathbb{N}$ there is an isomorphism of $R[G]$-modules*

$$\mathcal{H}^{\mathfrak{q}}(n)/(\mathcal{H}(n) + \mathcal{H}(\mathfrak{q}n)) \cong (R/\mathfrak{m}^d)(+) \oplus (R/\mathfrak{m}^d)(-).$$

Proof Let A be the image of $\mathcal{H}^{\mathfrak{q}}(n)$ in $H^1(F_\mathfrak{q}, T)$. Write A_f and A_{tr} for the intersections of A with $H^1_f(F_\mathfrak{q}, T)$ and $H^1_{\text{tr}}(F_\mathfrak{q}, T)$, where the former is uniquely determined and the latter depends on a choice of transverse submodule. Restriction at $\mathfrak{q}$ provides an isomorphism, via, for example, the exact sequences of Proposition 5.3.2, where this formula defines B

$$B = A/(A_f + A_{\text{tr}}) \cong \mathcal{H}^{\mathfrak{q}}(n)/(\mathcal{H}(n) + \mathcal{H}(\mathfrak{q}n)).$$

With respect to the decomposition, $H^1(F_q, T) = H^1_f(F_q, T) \oplus H^1_{tr}(F_q, T)$ for $x \in A$, we write $x_f \in A_f$ and $x_{tr} \in A_{tr}$ for the projections onto the finite and transverse submodules. Define a pairing on A by

$$[,]: A \times A \to R, \quad [x, y] = <x_f, y_{tr}>_q^{H.4}, \quad x, y \in A,$$

where $<,>_q^{H.4}$ is the local symmetric pairing on $H^1(F_q, T)$ given by H.4(c).

From Proposition 8.3.6, we have that A is maximal isotropic for the local pairing $<,>_q^{H.4}$ on $H^1(F_q, T)$. The dual of T is T itself, and furthermore the finite and transverse submodules $H^1_f(F_q, T)$ and $H^1_{tr}(F_q, T)$ are isotropic under the local Poitou-Tate pairing (by Proposition 4.5.4 or [54, proposition 1.3.2]) and hence they are isotropic for the $<,>_q^{H.4}$ pairing. We then have $[x, y] = -[y, x]$, for $x, y \in A$, for we have $<x, y>_q^{H.4} = 0$ because A, the finite and the transverse submodules are all isotropic for the H.4 pairing.

Suppose that $x \in A$ is in the kernel of this pairing $[,]$. Then we have $0 = <x_f, y_{tr}>_q^{H.4} = <x_f, y_f + y_{tr}>^{H.4} = <x_f, y>_q^{H.4}$ for every $y \in A$ and so $x_f \in A$ as A is maximal isotropic. As $x = x_f + x_{tr}$, it follows that we must have $x_{tr} \in A$ and hence $x \in A_f + A_{tr}$ and the image of x in $B = A/(A_f + A_{tr})$ is zero. Therefore, $[,]$ is induced by a non-degenerate alternating pairing

$$B \times B \to R.$$

It follows that there is an R-module D for which there are isomorphisms

$$B \cong \mathcal{H}^q(n)/(\mathcal{H}(n) + \mathcal{H}(qn)) \cong D \oplus D.$$

Since $\mathcal{H}^q(n)/\mathcal{H}(n)$ is a submodule of $H^1_s(F_q, T)$ which is a free R-module of rank 2 (Lemma 8.3.4), it follows that $\mathcal{H}^q(n)/(\mathcal{H}(n) + \mathcal{H}(qn))$ can be generated by 2 elements. Therefore, the R-module D is cyclic and of the form $R/\mathfrak{m}^d$ for some d.

The module $\mathcal{H}^q(n)/(\mathcal{H}(qn))$ is a submodule of $H^1_f(F_q, T)$ by the exact sequences of Theorem 5.3.2, where this latter $R[G]$-module is $R[G]$-isomorphic to T by Lemma 8.3.4. It follows from H.4(b) and Sect. 1.6 that T decomposes as as sum of 2 free R-modules of rank 1, namely, $T \cong T^+ \oplus T^-$. We then have that the $R[G]$-module $H^1_f(F_q, T)$ is isomorphic to $R(+) \oplus R(-)$ (Lemma 8.3.4). Hence, the $R[G]$-module $\mathcal{H}^q(n)/(\mathcal{H}(n) + \mathcal{H}(qn))$ is a quotient of a submodule of $R(+) \oplus R(-)$, and the result follows. □

Proposition 8.3.23 *Let* $nq \in \mathcal{M}_{0,F}$ *where* q *is a prime divisor of* F, *coprime to* n. *There are integers* $a^+, a^-, d \in \mathbb{N}$ *such that in this diagram of inclusions, the cokernel of each inclusion is cyclic of the given length, where* L *is the length of* R *and where* $\delta = +$ *or* $-$,

8.3 Kolyvagin Systems over Artinian Principal Ideal Rings

$$\begin{array}{ccc}
 & \mathcal{H}^{\mathsf{q}}(n)^\delta & \\
L - a^\delta \nearrow & & \nwarrow a^\delta + d \\
\mathcal{H}(n)^\delta & & \mathcal{H}(n\mathsf{q})^\delta \\
a^\delta \nwarrow & & \nearrow L - a^\delta - d \\
 & \mathcal{H}_{\mathsf{q}}(n)^\delta &
\end{array} \qquad (8.3.24)$$

The integer d is independent of δ.

Proof We have for $\delta = +, -$ that the four cokernels of the diagram $\mathcal{H}^{\mathsf{q}}(n)^\delta/\mathcal{H}(n)^\delta$, $\mathcal{H}^{\mathsf{q}}(n)^\delta/\mathcal{H}(n\mathsf{q})^\delta$, $\mathcal{H}(n)^\delta/\mathcal{H}_{\mathsf{q}}(n)^\delta$ and $\mathcal{H}(n\mathsf{q})^\delta/\mathcal{H}_{\mathsf{q}}(n)^\delta$ are all submodules of free R-modules of rank 1, namely, $H^1_{\mathrm{f}}(F_{\mathsf{q}}, T)^\delta$ and $H^1_{\mathrm{tr}}(F_{\mathsf{q}}, T)^\delta$ by Lemma 8.3.4 and H.5(b). Therefore, the four cokernels of the diagram are cyclic R-modules.

We then have that $\mathcal{H}(n)/\mathcal{H}_{\mathsf{q}}(n)$ is a direct sum of two cyclic R-modules which are isotypical components under $\mathrm{Gal}(F/k)$ and of lengths a^+, a^-, say where $0 \leq a^+, a^- \leq L$.

The relation between the lower left and upper left cokernels of the diagram follows from global duality (Proposition 5.3.2) and from Lemma 8.3.4; the relation between the lower right and upper right cokernels follows similarly.

The relation between the diagonally opposite cokernels $\mathcal{H}^{\mathsf{q}}(n)^\delta/\mathcal{H}(n)^\delta$ and $\mathcal{H}(n\mathsf{q})^\delta/\mathcal{H}_{\mathsf{q}}(n)^\delta$ as well as $\mathcal{H}^{\mathsf{q}}(n)^\delta/\mathcal{H}(n\mathsf{q})^\delta$ and $\mathcal{H}(n)^\delta/\mathcal{H}_{\mathsf{q}}(n)^\delta$ follows immediately from the preceding Proposition 8.3.22 where d is independent of $\delta = \pm$ by this same proposition. □

Proposition 8.3.25 *For $n\mathsf{q} \in \mathcal{M}_{0,F}$, where q is a prime divisor of F coprime to n, then there is a non-canonical isomorphism of R-modules*

$$\mathrm{res}_{F_{\mathsf{q}}/F}(\mathrm{St}(n)) \cong \mathrm{res}_{F_{\mathsf{q}}/F}(\mathrm{St}(n\mathsf{q})).$$

Proof For a^+, a^-, d as in Proposition 8.3.23 and $L = \mathrm{length}(R)$, we have that the two bottom inclusions in Proposition 8.3.23 give

$$\mathrm{length}_R(\mathcal{H}(n)) = \mathrm{length}_R(\mathcal{H}_{\mathsf{q}}(n)) + a^+ + a^-$$

and

$$\mathrm{length}_R(\mathcal{H}(n\mathsf{q})) = \mathrm{length}_R(\mathcal{H}_{\mathsf{q}}(n)) + (L - a^+ - d) + (L - a^- - d).$$

By Corollary 8.3.18, there is a decomposition $\mathcal{H}(n) \cong R^\epsilon \oplus M(n) \oplus M(n)$ where $\epsilon \in \{0, 1\}$ and M are a finitely generated R-module, and there is a similar decomposition of $\mathcal{H}(n\mathsf{q})$ with the same value of ϵ as this is independent of n. We obtain $\mathrm{length}_R(\mathcal{H}(n)) = L\epsilon + 2\lambda(n)$ and $\mathrm{length}_R(\mathcal{H}(n\mathsf{q})) = L\epsilon + 2\lambda(n\mathsf{q})$ where $\lambda(n) = \mathrm{length}\, M(n)$, $\lambda(n\mathsf{q}) = \mathrm{length}$

$M(n\mathfrak{q})$. Subtracting these two equations above gives us

$$2\lambda(n) - 2\lambda(n\mathfrak{q}) = a^+ + a^- - (L - a^+ - d) - (L - a^- - d) = 2(a^+ + a^- + d - L)$$

from which we obtain the relation

$$\lambda(n\mathfrak{q}) = \lambda(n) + L - a^+ - a^- - d. \tag{8.3.26}$$

The stub Selmer modules $\mathrm{St}(n)$, $\mathrm{St}(n\mathfrak{q})$ are cyclic R-modules (Definition 8.3.19), and hence their restrictions under $\mathrm{res}_{F_\mathfrak{q}/F}$ are also cyclic. Let $\mathfrak{m}^f$ be the annihilator ideal of $\mathrm{res}_{F_\mathfrak{q}/F}(\mathrm{St}(n))$ and $\mathfrak{m}^e$ be the annihilator ideal of $\mathrm{res}_{F_\mathfrak{q}/F}(\mathrm{St}(n\mathfrak{q}))$.

Using the exact sequences of Theorem 5.3.2 and referring to diagram (8.3.24) of Proposition 8.3.23, where $L = \mathrm{length}(R)$, we have that $\mathrm{res}_{F_\mathfrak{q}/F}(\mathcal{H}(n))$ is the sum of two cyclic R-modules of lengths a^+, a^- and $\mathrm{res}_{F_\mathfrak{q}/F}(\mathcal{H}(n\mathfrak{q}))$ is the sum of two cyclic R-modules of lengths $L - a^+ - d, L - a^- - d$.

Therefore, the annihilator of $\mathrm{res}_{F_\mathfrak{q}/F}(\mathrm{St}(n))$ is $\mathfrak{m}^f$ where

$$\lambda(n) + f = \max(a^+, a^-)$$

if this gives a non-negative value for f; otherwise, f is zero. Therefore, we have

$$f = \max(0, \max(a^+, a^-) - \lambda(n)).$$

Similarly, the annihilator of $\mathrm{res}_{F_\mathfrak{q}/F}(\mathrm{St}(n\mathfrak{q}))$ is $\mathfrak{m}^e$ where

$$\lambda(n\mathfrak{q}) + e = \max(L - a^+ - d, L - a^- - d)$$

if this gives a non-negative value for e; otherwise, e is zero. So we have

$$e = \max(0, \max(L - a^+ - d, L - a^- - d) - \lambda(n\mathfrak{q})).$$

By Equation (8.3.26), this gives

$$e = \max(0, \max(L - a^+ - d, L - a^- - d) - \lambda(n\mathfrak{q}))$$

$$= \max(0, \max(L - a^+ - d, L - a^- - d) - \lambda(n) - L + a^+ + a^- + d)$$

$$= \max(0, \max(a^+, a^-) - \lambda(n)) = f.$$

The annihilator ideals of these two localizations of stub Selmer modules are the same; hence, there is a non-canonical isomorphism of cyclic modules $\mathrm{res}_{F_\mathfrak{q}/F}(\mathrm{St}(n)) \cong \mathrm{res}_{F_\mathfrak{q}/F}(\mathrm{St}(n\mathfrak{q}))$.

8.3 Kolyvagin Systems over Artinian Principal Ideal Rings

[If a choice is made of a $\text{Gal}(F/k)$-equivariant finite-singular comparison isomorphism $\phi_{\mathfrak{q}}^{\text{fs}} : H_f^1(F_{\mathfrak{q}}, T/I_{n\mathfrak{q}}T) \longrightarrow H_s^1(F_{\mathfrak{q}}, T/I_{n\mathfrak{q}}T) \otimes_{\mathbb{Z}} \Gamma_{\mathfrak{q}}$, then it is easy to show that an isomorphism of R-modules $\text{res}_{F_{\mathfrak{q}}/F}(\text{St}(n)) \cong \text{res}_{F_{\mathfrak{q}}/F}(\text{St}(n\mathfrak{q}))$ is induced by the Kolyvagin system diagram (7.1.8).] □

Core vertices

(8.3.27) The aim of this subsection is to show that for a Kolyvagin system $\kappa \in \text{KS}(T, \mathcal{F}, \mathcal{L})$ where $(T, \mathcal{F}, \mathcal{L})$ is odd, then the isomorphism class of the submodule $\kappa_n R$ of $\mathcal{H}(n)$ is independent of n for all *core vertices* n under some mild hypotheses (Theorem 8.3.35).

We continue to assume (see (8.3.14)) that $(T^*, \mathcal{F}^*, \mathcal{L})$ as well as $(T, \mathcal{F}, \mathcal{L})$ satisfy the hypotheses H.0-H.5 of Sect. 8.1 where $T \cong T^*$ by Remark 8.1.11.

Definition 8.3.28 (i) We call $n \in \mathcal{M}_{0,F}$ a *core vertex* if $\rho(n) = 1$.

That $n \in \mathcal{M}_{0,F}$ be a core vertex is equivalent to $M(n) = 0$ and $\epsilon = 1$ in the decomposition $\mathcal{H}(n) \cong R^{\epsilon} \oplus M(n) \oplus M(n)$ of Corollary 8.3.18; that $n \in \mathcal{M}$ be a core vertex is also equivalent to $\mathcal{H}(n) \cong R$. That a core vertex exists evidently implies that $(T, \mathcal{F}, \mathcal{L})$ is odd.

Denote by $\mathcal{X}$ the graph whose vertices are the core vertices, and where we have an edge between vertices n and $n\mathfrak{q}$ for $n\mathfrak{q} \in \mathcal{M}_{0,F}$ and $\mathfrak{q}$ a prime.

(ii) If U denotes either T or $T/\mathfrak{m}T$, we write $\text{res}_{F_{\mathfrak{q}}/F}^s : H^1(F, U) \to H_s^1(F_{\mathfrak{q}}, U)$ for the composite of $\text{res}_{F_{\mathfrak{q}}/F} : H^1(F, U) \to H^1(F_{\mathfrak{q}}, U)$ with the projection $H^1(F_{\mathfrak{q}}, U) \to H_s^1(F_{\mathfrak{q}}, U)$ on to the singular quotient $H_s^1(F_{\mathfrak{q}}, U)$. Similarly, $\text{res}_{F_{\mathfrak{q}}/F}^s$ is defined for the variants $H_{\mathcal{F}_b^a(n)}^1(F, U)$.

Proposition 8.3.29 *Suppose that $(T, \mathcal{F}, \mathcal{L})$ is odd. Let n be a core vertex and $n\mathfrak{q} \in \mathcal{M}_{0,F}$ for a prime $\mathfrak{q}$ which is coprime to n. Then the following are equivalent:*

(1) $n\mathfrak{q}$ is a core vertex;
(2) $\text{res}_{F_{\mathfrak{q}}/F} : \mathcal{H}(n) \longrightarrow H^1(F_{\mathfrak{q}}, T)$ is injective;
(3) $\text{res}_{F_{\mathfrak{q}}/F} : \mathcal{H}(n\mathfrak{q}) \longrightarrow H^1(F_{\mathfrak{q}}, T)$ is injective;
(4) $\text{res}_{F_{\mathfrak{q}}/F} \overline{\mathcal{H}}(n) \neq 0$;
   Furthermore, these four conditions imply
(5) $\text{res}_{F_{\mathfrak{q}}/F} \overline{\mathcal{H}}(n\mathfrak{q}) \neq 0$.

Finally, if $n\mathfrak{q}$ is a core vertex, then there is an isomorphism

$$(\text{res}_{F_{\mathfrak{q}}/F}^s)^{-1} \circ \phi_{\mathfrak{q}}^{\text{fs}} \circ \text{res}_{F_{\mathfrak{q}}/F} : \mathcal{H}(n) \cong \mathcal{H}(n\mathfrak{q})$$

where $\phi_{\mathfrak{q}}^{\text{fs}}$ is a particular choice of finite singular comparison isomorphism and $\text{res}_{F_{\mathfrak{q}}/F}^s$ denotes the isomorphism obtained by restriction $\mathcal{H}(n\mathfrak{q}) \cong H_s^1(F_{\mathfrak{q}}, T)$.

Proof (2) ⇔ (3). The kernel of maps (2) and (3) is $\mathcal{H}_q(n)$; therefore, (2) and (3) are equivalent.

(2) ⇔ (4). Since n is a core vertex, which means that $\overline{\mathcal{H}}(n) \cong R/\mathfrak{m}$, we have that (4) is equivalent to $\mathrm{res}_{F_q/F} : \overline{\mathcal{H}}(n) \to H^1(F_q, \overline{T})$ being injective. By Lemma 8.1.16, we have $\overline{\mathcal{H}}_q(n) = \mathcal{H}_q(n)[\mathfrak{m}]$, and so one kernel is zero if and only if the other kernel is also zero.

(1) ⇔ (4). Since n is a core vertex, we must have $\rho^\delta(n) = 1$ and $\rho^{-\delta}(n) = 0$ for some choice of sign $\delta = \pm$. By Proposition 8.3.8, it follows that $n\mathfrak{q}$ is a core vertex if and only if $\mathrm{res}_{F_q/F}\overline{\mathcal{H}}(n)^\delta \neq 0$. This last condition is equivalent to (4) since $\rho^{-\delta}(n) = 0$ where $-\delta$ denotes the opposite sign to δ; we then have that $\overline{\mathcal{H}}(n)^\delta = \overline{\mathcal{H}}(n)$.

(1) ⇒ (5). From the implication (1) ⇒ (3) and that by Lemma 8.1.16, we have $\overline{\mathcal{H}}_q(n) = \mathcal{H}_q(n)[\mathfrak{m}]$; it follows that we have that $\mathrm{res}_{F_q/F} : \overline{\mathcal{H}}(n\mathfrak{q}) \longrightarrow H^1(F_q, \overline{T})$ is injective as its kernel is isomorphic to $\overline{\mathcal{H}}_q(n)$, which is zero by (2). As $n\mathfrak{q}$ is a core vertex, we have that $\overline{\mathcal{H}}(n\mathfrak{q}) \cong R/\mathfrak{m}$ and hence that $\mathrm{res}_{F_q/F}\overline{\mathcal{H}}(n\mathfrak{q}) \neq 0$.

For the final part, suppose that $n\mathfrak{q}$ is a core vertex. By Proposition 8.3.8, we must have that $\rho^{-\delta}(n\mathfrak{q}) = 1$, $\rho^\delta(n\mathfrak{q}) = 0$, for some choice of sign $\delta = \pm$, and $\mathcal{H}(n\mathfrak{q}) \cong R$. Hence, we have that $\mathrm{res}_{F_q/F}\mathcal{H}(n) \subseteq H^1_f(F_q, T)^\delta \cong R$. But since n is a core vertex, we have $\mathcal{H}(n) \cong R$. So (2) implies that $\mathrm{res}_{F_q/F}$ induces an isomorphism

$$\mathrm{res}_{F_q/F} : \mathcal{H}(n) \cong H^1_f(F_q, T)^\delta.$$

Analogously, by (3) the map $\mathrm{res}_{F_q/F} : \mathcal{H}(n\mathfrak{q}) \longrightarrow H^1(F_q, T)$ induces an isomorphism

$$\mathrm{res}^s_{F_q/F} : \mathcal{H}(n\mathfrak{q}) \cong H^1_s(F_q, T)^{-\delta}.$$

We may then select a finite-singular comparison isomorphism $\phi^{fs}_q : H^1_f(F_q, T) \to H^1_s(F_q, T)$ that interchanges the τ-eigenspaces in that $\phi^{fs}_q(H^1_f(F_q, T)^\eta) = H^1_s(F_q, T)^{-\eta}$ for $\eta = +$ and $\eta = -$ where evidently the groups $H^1_s(F_q, T)^{-\eta}$ for $\eta = \pm$ have the same block decompositions. This gives the isomorphism in the proposition. □

Proposition 8.3.30 *Suppose that $(T, \mathcal{F}, \mathcal{L})$ is odd. Let $\kappa \in \mathbf{KS}(T, \mathcal{F}, \mathcal{L})$. If n and m are core vertices that are connected in $\mathcal{X}$, then $\kappa_n \in \mathcal{H}(n)$ and $\kappa_m \in \mathcal{H}(m)$ generate isomorphic R-modules.*

Proof We may immediately reduce to the case where $m = n\mathfrak{q}$ and $\mathfrak{q} \in \mathcal{M}$ is a prime cycle. By the Kolyvagin system relations, we have $(\phi^{fs}_q \circ \mathrm{res}_{F_q/F})(\kappa_n) = \mathrm{res}^s_{F_q/F}(\kappa_{n\mathfrak{q}})$, where $\mathrm{res}^s_{F_q/F}$ denotes the isomorphism obtained by restriction $\mathcal{H}(n\mathfrak{q}) \cong H^1_s(F_q, T)$. By Proposition 8.3.29, there is the isomorphism which maps κ_n to $\kappa_{n\mathfrak{q}}$

$$(\mathrm{res}^s_{F_q/F})^{-1} \circ \phi^{fs}_q \circ \mathrm{res}_{F_q/F} : \mathcal{H}(n) \cong \mathcal{H}(n\mathfrak{q}).$$

This implies that $\kappa_n R \subseteq \mathcal{H}(n)$ and $\kappa_{n\mathfrak{q}} R \subseteq \mathcal{H}(n\mathfrak{q})$ are isomorphic R-modules. □

8.3 Kolyvagin Systems over Artinian Principal Ideal Rings 251

Proposition 8.3.31 *Assume that $\mathcal{L} = \mathcal{L}_L$ contains all but finitely primes $\mathfrak{q}$ of F which lie over primes of k which are inert and unramified in F/k and such that $\mathrm{Fr}_\mathfrak{q}$ acts trivially on T.*

Let $c_1, \ldots, c_n \in H^1(F, T)$ be non-zero classes such that c_i is in the ϵ_i isotypical component under τ where $\epsilon_1, \ldots, \epsilon_n$ is any sequence of elements in $\{+, -\}$. If $l > n$, where l is the residue characteristic of R, then there is a set of primes $\mathfrak{q} \in \mathcal{L}$ of positive Dirichlet density such that the restrictions $\mathrm{res}_{F_\mathfrak{q}/F} c_i$ are all non-zero.

Proof Note first that by definition that $\mathfrak{q} \in \mathcal{L}$, where $\mathcal{L} = \mathcal{L}_L$, implies that $I_\mathfrak{q} = 0$ and that $\mathrm{Fr}_\mathfrak{q}$ acts trivially on T (see Remark 7.1.19(iii)). Conversely, if $\mathfrak{q}$ is a non-Archimedean prime place of F such that $\mathfrak{q} \notin S(\mathcal{F})$ (see (7.1.3)) which lies over a place $\mathfrak{p}$ of k which is unramified and inert in F/k and that $\mathrm{Fr}_\mathfrak{q}$ acts trivially on T then $\mathrm{Fr}_\mathfrak{p}^2 = \mathrm{Fr}_\mathfrak{q}$ and the minimal polynomial $\det(1 - x.\mathrm{Fr}_\mathfrak{p}|T)$ of $\mathrm{Fr}_\mathfrak{p}$ acting on T equals $1 - x^2 \in R[x]$; hence $I_\mathfrak{q} = 0$. and $T/(\mathfrak{m}^L + (\mathrm{Fr}_\mathfrak{q} - 1)T)$ is a free R-module of rank 2 (see (7.3.4)) (see the argument of Remark 7.1.19(iii)).

Therefore, the hypothesis of the proposition that $\mathcal{L}$ contains all but finitely primes $\mathfrak{q}$ of F which lie over primes of k which are inert and unramified in F/k and such that $\mathrm{Fr}_\mathfrak{q}$ acts trivially on T is compatible with the hypothesis $\mathcal{L} = \mathcal{L}_L$ (given in (8.3.1)); furthermore, the former hypothesis says *grosso modo* that $\mathcal{L}$ is sufficiently large subject to the constraint $\mathcal{L} = \mathcal{L}_L$.

By Lemma 8.1.16, there is an isomorphism of R-modules $H^1(F, T)[\mathfrak{m}] \cong H^1(F, T/\mathfrak{m}T)$. Furthermore c_i then has a non-zero R-multiple in $(H^1(F, T)[\mathfrak{m}])^{\epsilon_i}$ for all i; we may therefore reduce to the case where R is a field and $T = T/\mathfrak{m}T$.

Let $F(T, \mu_{l^\infty})$ be the smallest galois extension of k containing F such that $G_{F(T, \mu_{l^\infty})}$ acts trivially on μ_{l^∞} and on T. By hypothesis H.2 we have $H^1(F(T, \mu_{l^\infty})/F, T/\mathfrak{m}T) = 0$. Let J be the Galois closure over k of $F(T, \mu_{lL})$. We have $J \subseteq F(T, \mu_{l^\infty})$, and so the restriction map

$$H^1(F, T/\mathfrak{m}T) \longrightarrow H^1(J, T/\mathfrak{m}T)^{\mathrm{Gal}(J/F)} \cong \mathrm{Hom}(G_J, T/\mathfrak{m}T)^{\mathrm{Gal}(J/F)} \quad (8.3.32)$$

is an injection by H.2 and Lemma 8.1.15. We identify the c_i with their images $\mathrm{Hom}(G_J, T/\mathfrak{m}T)$ under this restriction map. Let E_i be the smallest extension of J with $c_i(G_{E_i}) = 0$. Set $G_i = \mathrm{Gal}(E_i/J)$. Then G_i is a $\mathbb{F}_l$-vector space with a natural action of $\mathrm{Gal}(J/k)$, and we let $G_i^\pm$ be the eigenspaces for the action of the element τ, given by H.4. Note that for $\delta = \pm$, we have $c_i(G_i^\delta) \subseteq (T/\mathfrak{m}T)^{\delta\epsilon_i}$.

By hypotheses H.1, the module $T/\mathfrak{m}T$ does not have any $R[G_F]$-submodule different from 0 to $T/\mathfrak{m}T$. If for some i the map $c_i : G_i^+ \longrightarrow (T/\mathfrak{m}T)^{\epsilon_i}$ were trivial, then we would have $c_i(G_i) = c_i(G_i^-) \subseteq (T/\mathfrak{m}T)^{-\epsilon_i}$. But then $R \cdot c_i(G_i)$ would be a $R[G_F]$-submodule of $T/\mathfrak{m}T$ contained in $(T/\mathfrak{m}T)^{-\epsilon_i}$ and hence $c_i(G_i) = 0$. This would contradict the injectivity of the restriction map (8.3.32). Therefore, the maps $c_i : G_i^+ \longrightarrow (T/\mathfrak{m}T)^{\epsilon_i}$ are non-trivial for all i.

We have the injective group homomorphism formed from all the c_i

$$\prod_i c_i : G_J / \cap_i G_{E_i} \longrightarrow \prod_i T/\mathfrak{m}T.$$

Write $G_1 = G_J / \cap_i G_{E_i}$. It follows that G_1 inherits from $T/\mathfrak{m}T$ the structure of an $(R/\mathfrak{m})[\text{Gal}(F/k)]$-module and in particular has an action by τ, the element of hypothesis H.4, so that G_1 has the two eigencomponents $G_1^\pm$ under the action of τ.

Let $H_i = \{\gamma \in G_1^+ | c_i(\gamma) = 0 \text{ in } T/\mathfrak{m}T^{\epsilon_i}\}$. By the previous two paragraphs, $H_i \neq G_1^+$. Let μ be the Haar measure on G_1^+ normalized so that $\mu(G_1^+) = 1$. Then we have $\mu(H_i) = 1/|c_i(G_1^+)| \leq 1/l$ as $c_i(G_1^+)$ is a non-zero $\mathbb{F}_l$ vector space. So if $l > n$, we have

$$\mu(\bigcup_i H_i) \leq \frac{n}{l} < 1 = \mu(G_1^+).$$

Hence, there are elements $\eta \in G_1^+$ such that $c_i(\eta) \neq 0$ for all i.

Fix such an element $\eta \in G_1^+$ such that $c_i(\eta) \neq 0$ for all i. As $\eta \in G_1^+$ is in the $+$ eigenspace under the action of τ, we may write $\eta \equiv (\tau\sigma)^2 \mod \cap_i G_{E_i}$ for some $\sigma \in G_J$. By the Chebotarev density theorem, there is set $\mathfrak{S}$ of primes $\mathfrak{p}$ of k with positive Dirichlet density such that its Frobenius class $\text{Fr}_\mathfrak{p}$ in $\text{Gal}(E_1 \ldots E_n/k)$ is $\tau\sigma$, and at which the localizations of c_i are unramified. Then the images of the localizations of the c_i in $H_f^1(F_\mathfrak{q}, T/\mathfrak{m}T)$ are the evaluations of c_i at $\text{Fr}_\mathfrak{q}$, the Frobenius of $\mathfrak{q}$ where $\mathfrak{q}$ is the prime of F lying over $\mathfrak{p}$ where $\mathfrak{p}$ is inert F/k because the Frobenius $\text{Fr}_\mathfrak{p}$ at $\mathfrak{p}$ in F/k is equal to τ. The image of $\text{Fr}_\mathfrak{q}$ in $\text{Gal}(E_1 \ldots E_n/F)$ is equal to $(\tau\sigma)^2 = \eta$, and hence we have $c_i(\text{Fr}_\mathfrak{q}) \neq 0$ for all primes $\mathfrak{q}$ of F lying over a prime in $\mathfrak{S}$.

For any $\mathfrak{p} \in \mathfrak{S}$, the Frobenius class at $\mathfrak{p}$ in $\text{Gal}(J/k)$ is equal to the class of τ in $\text{Gal}(J/k)$ and which has order exactly two by hypothesis H.4. Then $\mathfrak{p}$ is unramified and inert in F/k and $\text{Fr}_\mathfrak{p}$ acting on T has characteristic polynomial $\det(1-x.\text{Fr}_\mathfrak{p}|T) = 1-x^2 \in R[x]$; therefore, the prime $\mathfrak{q}$ of F lying over $\mathfrak{p}$ is such that $\text{Fr}_\mathfrak{q}$ acts trivially on T. It follows from the hypothesis made on $\mathcal{L}$ (see the first paragraph of this proof), we may then select a place $\mathfrak{q}$ of F lying over a place $\mathfrak{p} \in \mathfrak{S}$ such that $c_i(\text{Fr}_\mathfrak{q}) \neq 0$ and $\mathfrak{q} \in \mathcal{L}$, and indeed there is a set of positive Dirichlet density of such prime places. $\square$

Corollary 8.3.33 *Suppose that $(T, \mathcal{F}, \mathcal{L})$ is odd and that $\mathcal{L} = \mathcal{L}_L$ contains all but finitely primes $\mathfrak{q}$ of F which lie over primes of k which are inert and unramified in F/k and such that $\text{Fr}_\mathfrak{q}$ acts trivially on T. Assume $l > 3$. Let $r(n) := \max(\rho^+(n), \rho^-(n))$. Let $n \in \mathcal{M}_{0,F}$ be such that $r(n) > 1$, and let $0 \neq c \in \mathcal{H}(n)$. Then there is a positive proportion of primes $\mathfrak{q} \in \mathcal{L}$ such that $\text{res}_{F_\mathfrak{q}/F}(c) \neq 0$ and $r(n\mathfrak{q}) < r(n)$.*

Proof Suppose first that $\rho^+(n), \rho^-(n) \geq 1$. By Proposition 8.3.31, there is a positive proportion of primes $\mathfrak{q} \in \mathcal{L}$ such that

(1) $\text{res}_{F_\mathfrak{q}/F}(\overline{\mathcal{H}}^+(n)) \neq 0$;

(2) $\operatorname{res}_{F_{\mathfrak{q}}/F}(\overline{\mathcal{H}}^{-}(n)) \neq 0$;
(3) $\operatorname{res}_{F_{\mathfrak{q}}/F}(c) \neq 0$.

The third condition can be ensured in the following way. There is $r \in R$ such that rc satisfies $0 \neq rc \in \mathcal{H}(n)[\mathfrak{m}]$. By Lemma 8.1.16, we have $\mathcal{H}(n)[\mathfrak{m}] = \overline{\mathcal{H}}(n)$, and so rc has non-zero projection in one of the eigenspaces, say $0 \neq rc^{\zeta} \in \overline{\mathcal{H}}^{\zeta}(n)$ for some $\zeta = +$ or $-$. Now we can choose $\mathfrak{q}$ such that $\operatorname{res}_{F_{\mathfrak{q}}/F}(rc^{\zeta}) \neq 0$, and this implies that $\operatorname{res}_{F_{\mathfrak{q}}/F}(c) \neq 0$.

By Proposition 8.3.8 and (1), (2) above and as $l > 3$, we have $\rho^{\pm}(n\mathfrak{q}) = \rho^{\pm}(n) - 1$ for both signs $\pm$, and hence $r(n\mathfrak{q}) = r(n) - 1$.

Suppose now that $\rho^{\delta}(n) = 0$ for some choice of sign $\delta = \pm$. Then we have $\rho^{-\delta}(n) > 1$. We consider primes $\mathfrak{q}$ in $\mathcal{L}$, again by Proposition 8.3.31, such that

(1) $\operatorname{res}_{F_{\mathfrak{q}}/F}(\overline{\mathcal{H}}^{-\delta}(n)) \neq 0$;
(2) $\operatorname{res}_{F_{\mathfrak{q}}/F}(c) \neq 0$.

By Proposition 8.3.8, we then have $\rho^{-\delta}(n\mathfrak{q}) = \rho^{-\delta}(n) - 1 \geq 1$ and $\rho^{\delta}(n\mathfrak{q}) = 1$, and hence $r(n\mathfrak{q}) < r(n)$. □

Proposition 8.3.34 *Suppose that $(T, \mathcal{F}, \mathcal{L})$ is odd and that $\mathcal{L} = \mathcal{L}_L$ contains all but finitely primes $\mathfrak{q}$ of F which lie over primes of k which are inert and unramified in F/k and such that $\operatorname{Fr}_{\mathfrak{q}}$ acts trivially on T. If $l > 4$ then the graph $\mathcal{X}$ is connected.*

Proof For two distinct core vertices n and m, define $f(n, m) = \nu(\operatorname{lcm}(n, m)/\gcd(n, m))$, where $\nu(t)$ denotes the number of prime divisors of a cycle t on F. Evidently $f(n, m)$ is the number of prime divisors of F dividing one of n or m but not both. We will show by induction on f that $\mathcal{X}$ is connected. We take $n \neq m$ to be distinct squarefree cycles which are core vertices. By interchanging n, m if necessary, we may suppose that there is a prime $\mathfrak{q}$ with $\mathfrak{q} | m$ but $\mathfrak{q}$ does not divide n.

Since n is a core vertex, $\rho^{\delta}(n) = 1$ and $\rho^{-\delta}(n) = 0$ for some sign $\delta = \pm$. By Proposition 8.3.8, we have $\rho^{\delta}(n\mathfrak{q}) \in \{0, 2\}$ and $\rho^{-\delta}(n\mathfrak{q}) = 1$. If $\rho^{\delta}(n\mathfrak{q}) = 0$, then $n\mathfrak{q}$ is also a core vertex, and then $m, n\mathfrak{q}$ are connected in $\mathcal{X}$ by the induction hypothesis applied to the pair of cycles $(n\mathfrak{q}, m)$.

So assume $\rho^{\delta}(n\mathfrak{q}) = 2$. We choose a prime $\mathfrak{r} \in \mathcal{L}$ such that

(1) $\operatorname{res}_{F_{\mathfrak{r}}/F}(\overline{\mathcal{H}}^{+}(n\mathfrak{q})) \neq 0$;
(2) $\operatorname{res}_{F_{\mathfrak{r}}/F}(\overline{\mathcal{H}}^{-}(n\mathfrak{q})) \neq 0$;
(3) $\operatorname{res}_{F_{\mathfrak{r}}/F}(\overline{\mathcal{H}}^{\delta}(n)) \neq 0$;
(4) $\operatorname{res}_{F_{\mathfrak{r}}/F}(\overline{\mathcal{H}}(m)) \neq 0$.

The existence of such an $\mathfrak{r}$ follows from Proposition 8.3.31 and that $l > 4$.

Then Proposition 8.3.8, together with (1) and (2), implies that $n\mathfrak{r}\mathfrak{q}$ is a core vertex. And (3) and (4) with Proposition 8.3.29 imply, respectively, that $n\mathfrak{r}$ and $m\mathfrak{r}$ are core vertices. This means that we have a two-step path $n, n\mathfrak{r}, n\mathfrak{q}\mathfrak{r}$ and a one-step path $m, m\mathfrak{r}$ in $\mathcal{X}$. Then $m\mathfrak{r}, n\mathfrak{q}\mathfrak{r}$ are connected in $\mathcal{X}$ by the induction hypothesis applied to the pair $(n\mathfrak{q}\mathfrak{r}, m\mathfrak{r})$. $\square$

Theorem 8.3.35 *Suppose that $(T, \mathcal{F}, \mathcal{L})$ is odd and that $\mathcal{L} = \mathcal{L}_L$ contains all but finitely primes $\mathfrak{q}$ of F which lie over primes of k which are inert and unramified in F/k and such that $\mathrm{Fr}_\mathfrak{q}$ acts trivially on T. Let $\mathcal{H}$ be the Selmer sheaf corresponding to $(T, \mathcal{F}, \mathcal{L})$ (see (7.1.18)). Suppose that $\kappa \in \mathbf{KS}(T, \mathcal{F}, \mathcal{L})$, and $l > 4$. Then there is an integer $\sigma(\kappa)$ with $0 \le \sigma(\kappa) \le \mathrm{length}(R)$ such that the stalk $\mathfrak{m}^{\sigma(\kappa)}\mathcal{H}(n)$ of the sheaf $\mathfrak{m}^{\sigma(\kappa)}\mathcal{H}$ at any core vertex n is isomorphic to the submodule $\kappa_n R$ of $\mathcal{H}(n)$.*

Proof The equality of submodules of $\mathcal{H}(n)$ for any core vertex n given by $\kappa_n R = \mathfrak{m}^{\sigma(\kappa)}\mathcal{H}(n)$ follows directly from Proposition 8.3.34, Proposition 8.3.30 and that $\mathcal{H}(n) \cong R$ for a core vertex n (Definition 8.3.28). $\square$

Proposition 8.3.36 *Suppose that $(T, \mathcal{F}, \mathcal{L})$ is odd and that $\mathcal{L} = \mathcal{L}_L$ contains all but finitely primes $\mathfrak{q}$ of F which lie over primes of k which are inert and unramified in F/k and such that $\mathrm{Fr}_\mathfrak{q}$ acts trivially on T. Suppose that $\kappa \in \mathbf{KS}(T, \mathcal{F}, \mathcal{L})$ and $l > 4$. If $\sigma(\kappa)$ is the integer given by Theorem 8.3.35, then we have $\sigma(\kappa) = \mathrm{length}(R)$ if and only if $\kappa = 0$.*

Proof As $\sigma(\kappa) = \mathrm{length}(R)$, we have that $\kappa_n = 0$ for all core vertices $n \in \mathcal{M}_{0,F}$. We have to show that $\kappa_n = 0$ for all $n \in \mathcal{M}_{0,F}$.

We argue by induction on $r(n) = \max(\rho^+(n), \rho^-(n))$ for $n \in \mathcal{M}_{0,F}$. Indeed, if $\kappa_n \ne 0$ and $r(n) > 1$, we could choose a prime $\mathfrak{q} \in \mathcal{L}$, by Corollary 8.3.33, such that $\mathrm{res}_{F_\mathfrak{q}/F}(\kappa_n) \ne 0$ and such that $r(n\mathfrak{q}) < r(n)$. But then we have the Kolyvagin system relation

$$\mathrm{res}^s_{F_\mathfrak{q}/F}(\kappa_{n\mathfrak{q}}) = \phi^{\mathrm{fs}}_\mathfrak{q}(\mathrm{res}_{F_\mathfrak{q}/F}(\kappa_n))$$

where $\mathrm{res}^s_{F_\mathfrak{q}/F}$ denotes the isomorphism obtained by restriction, by Proposition 8.3.29, $\mathcal{H}(n\mathfrak{q}) \cong H^1_s(F_\mathfrak{q}, T)$. This relation would imply that $\mathrm{res}_{F_\mathfrak{q}/F}(\kappa_{n\mathfrak{q}}) \ne 0$, which would contradict the induction hypothesis. $\square$

The Sheaf of Stub Selmer Modules

(8.3.37) Let $\mathcal{H}$ be the Selmer sheaf attached to the Selmer triple $(T, \mathcal{F}, \mathcal{L})$ (see (7.1.18)). Then $\mathcal{H}$ depends on a choice of finite-singular comparison maps. Because of this, the restriction of $\mathcal{H}$ to the stub Selmer modules is not usually sheaf unlike for the stub Selmer sheaf of [54]. For this reason, we have to select carefully the finite-singular comparison maps in order to match up the images of the vertex to edge maps of the simplicial sheaf $\mathcal{H}$ in order to define the sheaf of stub Selmer modules as a subsheaf of $\mathcal{H}$.

8.3 Kolyvagin Systems over Artinian Principal Ideal Rings

Definition 8.3.38 The *sheaf of stub Selmer modules* $\text{St}(\mathcal{H})$ is the subsheaf of $\mathcal{H}$ where the finite singular comparison isomorphisms $\phi_\mathfrak{q}^{\text{fs}}$ are selected as follows.

Recall that the graph G has vertices corresponding to the elements of $\mathcal{M}_{0,F}$ and an edge joining $n, n\mathfrak{q} \in \mathcal{M}_{0,F}$ whenever $\mathfrak{q} \in \mathcal{M}_{0,F}$ is a prime cycle. The modules $\mathcal{H}(n)$ for $n \in \mathcal{M}_{0,F}$ and $\mathcal{H}(e)$ for an edge e joining $n, n\mathfrak{q} \in \mathcal{M}_{0,F}$ as well as the vertex to edge maps are defined in (7.1.18).

For the stub Selmer sheaf, then we put:

- $\text{St}(\mathcal{H})(n) = \text{St}(n) \otimes \Gamma_n = \text{St}(n) \subseteq \mathcal{H}(n)$ if $n \in \mathcal{M}_{0,F}$;
- If e is the edge of G joining n and $n\mathfrak{q}$, then $\text{St}(\mathcal{H})(e)$ is the image of $\text{St}(n\mathfrak{q})$ in $\mathcal{H}(e) = H_s^1(F_\mathfrak{q}, T) \otimes \Gamma_{n\mathfrak{q}}$ under the vertex to edge map of $\mathcal{H}$ which is is localization at $\mathfrak{q}$ followed by the projection onto the singular quotient, $\text{St}(\mathcal{H})(n\mathfrak{q}) \to \text{St}(\mathcal{H})(e)$.

The vertex to edge maps of $\text{St}(\mathcal{H})$ is the restrictions of those of $\mathcal{H}$, namely:

- The map $\text{St}(f_{n\mathfrak{q},e}) : \text{St}(\mathcal{H})(n\mathfrak{q}) \to \text{St}(\mathcal{H})(e)$ is localization at $\mathfrak{q}$ followed by the projection onto the singular quotient;
- The map $\text{St}(f_{n,e}) : \text{St}(\mathcal{H})(n) \to \text{St}(\mathcal{H})(e)$ is the localization at $\mathfrak{q}$ followed by a finite singular comparison isomorphism $\phi_\mathfrak{q}^{\text{fs}}$ (see (7.1.18)) and is obtained by restriction of the composite map

$$H^1_{\mathcal{F}(n)}(F, T/I_n T) \otimes \Gamma_n \to H^1_f(F_\mathfrak{q}, T/I_{n\mathfrak{q}}T) \otimes \Gamma_n \xrightarrow{\phi_\mathfrak{q}^{\text{fs}}} H^1_s(F_\mathfrak{q}, T/I_{n\mathfrak{q}}T) \otimes \Gamma_{n\mathfrak{q}}$$

to $\text{St}(\mathcal{H})(n)$. Here the first map is localization at $\mathfrak{q}$, and the second map $\phi_\mathfrak{q}^{\text{fs}}$ is a finite singular comparison isomorphism. The isomorphism $\phi_\mathfrak{q}^{\text{fs}}$ is here chosen, so that the image of $\text{St}(\mathcal{H})(n) \to \text{St}(\mathcal{H})(e)$ coincides with the image of $\text{St}(f_{n\mathfrak{q},e}) : \text{St}(\mathcal{H})(n\mathfrak{q}) \to \text{St}(\mathcal{H})(e)$ where both these images are cyclic R-modules of the same length by Proposition 8.3.25.

The vertex to edge map $\text{St}(f_{n,e})$ is well defined by a suitable choice of $\phi_\mathfrak{q}^{\text{fs}}$. Both vertex to edge maps $\text{St}(f_{n,e})$ and $\text{St}(f_{n\mathfrak{q},e})$ are surjective because by construction the image $\text{St}(\mathcal{H})(e)$ of the composite map

$$\text{St}(n) \xrightarrow{\text{res}_{F_\mathfrak{q}/F}} H^1_f(F_\mathfrak{q}, T) \xrightarrow{\phi_\mathfrak{q}^{\text{fs}}} H^1_s(F_\mathfrak{q}, T)$$

is equal to the image of

$$\text{St}(n\mathfrak{q}) \xrightarrow{\text{res}_{F_\mathfrak{q}/F}} H^1_s(F_\mathfrak{q}, T).$$

8.3.39. Bibliographical Remarks. Parts of this section and the next Sect. 8.4 are partially based on the references, which are restricted to the ground field $\mathbb{Q}$, [30, §§1.5, 1.6] and [82].

8.4 Kolyvagin Systems over Discrete Valuation Rings

(8.4.1) In this section, we show an essential connection between Kolyvagin systems and the stub Selmer module. Namely, we demonstrate that a Kolyvagin system, taken to be a global section of the Selmer sheaf and under hypotheses, is in fact a global section of the subsheaf of stub Selmer modules (see Proposition 8.4.14).

(8.4.2) The notation is the same as in the previous section (8.3.1) with the essential difference that the coefficient ring R is now a mixed characteristic discrete valuation ring. We then have that

k is a global field equipped with the special set of places ∞;
F is an imaginary quadratic field extension of k with respect to ∞;
R is a coefficient ring (see (4.1.2)) which is a mixed characteristic complete discrete valuation ring with local parameter π and maximal ideal $\mathfrak{m}$ and where the residue characteristic l of R is different from 2 and the characteristic of k;
T is an R-representation of G_F;
$(T, \mathcal{F}, \mathcal{L})$ is a Selmer triple on F of the Heegner point and CM point type given by Definition 7.3.3, (7.3.4), (7.3.2) and for the above imaginary quadratic extension F/k and the integers $d_\mathfrak{q}$ for all $\mathfrak{q} \in \mathcal{L}$, defined in (7.3.4).

We assume for the rest of this section Sect. 8.4 that

- $(T, \mathcal{F}, \mathcal{L})$ is such that both $(T^*, \mathcal{F}^*, \mathcal{L})$ and $(T, \mathcal{F}, \mathcal{L})$ satisfy the hypotheses H.0-H.5 of Sect. 8.1 where $T \cong T^*$ by Remark 8.1.11;
- $\tau \in G_k$ is the element given by hypothesis H.4(b);
- $\mathcal{L}$ contains all but finitely many primes of F lying over primes of k unramified and inert in F/k;
- $\kappa \in \mathbf{KS}(T, \mathcal{F}, \mathcal{L})$ is a Kolyvagin system for $(T, \mathcal{F}, \mathcal{L})$.

For any $i \in \mathbb{N}$, define

$$R^{(i)} = R/\mathfrak{m}^i, \ T^{(i)} = T/\mathfrak{m}^i T, \ \mathcal{L}^{(i)} = \mathcal{L}_i(T), \mathcal{M}_{0,F}^{(i)} = \mathcal{M}_{0,F}(\mathcal{L}^{(i)})$$

where the last means that $\mathcal{M}_{0,F}^{(i)}$ is the set of squarefree cycles formed of primes $\mathfrak{q}$ of $\mathcal{L}^{(i)}$ lying over primes $\mathfrak{p}$ of k which are inert and unramified in F/k (see (7.3.4)).

Remarks 8.4.3

(i) This Selmer triple $(T, \mathcal{F}, \mathcal{L})$ on F of Heegner point and CM point type Sect. 7.3 with imaginary quadratic extension F/k then has the integers $d_\mathfrak{q}$ for all $\mathfrak{q} \in \mathcal{L}$ defined in (7.3.4).

8.4 Kolyvagin Systems over Discrete Valuation Rings

Associated with these $(T, \mathcal{F}, \mathcal{L})$, $d_\mathfrak{q}$, there is an abelian variety J/k, where $T = T_l(J) \otimes_{\mathbb{Z}_l} R$, and the uniquely determined objects $\mathcal{L}_i$, Γ_n, $\mathcal{M}_{i,F}$, I_n which are constructed exactly as in Sect. 7.1 or (7.3.4) from these $(T, \mathcal{F}, \mathcal{L})$, F/k and $d_\mathfrak{q}$.

(ii) The sets $\mathcal{L}_i$ are the subsets of $\mathcal{L}$ defined in (7.1.4), for all $i \in \mathbb{N}$ where $\mathcal{L}_0 = \mathcal{L}$. Also $\mathcal{M}_{i,F}$, for $i \in \mathbb{N}$ is the set of squarefree cycles which are products of distinct primes q of $\mathcal{L}_i$ lying over primes $\mathfrak{p}$ of k which are inert and and unramified in F/k.

(iii) As T is $T_l \otimes_{\mathbb{Z}_l} R$ where T_l is the l-adic Tate module of the abelian variety J/k mentioned above, the hypothesis H.0 shows that J/k is an elliptic curve.

(iv) The hypothesis of (8.4.2) that $\mathcal{L}$ contains all but finitely many primes of F lying over primes of k unramified and inert in F/k corresponds to, and replaces, the hypothesis of [30, §1.6] that $\mathcal{L}_s(T) \subseteq \mathcal{L}$ for all sufficiently large s in the notation of [30].

(v) At the end of the last section Sect. 8.3, some results have the assumption that $\mathcal{L} = \mathcal{L}_L$ contains all but finitely primes q of F which lie over primes of k which are inert and unramified in F/k and such that $\mathrm{Fr}_\mathfrak{q}$ acts trivially on T (see Proposition 8.3.31 et seq.). For this reason, the hypothesis that $\mathcal{L}$ contains all but finitely many primes of F lying over primes of k unramified and inert in F/k is included as an hypothesis in this section Sect. 8.4 in (8.4.2).

By definition (see (7.4.4) and remark 7.3.6(i)), the sets $\mathcal{L}_i$ for $i \geq 1$ only contain primes of F which lie over primes of k which are inert and unramified in F/k. Here by definition $\mathcal{L}_0 = \mathcal{L}$ is any set of primes of F disjoint from the finite set $S(\mathcal{F})$ (see Definition 7.3.3). Therefore, this hypothesis on $\mathcal{L}$ in (8.4.2) is compatible with the definition of the sets $\mathcal{L}_i$ for $i \geq 1$.

(8.4.4) If M is an R-module with an action by τ where τ restricted to M has order 2, denote by M^+, M^- the isotypical components of M on which τ acts like $+1, -1$, respectively. Write

$$\mathcal{H}_b^a(c) = H^1_{\mathcal{F}_b^a(c)}(F, T), \overline{\mathcal{H}}_b^a(c) = H^1_{\mathcal{F}_b^a(c)}(F, T/\mathfrak{m}T)$$

$$\mathcal{H}_b^{(i),a}(c) = H^1_{\mathcal{F}_b^a(c)}(F, T/\mathfrak{m}^i T).$$

If anyone of a, b, c is equal to 1, then it will be omitted from the notation.

Kolyvagin Systems and the Stub Selmer Module

(8.4.5) In this subsection, we show that for a Kolyvagin system which by definition is a global section of the Selmer sheaf of $(T, \mathcal{F}, \mathcal{L})$, is in fact a global section of the sheaf of stub Selmer modules (Proposition 8.4.14). This is applied in the second part of this section to give one of the main results Theorem 8.4.24.

Proposition 8.4.6 *The Selmer triple $(T^{(i)}, \mathcal{F}, \mathcal{L}^{(i)})$ satisfies the hypotheses H.0-H.5.*

Proof It is readily checked that these hypotheses H.0-H.5 persist under the base change $R \to R^{(i)}$ of the coefficient ring, so the result follows as $(T, \mathcal{F}, \mathcal{L})$ is a Selmer triple which satisfies hypotheses H.0-H.5 (see remark 8.1.3(i)). □

Proposition 8.4.7 *For $n \in \mathcal{M}_{0,F}^{(i)}$, we have the decomposition of R-modules*

$$\mathcal{H}^{(i)}(n) \cong R^{(i),\epsilon} \oplus M^{(i)}(n) \oplus M^{(i)}(n) \tag{8.4.8}$$

in which the integer $\epsilon \in \{0, 1\}$ is independent of i and n.

Proof The decomposition follows from Proposition 8.4.6 and Corollary 8.3.18. That the integer $\epsilon \in \{0, 1\}$ is independent of i and n follows from Corollary 8.3.18 or Proposition 8.3.15(iii). □

Definition 8.4.9 In view of the decomposition (8.4.8) of $\mathcal{H}^{(i)}(n) = H^1_{\mathcal{F}(n)}(F, T^{(i)})$, put

$$\lambda^{(i)}(n) = \text{length}_R(M^{(i)}(n)), \ \text{St}^{(i)}(n) = \mathfrak{m}^{\lambda^{(i)}(n)} \mathcal{H}^{(i)}(n) \tag{8.4.10}$$

where $\text{St}^{(i)}(n)$ is the stub Selmer module for $T^{(i)}$ (see Definition 8.3.19).

We say that $(T, \mathcal{F}, \mathcal{L})$ is *odd* (resp., *even*) if $(T^{(i)}, \mathcal{F}, \mathcal{L}^{(i)})$ is odd (resp., even) for all i where this parity is independent of i by Proposition 8.4.7 (cf. Definition 8.3.20).

Proposition 8.4.11 *The reduction mod $\mathfrak{m}^i$ of the Kolyvagin system κ is a Kolyvagin system $\kappa^{(i)} \in \mathbf{KS}(T^{(i)}, \mathcal{F}, \mathcal{L}^{(i)})$.*

Proof This follows from the given Kolyvagin system κ and the functorial properties of Kolyvagin systems of (7.1.11). □

Proposition 8.4.12 *Suppose that $1 \leq i \leq h$ where $i, h \in \mathbb{N}$, $n \in \mathcal{M}_{0,F}^{(h)}$ and $\text{St}^{(i)}(n) \neq 0$. Then in the R-module decomposition for all $i \leq j \leq h$ provided by Proposition 8.4.7*

$$\mathcal{H}^{(j)}(n) \cong R^{(j),\epsilon} \oplus M^{(j)}(n) \oplus M^{(j)}(n) \tag{8.4.13}$$

we have:

(i) *$\epsilon = 1$;*
(ii) *$\text{St}^{(j)}(n) \neq 0$ for all j such that $i \leq j \leq h$;*
(iii) *$M^{(i)}(n) \cong M^{(j)}(n)$ for all $i \leq j \leq h$;*
(iv) *if $h \geq 2i - 1$ and $n \in \mathcal{M}_{0,F}^{(h)}$, then the image of the natural map $H^1_{\mathcal{F}(n)}(F, T^{(h)}) \to H^1_{\mathcal{F}(n)}(F, T^{(i)})$ is a free $R^{(i)}$-module of rank 1 and, in particular, $\kappa_n^{(i)}$ belongs to a non-zero free $R^{(i)}$-submodule of $\mathcal{H}^{(i)}(n)$.*

8.4 Kolyvagin Systems over Discrete Valuation Rings

Proof In the decomposition $H^1_{\mathcal{F}(n)}(F, T^{(k)}) \cong R^{(k),\epsilon} \oplus M^{(k)}(n) \oplus M^{(k)}(n)$, the integer $\epsilon \in \{0, 1\}$ is independent of $k \in \mathbb{N}$ and n and $M^{(k)}(n)$ is a finite $R^{(k)}$-module depending on k and n.

Plainly $\mathrm{St}^{(i)}(n) \neq 0$ if and only if $\mathrm{length}_R(M^{(i)}(n)) < i$ and $\epsilon = 1$. Therefore (i) holds. As $H^1_{\mathcal{F}(n)}(F, T^{(i)})$ can be identified with $H^1_{\mathcal{F}(n)}(F, T^{(h)})[\mathfrak{m}^i]$ (by Lemma 8.1.16) and because $\mathrm{length}_R(M^{(i)}(n)) < i$, we must have that $\mathrm{length}_R(M^{(h)}(n)) < i$. It follows that $\mathrm{St}^{(h)}(n) \neq 0$ for $h \geq i$ so that (ii) holds.

Under the identification $H^1_{\mathcal{F}(n)}(F, T^{(i)}) \cong H^1_{\mathcal{F}(n)}(F, T^{(h)})[\mathfrak{m}^i]$ of Lemma 8.1.16, as $\epsilon = 1$, it follows that there is an isomorphism

$$M^{(i)}(n) \cong M^{(h)}(n)[\mathfrak{m}^i].$$

As $\mathrm{length}(M^{(i)}(n)) < i$, it follows that $M^{(i)} \cong M^{(j)}$ for all $i \leq j \leq h$ and (iii) holds.

Suppose finally that $h \geq 2i - 1$. Under the identification $H^1_{\mathcal{F}(n)}(F, T^{(i)}) \cong H^1_{\mathcal{F}(n)}(F, T^{(h)})[\mathfrak{m}^i]$ of Lemma 8.1.16, the map $H^1_{\mathcal{F}(n)}(F, T^{(h)}) \to H^1_{\mathcal{F}(n)}(F, T^{(i)})$ may be identified with multiplication by π^{h-i} and as this map annihilates $M^{(h)}(n) \cong M^{(i)}(n)$ (by (iii)) it follows that its image is isomorphic to $R^{(i)}$. As $\kappa_n^{(i)}$ is the image of $\kappa_n^{(h)}$ under this map, it also follows that $\kappa_n^{(i)}$ lies in a submodule isomorphic to $R^{(i)}$ of $H^1_{\mathcal{F}(n)}(F, T^{(i)})$. □

Proposition 8.4.14 *Suppose that $\mathcal{L}$ contains all but finitely primes $\mathfrak{q}$ of F which lie over primes of k which are inert and unramified in F/k. If $i \geq 1$ and $n \in \mathcal{M}_{0,F}^{(2i-1)}$, then we have $\kappa_n^{(i)} \in \mathrm{St}^{(i)}(n) \otimes \Gamma_n$.*

Proof Let $n \in \mathcal{M}_{0,F}^{(2i-1)}$. Put (as in Definition 8.3.7)

$$\rho(n) = \dim_{R/\mathfrak{m}} \overline{\mathcal{H}}(n)$$

$$\rho^\delta(n) = \dim_{R/\mathfrak{m}} \overline{\mathcal{H}}(n)^\delta \qquad (\delta = \pm).$$

Fix a generator for the cyclic group $\Gamma_\mathfrak{q}$ for every $\mathfrak{q} \in \mathcal{M}_{0,F}^{(2i-1)}$, so that we make the identification $H^1_{\mathcal{F}(n)}(F, T^{(i)}) \otimes \Gamma_n \cong H^1_{\mathcal{F}(n)}(F, T^{(i)})$.

From Proposition 8.4.7, for $n \in \mathcal{M}_{0,F}^{(i)}$, we have the decomposition of R-modules

$$\mathcal{H}^{(i)}(n) \cong R^{(i),\epsilon} \oplus M^{(i)}(n) \oplus M^{(i)}(n)$$

in which the integer $\epsilon \in \{0, 1\}$ is independent of i and n.

Principal Case. First suppose $\mathrm{St}^{(i)}(n) \neq 0$ (see (8.4.10)).

In this case, we proceed by induction on the integer $i \geq 1$. If $i = 1$, then evidently we have $\kappa_n^{(1)} \in \mathrm{St}^{(1)}(n) \otimes \Gamma_n$ for then $\mathrm{St}^{(1)}(n) = \overline{\mathcal{H}}(n)$.

We may then assume that $i \geq 2$. We must have that $\epsilon = 1$ and $\lambda^{(i)}(n) < i$. Let $j = \lambda^{(i)}(n)$ so that $j < i$. By the induction hypothesis, we have $\kappa_n^{(j)} \in \mathrm{St}^{(j)}(n)$. By Lemma 8.1.16, we have an isomorphism of R-modules $M^{(j)} \cong M^{(i)}[\mathfrak{m}^j] = M^{(i)}$, so that $\lambda^{(j)}(n) = \lambda^{(i)}(n) = j$. This implies that $\mathrm{St}^{(j)}(n) = 0$, and so $\kappa_n^{(j)} = 0$. Again from Lemma 8.1.16, this is equivalent to $\pi^{i-j}\kappa_n^{(i)} = 0$. Now as $n \in \mathcal{M}_{0,F}^{(2i-1)}$, by Proposition 8.4.12(iv), we have that $\kappa_n^{(i)}$ belongs to a non-zero free $R^{(i)}$-submodule of $\mathcal{H}^{(i)}(n)$ and therefore $\kappa_n^{(i)}$ is divisible by π^j in $\mathcal{H}^{(i)}(n)$. Therefore, we have that

$$\kappa_n^{(i)} \in \mathrm{St}^{(i)}(n) = \mathfrak{m}^{\lambda^{(i)}(n)} \mathcal{H}^{(i)}(n) = \mathfrak{m}^j \mathcal{H}^{(i)}(n)$$

proving the result in this case.

It remains to show that if $\mathrm{St}^{(i)}(n) = 0$, then $\kappa_n^{(i)} = 0$, and this is divided into three sub-cases.

Case 1. Suppose that $\rho(n) \leq 1$ and $\mathrm{St}^{(i)}(n) = 0$.

If $\rho(n) = 1$, then we have $\mathcal{H}^{(i)}(n) \cong R^{(i)} \cong \mathrm{St}^{(i)}(n) \cong 0$ which is impossible, so that we must have $\rho(n) = 0$. But then it follows that $\kappa_n^{(i)} \in \mathcal{H}^{(i)}(n) = 0 \subseteq \mathrm{St}^{(i)}(n)$ as required.

Case 2: Suppose that $\rho(n)^+$ and $\rho(n)^-$ are both non-zero and $\mathrm{St}^{(i)}(n) = 0$.

In this case, we proceed by induction on the integer $\rho(n)$. We may have, as $\rho(n)^+$ and $\rho(n)^-$ are both non-zero, that $\rho(n) > 1$. Using Lemma 8.1.16, we may identify $H^1_{\mathcal{F}(n)}(F, T^{(i)})[\mathfrak{m}] \cong H^1_{\mathcal{F}(n)}(F, T/\mathfrak{m}T)$.

Suppose that $\kappa_n^{(i)} \neq 0$; then we have that $\kappa_n^{(i)}$ has some non-zero R-multiple which has at least one τ-eigencomponent $d \in (\mathcal{H}^{(i)}(n)[\mathfrak{m}])^\zeta$ which is non-zero where ζ is either $+$ or $-$. By Proposition 8.3.31, we may choose a prime $\mathfrak{q} \in \mathcal{L}^{(2i-1)}$ at which both

(a) $\mathrm{res}_{F_\mathfrak{q}/F}(d) \neq 0$

(b) the localization map $H^1_{\mathcal{F}(n)}(F, T/\mathfrak{m}T)^{-\zeta} \xrightarrow{\mathrm{res}_{F_\mathfrak{q}/F}} H^1(F_\mathfrak{q}, T/\mathfrak{m}T)$ is non-zero. Both eigenspaces $\overline{\mathcal{H}}(n)^\pm$ are non-zero by hypothesis. Therefore, we have $\mathrm{res}_{F_\mathfrak{q}/F}(\overline{\mathcal{H}}(n)^\delta) \neq 0$ for $\delta = +$ and $\delta = -$. By Proposition 8.3.8, we have $\rho(n\mathfrak{q}) = \rho(n) - 2$, and so by the induction hypothesis, we have $\kappa_{n\mathfrak{q}}^{(i)} \in \mathrm{St}^{(i)}(n\mathfrak{q})$. Because $\mathrm{St}^{(i)}(n) = 0$, Proposition 8.3.25 implies that $\mathrm{res}_{F_\mathfrak{q}/F}(\mathrm{St}^{(i)}(n\mathfrak{q})) = 0$ and hence $\mathrm{res}_{F_\mathfrak{q}/F}(\kappa_{n\mathfrak{q}}^{(i)}) = 0$. But then the Kolyvagin system relations given by the diagram (7.1.8) imply that $\mathrm{res}_{F_\mathfrak{q}/F}(\kappa_n^{(i)}) = 0$, contradicting the choice of $\mathfrak{q}$. Hence, we have $\kappa_n^{(i)} = 0 \in \mathrm{St}^{(i)}(n)$.

Case 3: Suppose that one of $\rho^\pm(n)$ is equal to zero and $\mathrm{St}^{(i)}(n) = 0$.

We again argue by induction on the integer $\rho(n)$. From Case 1, we may assume that $\rho(n) > 1$. Suppose $\rho^{-\zeta}(n) = 0$ where ζ is either $+$ or $-$ and $-\zeta$ is the opposite sign; we then have $\rho^\zeta(n) > 1$.

Suppose that $\kappa_n^{(i)} \neq 0$; we may choose a non-zero R-multiple of $\kappa_n^{(i)}$, $d \in \mathcal{H}^{(i)}(n)[\mathfrak{m}]^\zeta$, and by Proposition 8.3.31, a prime $\mathfrak{q} \in \mathcal{L}^{(2i-1)}$ for which $\mathrm{res}_{F_\mathfrak{q}/F}(d) \neq 0$. By Proposition 8.3.8, $\rho^\pm(n\mathfrak{q})$ are both nonzero and $\rho(n\mathfrak{q}) = \rho(n)$. Thus, by Case 2 applied to $n\mathfrak{q}$, we have that $\kappa_{n\mathfrak{q}}^{(i)} \in \mathrm{St}^{(i)}(n\mathfrak{q})$. Because $\mathrm{St}^{(i)}(n) = 0$, Proposition 8.3.25 implies that we have $\mathrm{res}_{F_\mathfrak{q}/F}(\mathrm{St}^{(i)}(n\mathfrak{q})) = 0$ and hence $\mathrm{res}_{F_\mathfrak{q}/F}(\kappa_{n\mathfrak{q}}^{(i)}) = 0$. But the Kolyvagin system relations given by the diagram (7.1.8) imply that $\mathrm{res}_{F_\mathfrak{q}/F}(\kappa_{n\mathfrak{q}}^{(i)}) \neq 0$. This contradiction shows that $\kappa_n^{(i)} = 0$ and hence $\kappa_n^{(i)} = 0 \in \mathrm{St}^{(i)}(n)$. □

8.5 Selmer Groups over Artinian Principal Ideal Rings

(8.5.1) In this section, we demonstrate two main results (Theorem 8.5.16 and Corollary 8.5.40) on the decomposition of Selmer groups for Heegner point and CM point Kolyvagin systems obtained from a Heegner system where the coefficient ring is an artinian principal ideal ring.

(8.5.2) The notation is almost exactly the same in this Sect. 8.5 as in the previous section Sect. 8.4, (8.4.2), and the only differences are that the coefficient ring R is now assumed to be an artinian ring which is quotient of a mixed characteristic discrete valuation ring and that $\mathcal{L}$ satisfies a corresponding different hypothesis. To be precise, we have that

k is a global field equipped with the special set of places ∞;
F is an imaginary quadratic field extension of k with respect to ∞;
R is an artinian principal ideal ring (see (4.1.2)) which is quotient of a mixed characteristic discrete valuation ring where R has local parameter π and maximal ideal $\mathfrak{m}$ and where the residue characteristic l of R is different from 2 and the characteristic of k;
T is an R-representation of G_F;
$(T, \mathcal{F}, \mathcal{L})$ is a Selmer triple on F of the Heegner point and CM point type given by Definition 7.3.3, (7.3.4), (7.3.2) and for the above imaginary quadratic extension F/k and the integers $d_\mathfrak{q}$ for all $\mathfrak{q} \in \mathcal{L}$, defined in (7.3.4);

We assume for the rest of this section Sect. 8.5 that

- $(T, \mathcal{F}, \mathcal{L})$ is such that both $(T^*, \mathcal{F}^*, \mathcal{L})$ and $(T, \mathcal{F}, \mathcal{L})$ satisfy the hypotheses H.0-H.5 of Sect. 8.1 where $T \cong T^*$ by Remark 8.1.11;
- $\tau \in G_k$ is the element given by hypothesis H.4(b);
- $\mathcal{L}$ satisfies $\mathcal{L} = \mathcal{L}_L$, where L is the length of R, and that $\mathcal{L}$ contains all but finitely primes $\mathfrak{q}$ of F which lie over primes of k which are inert and unramified in F/k and such that $\mathrm{Fr}_\mathfrak{q}$ acts trivially on T;
- $\kappa \in \mathbf{KS}(T, \mathcal{F}, \mathcal{L})$ is a Kolyvagin system for $(T, \mathcal{F}, \mathcal{L})$.

For any $i \in \mathbb{N}$, define

$$R^{(i)} = R/\mathfrak{m}^i, \quad T^{(i)} = T/\mathfrak{m}^i T, \quad \mathcal{L}^{(i)} = \mathcal{L}_i(T), \quad \mathcal{M}_{0,F}^{(i)} = \mathcal{M}_{0,F}(\mathcal{L}^{(i)})$$

where the last means that $\mathcal{M}_{0,F}^{(i)}$ is the set of squarefree cycles formed of primes q of $\mathcal{L}^{(i)}$ lying over primes p of k which are inert and and unramified in F/k (see (7.3.4)).

Definition 8.5.3 Let $\mathcal{S}$ be a simplicial sheaf on a graph G with values in the category of R-modules as in Sect. 7.1 Definition 7.1.17. If v, w are vertices of G a *path* from v to w is a sequence of vertices $v = u_1, u_2, \ldots, u_n = w$ of G for some $n \geq 1$ beginning at v and ending at w such that u_i, u_{i+1} is joined by an edge of G for all i.

The sheaf $\mathcal{S}$ is said to be *locally cyclic* if all R-modules $\mathcal{S}(v), \mathcal{S}(e)$ are cyclic for all vertices v and edges e of G and all maps $f_{v,e} : \mathcal{S}(v) \to \mathcal{S}(e)$ are surjective whenever e is an edge with v as an endpoint.

If the sheaf $\mathcal{S}$ is locally cyclic, then a *surjective path* $P = v_1, \ldots, v_n$ is a path in G such that if v_i and v_{i+1} are joined by an edge e_i then $f_{v_{i+1},e_i} : \mathcal{S}(v_{i+1}) \to \mathcal{S}(e)$ is an isomorphism. If P is a surjective path from v_1 to v_n, then there is a surjective homomorphism of R-modules $\psi_P : \mathcal{S}(v_1) \to \mathcal{S}(v_n)$ given by

$$\psi_P = f_{v_n,e_{n-1}}^{-1} \circ f_{v_{n-1},e_{n-1}} \circ f_{v_{n-1},e_{n-2}}^{-1} \circ f_{v_{n-2},e_{n-2}} \circ \ldots \circ f_{v_2,e_1}^{-1} \circ f_{v_1,e_1}$$

where all maps f_{v_{i+1},e_i} are by definition isomorphisms.

A vertex v is a *hub* of the locally cyclic sheaf $\mathcal{S}$ if for every vertex w there is a surjective path from v to w.

Proposition 8.5.4 *Let* $\mathrm{St}(\mathcal{H})$ *be the stub Selmer sheaf (Definition 8.3.38) and suppose that* $n \in \mathcal{M}_{0,F}$ *is a hub of* $\mathrm{St}(\mathcal{H})$. *Then we have:*

(i) $\mathrm{St}(\mathcal{H})$ *is locally cyclic;*
(ii) *The map* $\Psi_n : \Gamma(\mathrm{St}(\mathcal{H})) \to \mathrm{St}(\mathcal{H})(n)$ *defined by* $\kappa \mapsto \kappa_n$ *is injective;*
(iii) *If* $\kappa \in \Gamma(\mathrm{St}(\mathcal{H}))$ *and if* u *is a vertex such that* $\kappa_u \neq 0$ *and* κ_u *generates* $\mathfrak{m}^i \mathrm{St}(\mathcal{H})(u)$ *for some* $i \in \mathbb{N}$ *then* κ_w *generates* $\mathfrak{m}^i \mathrm{St}(\mathcal{H})(w)$ *for every vertex* w.

Proof

(i) It follows from the Definition 8.3.38 of $\mathrm{St}(\mathcal{H})$ that all R-modules $\mathrm{St}(\mathcal{H})(v)$ and $\mathrm{St}(\mathcal{H})(e)$ are cyclic and all the maps $f_{v,e}$ are surjective (see the discussion in Definition 8.3.38) for all vertices $e \in \mathcal{M}_{0,F}$ and all edges e, where an edge joins two vertices of the form n, $n\mathfrak{q} \in \mathcal{M}_{0,F}$ whenever $\mathfrak{q} \in \mathcal{M}_{0,F}$ is a prime cycle.
(ii) For every vertex w, let P_w be a surjective path joining n to w. If $\kappa \in \Gamma(\mathrm{St}(\mathcal{H}))$, then $\kappa_w = \psi_{P_w}(\kappa_n)$ where n is the given hub. As this holds for any vertex w, it follows that the map $\Psi_n : \Gamma(\mathrm{St}(\mathcal{H})) \to \mathrm{St}(\mathcal{H})(n)$, $\kappa \mapsto \kappa_n$, is injective.
(iii) This follows from the surjectivity of the maps ψ_{P_w} of the proof of part (ii) and Nakayama's lemma. □

8.5 Selmer Groups over Artinian Principal Ideal Rings

Proposition 8.5.5 *Suppose that $(T, \mathcal{F}, \mathcal{L})$ is odd (Definition 8.4.9) and $l > 4$. Then the simplicial sheaf $\mathrm{St}(\mathcal{H})$ (Definition 8.3.38) is locally cyclic, and its underlying graph G is connected, and every $n \in \mathcal{M}_{0,F}$ with $\rho(n) = 1$ (equivalently, n is a core vertex of $\mathcal{H}$ (Definition 8.3.28)) is a hub of $\mathrm{St}(\mathcal{H})$. In particular, $\mathrm{St}(\mathcal{H})$ has a hub.*

Proof The graph G, on which the simplicial sheaf $\mathrm{St}(\mathcal{H})$ is based (Definition 8.3.38), contains the subgraph $\mathcal{X}$ whose vertices are the core vertices (Definition 8.3.28). By Proposition 8.3.34 if $(T, \mathcal{F}, \mathcal{L})$ is odd (Definition 8.3.20), $l > 4$, then the graph $\mathcal{X}$ is connected.

It is already shown in Proposition 8.5.4(i) that $\mathrm{St}(\mathcal{H})$ is locally cyclic (Definition 8.5.3).

It remains to show that if $n \in \mathcal{M}_{0,F}$ and n are a core vertex, then n is a hub of $\mathrm{St}(\mathcal{H})$ (see Definition 8.5.3). Fix such a core vertex n, where the existence of such a core vertex follows obviously from Corollary 8.3.33 or Proposition 8.3.34. Let m be any other cycle in $\mathcal{M}_{0,F}$. We show by induction on $r(m) := \max(\rho^+(m), \rho^-(m))$, with the notation of Definition 8.3.7, that there is a surjective path from n to m (see Definition 8.5.3).

The cycle n is a core vertex and if m is another core vertex, then there is a path in the subgraph $\mathcal{X}$ of G from n to m as $\mathcal{X}$ is connected by Proposition 8.3.34.

Suppose that $m = nq$ where m, n are core vertices and $q \in \mathcal{L}$ is a prime coprime to n. Let e be the edge joining n and nq in $\mathcal{X}$. Then we have $H^1_{\mathcal{F}(n)}(F, \overline{T}) \cong H^1_{\mathcal{F}(n)}(F, T)[m]$ and $H^1_f(F_q, \overline{T}) \cong H^1_f(F_q, T)[m]$. The R-module $H^1_{\mathcal{F}(n)}(F, T)$ is a single eigenspace under the action of $\mathrm{Gal}(F/k)$ so let $\delta = \pm$ be the sign of this representation. Then $H^1_{\mathcal{F}(n)}(F, T)^\delta$ and $H^1_f(F_q, T)^\delta$ are both free R-modules of rank 1.

By Proposition 8.3.29, the restriction maps

$$\mathcal{H}(n) \to H^1(F_q, T), \quad \mathcal{H}(nq) \to H^1(F_q, T)$$

are injective. It follows that the restriction homomorphisms of the free R-modules of rank 1

$$\mathcal{H}(n)^\delta \to H^1_f(F_q, T)^\delta, \quad \mathcal{H}(nq)^{-\delta} \to H^1_s(F_q, T)^{-\delta}$$

are isomorphisms. Hence, as $\mathcal{X}$ is connected, we have that all vertex to edge maps of $\mathcal{X}$ are isomorphisms (see Definition 8.3.38). Therefore, if n and m are core vertices, then there is a surjective path in $\mathcal{X}$ from n to m, and indeed every path in $\mathcal{X}$ is a surjective path.

Suppose now that $m \in \mathcal{M}_{0,M}$ where m is not a core vertex, that is to say $\rho(m) = \rho^+(n) + \rho^-(n) > 1$. We have to show that there is a surjective path from n to m of the simplicial sheaf $\mathrm{St}(\mathcal{H})$. As $\rho(m) > 1$ and $(T, \mathcal{F}, \mathcal{L})$ is odd and so $\rho(m)$ is odd, we have that $\rho^\delta(m) \geq 2$ for some $\delta = \pm$ and hence that $r(m) > 1$.

Then at least one eigencomponent $H^1_{\mathcal{F}(m)}(F, T)^\delta$ for $\delta = \pm$ is non-zero. Indeed as $(T, \mathcal{F}, \mathcal{L})$ is odd, $H^1_{\mathcal{F}(m)}(F, T)^\delta$ for some $\delta = \pm$ contains a copy of $R(\delta)$ which is an indecomposable R-representation of $\mathrm{Gal}(F/k)$ for some $\delta = \pm$ and which is a free R-

module of rank 1. We may then select $c \neq 0$ such that $R(\delta) \cong cR \subseteq H^1_{\mathcal{F}(m)}(F,T)^\delta$ for some $\delta = \pm$.

By Corollary 8.3.33, we may select a prime $q \in \mathcal{L}$ such that $\mathrm{res}_{F_q/F}(c) \neq 0$ and $r(mq) < r(m)$. By the induction hypothesis, there is a surjective path in G from n to mq.

We then have that the localization map $\mathfrak{m}^{L-1} H^1_{\mathcal{F}(m)}(F,T)^\delta \to H^1_f(F_q, T)^\delta$ is non-zero where $L = \mathrm{length}(R)$.

The cyclic R-module $\mathrm{St}(\mathcal{H})(m)$ is a single eigenspace of $\mathrm{Gal}(F/k)$, and let $\delta = \pm$ be the corresponding character. If $\mathrm{St}(\mathcal{H})(m) = 0$, then there is nothing to prove as there is an evident surjective path from mq to m in $\mathrm{St}(\mathcal{H})$, so we may assume that $\mathrm{St}(\mathcal{H})(m)$ is non-zero.

Let e be the path of the sheaf $\mathrm{St}(\mathcal{H})$ joining m and mq. The vertex to edge maps of $\mathrm{St}(\mathcal{H})$ are the following (see Definition 8.3.38)

- The map $\mathrm{St}(f_{mq,e}) : \mathrm{St}(\mathcal{H})(mq) \to \mathrm{St}(\mathcal{H})(e)$ is localization at q followed by the projection onto the singular quotient;
- The map $\mathrm{St}(f_{m,e}) : \mathrm{St}(\mathcal{H})(m) \to \mathrm{St}(\mathcal{H})(e)$ is localization at q followed by a finite singular comparison isomorphism ϕ_q^{fs} (see (7.1.18)) and is obtained by the restriction to $\mathrm{St}(\mathcal{H})(m)$ of the composite map

$$H^1_{\mathcal{F}(m)}(F, T/I_m T) \otimes \Gamma_m \to H^1_f(F_q, T/I_{mq}T) \otimes \Gamma_m \xrightarrow{\phi_q^{\mathrm{fs}}} H^1_s(F_q, T/I_{mq}T) \otimes \Gamma_{mq}$$

The maps here are explained in detail in Definition 8.3.38. The finite singular comparison isomorphism ϕ_q^{fs} is here to chosen so that the image of $\mathrm{St}(\mathcal{H})(m) \to \mathrm{St}(\mathcal{H})(e)$ coincides with the image of $\mathrm{St}(f_{mq,e}) : \mathrm{St}(\mathcal{H})(mq) \to \mathrm{St}(\mathcal{H})(e)$ where both these images are cyclic R-modules of the same length by Proposition 8.3.25. Furthermore, as $\mathrm{St}(\mathcal{H})(mq)$ and $\mathrm{St}(\mathcal{H})(m)$ are in opposite eigenspaces under the action of $\mathrm{Gal}(F/k)$ the isomorphism ϕ_q^{fs} has to be chosen, so that it interchanges opposite eigenspaces.

As q is chosen so that the localization map $\mathfrak{m}^{L-1} H^1_{\mathcal{F}(m)}(F,T)^\delta \to H^1_f(F_q, T)^\delta$ is non-zero, it follows that $H^1_{\mathcal{F}(m)}(F,T)^\delta \to H^1_f(F_q, T)^\delta$ is surjective, because $H^1_f(F_q, T)^\delta$ is a free R-module of rank 1.

Therefore, we have that $\mathfrak{m}^{\lambda(m)} H^1_{\mathcal{F}(m)}(F,T) \to \mathfrak{m}^{\lambda(m)} H^1_f(F_q, T)$ is also surjective. These two R-modules $\mathfrak{m}^{\lambda(m)} H^1_{\mathcal{F}(m)}(F,T), \mathfrak{m}^{\lambda(m)} H^1_f(F_q, T)$ are cyclic of length $\max(0, L - \lambda(m))$; hence $\mathfrak{m}^{\lambda(m)} H^1_{\mathcal{F}(m)}(F,T) \to \mathfrak{m}^{\lambda(m)} H^1_f(F_q, T)$ is an isomorphism. Therefore, if e is the edge joining m and mq, we have that the map $\mathrm{St}(\mathcal{H})(m) \to \mathrm{St}(\mathcal{H})(e)$ is an isomorphism. Therefore, the path from mq to m is a surjective path, so by composition, we obtain a surjective path from n to m. □

Definition 8.5.6 The Selmer triple $(T, \mathcal{F}, \mathcal{L})$ is said to be *liftable to a discrete valuation ring* if there is a complete mixed characteristic discrete valuation ring D where R is a quotient of D and a Selmer triple $(\widetilde{T}, \widetilde{\mathcal{F}}, \widetilde{\mathcal{L}})$ over D such that the following conditions hold:

8.5 Selmer Groups over Artinian Principal Ideal Rings

- The Selmer triple $(\widetilde{T}, \widetilde{\mathcal{F}}, \widetilde{\mathcal{L}})$ over D is of Heegner point and CM point type as in (7.3.4) and we have $T \cong \widetilde{T} \otimes_D R$, $\mathcal{F} = \widetilde{\mathcal{F}} \otimes_D R$, $\mathcal{L} = \widetilde{\mathcal{L}}_i(\widetilde{T})$ where i is the length of R (see (8.4.2));
- $\widetilde{\mathcal{L}}$ contains all but finitely many primes of F lying over primes of k unramified and inert in F/k;
- Both $(\widetilde{T}^*, \widetilde{\mathcal{F}}^*, \widetilde{\mathcal{L}}^*)$ and $(\widetilde{T}, \widetilde{\mathcal{F}}, \widetilde{\mathcal{L}})$ satisfy the hypotheses H.0-H.5 of Sect. 8.1 where $\widetilde{T} \cong \widetilde{T}^*$ (by Remark 8.1.11).

Proposition 8.5.7 *Suppose that $(T, \mathcal{F}, \mathcal{L})$ over R is odd (Definitions 8.3.20 and 8.4.9) and $l > 4$. Then we have:*

(i) *The R-module $\Gamma(\operatorname{St}(\mathcal{H}))$ is cyclic.*
(ii) *Suppose that $\kappa \in \Gamma(\operatorname{St}(\mathcal{H}))$, $m \in \mathcal{M}_{0,F}$ and $\kappa_m \neq 0$. Let $j \geq 0$ be such that κ_m generates $\mathfrak{m}^j \operatorname{St}(\mathcal{H})(m)$. Then κ_r generates $\mathfrak{m}^j \operatorname{St}(\mathcal{H})(r)$ for every $r \in \mathcal{M}_{0,F}$, so that we have*

$$\kappa_r \in \mathfrak{m}^{j+\lambda(r)} \mathcal{H}(r) \text{ for every } r \in \mathcal{M}_{0,F}.$$

(iii) *Suppose that $(T, \mathcal{F}, \mathcal{L})$ is liftable to a Selmer triple $(\widetilde{T}, \widetilde{\mathcal{F}}, \widetilde{\mathcal{L}})$ over a complete mixed characteristic discrete valuation ring D and $i \in \mathbb{N}$ is the length of R. Then there is an isomorphism of R-modules*

$$\mathbf{KS}(T, \mathcal{F}, \mathcal{L}^{(2i-1)}) \cong \Gamma(\operatorname{St}(\mathcal{H})(T, \mathcal{F}, \widetilde{\mathcal{L}}^{(2i-1)}))$$

and $\mathbf{KS}(T, \mathcal{F}, \widetilde{\mathcal{L}}^{(2i-1)}))$ is a cyclic R-module.
   Let $\kappa \in \mathbf{KS}(T, \mathcal{F}, \widetilde{\mathcal{L}}^{(2i-1)}))$ and $m \in \mathcal{M}_{0,F}^{(2i-1)}$ be such that $\kappa_m \neq 0$. Let $j \geq 0$ be such that κ_m generates $\mathfrak{m}^j \operatorname{St}(\mathcal{H}(T, \mathcal{F}, \widetilde{\mathcal{L}}^{(2i-1)}))(m)$. Then κ_r generates $\mathfrak{m}^j \operatorname{St}(\mathcal{H}(T, \mathcal{F}, \widetilde{\mathcal{L}}^{(2i-1)}))(r)$ for every $r \in \mathcal{M}_{0,F}^{(2i-1)}$.

Proof

(i) As $\operatorname{St}(\mathcal{H})$ has a hub $n \in \mathcal{M}_{0,F}$ by Proposition 8.5.5, it follows from Proposition 8.5.4 that the homomorphism of R-modules $\Gamma(\operatorname{St}(\mathcal{H})) \to \operatorname{St}(\mathcal{H})(n) \cong R$, $\kappa \mapsto \kappa_n$, is injective. Hence, there is a non-canonical isomorphism of $\Gamma(\operatorname{St}(\mathcal{H}))$ with an ideal of R.
(ii) By Proposition 8.5.5, $\operatorname{St}(\mathcal{H})$ has a hub $n \in \mathcal{M}_{0,F}$. By Proposition 8.5.4(iii), we have that if $\kappa \in \Gamma(\operatorname{St}(\mathcal{H}))$ and if u is a vertex such that $\kappa_u \neq 0$ and κ_u generates $\mathfrak{m}^i \operatorname{St}(\mathcal{H})(v)$ for some $i \in \mathbb{N}$ then κ_w generates $\mathfrak{m}^i \operatorname{St}(\mathcal{H})(w)$ for every vertex w. As $\operatorname{St}(\mathcal{H})(v)$ is a cyclic R-module, it follows that if $\kappa \in \Gamma(\operatorname{St}(\mathcal{H}))$ then κ_u generates $\mathfrak{m}^i \operatorname{St}(\mathcal{H})(v)$ for some $i \in \mathbb{N}$ and the result follows.

(iii) Assume that $(\widetilde{T}, \widetilde{\mathcal{F}}, \widetilde{\mathcal{L}})$ is a lifting of $(T, \mathcal{F}, \mathcal{L})$ to D and $\mathfrak{n}$ is the maximal ideal of D; in particular, R has length i, and is a quotient of D. There is a core vertex n of $\mathrm{St}(\mathcal{H}(\widetilde{T}^{(2i-1)}, \widetilde{\mathcal{F}}^{(2i-1)}, \widetilde{\mathcal{L}}^{(2i-1)}))$ (by Corollary 8.3.33 or Proposition 8.3.34), where $(\widetilde{T}^{(2i-1)}, \widetilde{\mathcal{F}}^{(2i-1)}, \widetilde{\mathcal{L}}^{(2i-1)})$ is the induced Selmer triple over $D/\mathfrak{n}^{2i-1}$ and $\mathrm{St}(\mathcal{H}(\widetilde{T}^{(2i-1)}, \widetilde{\mathcal{F}}^{(2i-1)}, \widetilde{\mathcal{L}}^{(2i-1)}))$ is the corresponding stub Selmer sheaf. Then n is a hub of $\mathrm{St}(\mathcal{H}(\widetilde{T}^{(2i-1)}, \widetilde{\mathcal{F}}^{(2i-1)}, \widetilde{\mathcal{L}}^{(2i-1)}))$ by Proposition 8.5.5. As $n \in \mathcal{M}_{0,F}^{(2i-1)} \subseteq \mathcal{M}_{0,F}^{(i)}$, we have that n is plainly also a hub of $\mathrm{St}(\mathcal{H}(T^{(i)}, \mathcal{F}^{(i)}, \widetilde{\mathcal{L}}^{(2i-1)}))$, where $(T^{(i)}, \mathcal{F}^{(i)}, \widetilde{\mathcal{L}}^{(2i-1)})$ is a the induced Selmer triple over R and $T^{(i)} = T$, $\mathcal{F}^{(i)} = \mathcal{F}$.

The Kolyvagin systems in $\mathbf{KS}(T, \mathcal{F}, \widetilde{\mathcal{L}}^{(2i-1)})$ are by definition Kolyvagin systems for $(T, \mathcal{F})$ but where the vertices run over the elements of the subset $\mathcal{M}_{0,F}^{(2i-1)}$ of $\mathcal{M}_{0,F}^{(i)}$. By Proposition 8.4.14, we have for any $\kappa \in \mathbf{KS}(T, \mathcal{F}, \widetilde{\mathcal{L}}^{(2i-1)})$ that $\kappa_m \in \mathrm{St}^{(i)}(m) \otimes \Gamma_m$ for all $m \in \mathcal{M}_{0,F}^{(2i-1)}$. Hence, we obtain an injective map of R-modules

$$\mathbf{KS}(T, \mathcal{F}, \widetilde{\mathcal{L}}^{(2i-1)}) \longrightarrow \Gamma(\mathrm{St}^{(i)}(\mathcal{H})(T, \mathcal{F}, \widetilde{\mathcal{L}}^{(2i-1)})).$$

This map is evidently surjective as an element of $\Gamma(\mathrm{St}^{(i)}(\mathcal{H})(T, \mathcal{F}, \widetilde{\mathcal{L}}^{(2i-1)}))$ belongs to $\mathbf{KS}(T, \mathcal{F}, \widetilde{\mathcal{L}}^{(2i-1)})$.

By Proposition 8.5.4 and as n is a hub, we obtain that the map $\Gamma(\mathrm{St}(\mathcal{H})) \to \mathrm{St}(\mathcal{H})(n)$ defined by $\kappa \mapsto \kappa_n$ is injective. Hence we obtain an injection of R-modules

$$\mathbf{KS}(T, \mathcal{F}, \widetilde{\mathcal{L}}^{(2i-1)}) \cong \Gamma(\mathrm{St}^{(i)}(\mathcal{H})(T, \mathcal{F}, \widetilde{\mathcal{L}}^{(2i-1)})) \to \mathrm{St}(\mathcal{H})(n).$$

As $\mathrm{St}(\mathcal{H})(n)$ is a cyclic R-module, it follows that $\mathbf{KS}(T, \mathcal{F}, \widetilde{\mathcal{L}}^{(2i-1)})$ is also a cyclic R-module, as required.

The last part, where $\mathrm{St}(\mathcal{H}(T, \mathcal{F}, \widetilde{\mathcal{L}}^{(2i-1)}))(r)$ denotes the stub Selmer module at r of the Selmer triple $(T, \mathcal{F}, \widetilde{\mathcal{L}}^{(2i-1)})$, follows from this and part (ii). □

Proposition 8.5.8 *Suppose that $\kappa \in \Gamma(\mathrm{St}(\mathcal{H}))$, the Selmer triple $(T, \mathcal{F}, \mathcal{L})$ over R is odd (Definition 8.4.9) and $l > 4$. Then the following are equivalent:*

(i) *There is $m \in \mathcal{M}$ such that $\mathrm{St}(\mathcal{H})(m) \neq 0$ and κ_m generates $\mathrm{St}(\mathcal{H})(m)$.*
(ii) *The element κ_m generates $\mathrm{St}(\mathcal{H})(m)$ for all $m \in \mathcal{M}$.*
(iii) *The image of κ in $\Gamma(\mathrm{St}(\mathcal{H}))(T/\mathfrak{m}T, \mathcal{F}^{(1)}, \mathcal{L}^{(1)})$, by taking the reduction of κ mod $\mathfrak{m}$, is non-zero.*

Proof (i) ⇔ (ii). The equivalence of (i) and (ii) follows from Proposition 8.5.7(ii).

(iii) ⇒ (i). Let $m \in \mathcal{M}_{0,F}$ and $L = \mathrm{length}(R)$. The image of κ_m in $H^1_{\mathcal{F}(m)}(F, T/\mathfrak{m}T)$ is non-zero if and only if $\mathfrak{m}^{L-1}\kappa_m \neq 0$ by Lemma 8.1.16. Therefore, if the image of κ_m in $H^1_{\mathcal{F}(m)}(F, T/\mathfrak{m}T)$ is non-zero, we have that κ_m generates $\mathrm{St}(\mathcal{H})(m)$ and furthermore that $\mathrm{St}(\mathcal{H})(m)$ is a free R-module of rank 1. Therefore (iii) implies (i).

8.5 Selmer Groups over Artinian Principal Ideal Rings

(ii) $\Rightarrow$ (iii) Suppose (ii) holds so that the element κ_m generates $\mathrm{St}(\mathcal{H})(m)$ for all $m \in \mathcal{M}$. Let $m \in \mathcal{M}_{0,F}$ be a core vertex (such a vertex exists by Corollary 8.3.33 or Proposition 8.3.34). Then $\mathrm{St}(\mathcal{H}(m))$ is a free R-module of rank 1 and hence $\mathfrak{m}^{L-1} \kappa_m \neq 0$. Therefore, the image of κ_m in $H^1_{\mathcal{F}(m)}(F, T/\mathfrak{m}T)$ is non-zero. So we have that (ii) implies (iii). □

Definition 8.5.9 The element $\kappa \in \mathbf{KS}(T)$ (respectively, $\kappa \in \Gamma(\mathrm{St}(\mathcal{H}))$) is said to be *primitive* if the image of κ in $\mathbf{KS}(T/\mathfrak{m}T)$ (respectively, in $\Gamma(\mathrm{St}(\mathcal{H}))(T/\mathfrak{m}T, \mathcal{F}^{(1)}, \mathcal{L}^{(1)})$) is non-zero.

Here $\Gamma(\mathrm{St}(\mathcal{H}))(T/\mathfrak{m}T, \mathcal{F}^{(1)}, \mathcal{L}^{(1)})$ means the module of global sections of the stub Selmer sheaf for the Selmer triple $(T/\mathfrak{m}T, \mathcal{F}^{(1)}, \mathcal{L}^{(1)})$.

[For Kolyvagin systems over a discrete valuation ring, primitivity is correspondingly defined in Definition 8.6.8 in the next section].

Proposition 8.5.10 *Suppose that $\kappa \in \Gamma(\mathrm{St}(\mathcal{H}))$ is primitive, $(T, \mathcal{F}, \mathcal{L})$ over R is odd (Definition 8.4.9) and $l > 4$. We have the decomposition $H_{\mathcal{F}}(F, T) \cong R \oplus M(1) \oplus M(1)$ (Proposition 8.3.15). Put $\lambda(1) = \mathrm{length}_R(M(1))$ and $L = \mathrm{length}(R)$. If $\kappa_1 \neq 0$, then we have*

$$\lambda(1) = L - \mathrm{length}_R(R\kappa_1).$$

If $\kappa_1 = 0$ then $\lambda(1) \geq L$.

Proof By Proposition 8.5.8, κ_1 generates $\mathrm{St}(\mathcal{H})(1) = \mathfrak{m}^{\lambda(1)} H_{\mathcal{F}}(F, T)$. We then have evidently if $\kappa_1 \neq 0$ that $\mathrm{length}_R(R\kappa_1) = \mathrm{length}(\mathrm{St}(\mathcal{H})(1)) = L - \lambda(1)$. □

Definition 8.5.11 If $\kappa \in \mathbf{KS}(T)$ and $r \in \mathbb{N}$ put, where $L = \mathrm{length}(R)$,

$$\partial^r(\kappa) = \min\{L - \mathrm{length}(R\kappa_m) | \ m \in \mathcal{M}_{0,F}, \nu(m) = r\}.$$

Here $\nu(m)$ denotes the number of distinct prime divisors of the cycle $m \in \mathcal{M}_{0,F}$.

The *elementary divisors* of κ are defined by

$$e_i(\kappa) = \partial^i(\kappa) - \partial^{i+1}(\kappa) \qquad (i \geq 0).$$

Lemma 8.5.12 *Suppose that $(T, \mathcal{F}, \mathcal{L})$ is odd (Definition 8.3.20) and $m \in \mathcal{M}_{0,F}$. Then there is a character $\psi = \pm$ of $G = \mathrm{Gal}(F/k)$ uniquely determined by m and an $R[G]$-module M such that there is a decomposition of $R[G]$-modules (Proposition 8.3.15)*

$$\mathcal{H}(m) \cong R(\psi) \oplus M \oplus M.$$

For any prime place $q \in \mathcal{M}_{0,F}$ which is coprime to m, we then have

$$\mathcal{H}(mq) \cong R(-\psi) \oplus N \oplus N.$$

for some $R[G]$-module N. The element ψ is called the **character** of $\mathcal{H}(m)$.

Proof In the decomposition for any $m \in \mathcal{M}_{0,F}$ given by

$$\mathcal{H}(m) \cong R(\psi) \oplus M \oplus M.$$

we have that in $\mathcal{H}(m)[\mathfrak{m}]$ the module $(R/\mathfrak{m})(\psi)$ occurs an odd number of times and $(R/\mathfrak{m})(-\psi)$ occurs an even number of times, so that the character $\psi = \pm$ is uniquely determined by m.

From Proposition 8.3.8 and with the notation of that proposition we have

$$\rho^\delta(nq) = \rho^\delta(n) \pm 1.$$

It follows that for any prime place $q \in \mathcal{M}_{0,F}$ which is coprime to m, we then have that in the $R[G]$-module decomposition of $\mathcal{H}(mq)[\mathfrak{m}]$ that the module $(R/\mathfrak{m})(\psi)$ occurs an even number of times and $(R/\mathfrak{m})(-\psi)$ occurs an odd number of times and the result follows.

$\square$

Lemma-Definition 8.5.13 *Under the hypotheses of Lemma 8.5.12, for any $m \in \mathcal{M}_{0,F}$, we have the $R[G]$-module decomposition $\mathcal{H}(m) \cong R(\psi) \oplus M(m) \oplus M(n)$, and we define*

$$\lambda^\psi(m) = \text{length}_R(M(m)^\psi)$$

for any character $\psi = \pm$ of $\text{Gal}(F/k)$.

With the notation of the diagram (8.3.24) and where $L = \text{length}(R)$, let $n, q \in \mathcal{M}_{0,F}$ where q is a prime divisor coprime to n. Then we have the following relations, where χ is the character of $\mathcal{H}(n)$,

$$\lambda^\chi(nq) = \lambda^\chi(n) + L - a^\chi - d/2$$

$$\lambda^{-\chi}(nq) = \lambda^{-\chi}(n) - a^{-\chi} - d/2$$

$$\lambda(nq) = \lambda(n) + L - a^+ - a^- - d.$$

Proof This is straightforward from the diagram (8.3.24) and is a variant of the equation (8.3.26) in the proof of Proposition 8.3.25 where the character of $\mathcal{H}(nq)$ is the contragredient of the character of $\mathcal{H}(n)$ by Lemma 8.5.12. The third relation is a restatement of (8.3.26).

$\square$

8.5 Selmer Groups over Artinian Principal Ideal Rings

(8.5.14) In order to state the next Theorem 8.5.16, for $n \in \mathcal{M}_{0,F}$, the $R[G]$-module $\mathcal{H}(n)$ decomposes as a sum of cyclic $R[G]$-modules where χ is the character of $\mathcal{H}(n)$

$$\mathcal{H}(n) = R(\chi) \oplus \bigoplus_{i \geq 1} \left(\frac{R}{\mathfrak{m}^{d_i(n,(-1)^i\chi)}} ((-1)^i \chi) \right)^2 \tag{8.5.15}$$

where each component $R/\mathfrak{m}^{d_i(n,(-1)^i\chi)}((-1)^i \chi)$ is in the $(-1)^i \chi$-eigenspace for all i and where we may assume that, by permuting the components and writing $d_i = d_i(n, (-1)^i\chi)$,

$$d_1 \geq d_3 \geq d_5 \geq \ldots$$

$$d_2 \geq d_4 \geq d_6 \geq \ldots .$$

This uniquely determines the integers $d_i(n, (-1)^i \chi) \in \mathbb{N}$ in the decomposition (8.5.15).

Theorem 8.5.16 *Suppose that $(T, \mathcal{F}, \mathcal{L})$ is odd (Definition 8.3.20) Assume that $(T, \mathcal{F}, \mathcal{L})$ is liftable to a mixed characteristic discrete valuation ring D with maximal ideal $\mathfrak{n}$ (Definition 8.5.6), and let $(\widetilde{T}, \widetilde{\mathcal{F}}, \widetilde{\mathcal{L}})$ be the corresponding Selmer triple over D and put $L = \text{length}(R)$. Let $(\widetilde{T}^{(2L-1)}, \widetilde{\mathcal{F}}^{(2L-1)}, \widetilde{\mathcal{L}}^{(2L-1)})$ be the induced Selmer triple over $D/\mathfrak{n}^{2L-1}$. Assume that $\mathcal{L} = \widetilde{\mathcal{L}}^{(2L-1)}$, so that $\mathcal{M}_{0,F} = \mathcal{M}_{0,F}(\widetilde{\mathcal{L}}^{(2L-1)})$.*

Suppose that $\kappa \in \mathbf{KS}(T, \mathcal{F}, \mathcal{L})$ and $m_0 \in \mathcal{M}_{0,F}$ are such that $\kappa_{m_0} \neq 0$. Let $j \in \mathbb{N}$ be the unique integer such that κ_{m_0} generates $\mathfrak{m}^j \text{St}(m_0)$. Let ψ be the character of $\mathcal{H}$. We have then

$$\min_{\mathfrak{q}_1 \ldots \mathfrak{q}_r} \lambda(\mathfrak{q}_1 \ldots \mathfrak{q}_r) = \lambda(1) - \sum_{i=1}^{r} d_i(1, (-1)^i\psi). \tag{8.5.17}$$

Here the minimum runs over all cycles $\mathfrak{q}_1 \ldots \mathfrak{q}_r \in \mathcal{M}_{0,F}$ which are products of r distinct primes in $\mathcal{M}_{0,F}$. Furthermore, for a cycle $\mathfrak{q}_1 \ldots \mathfrak{q}_r \in \mathcal{M}_{0,F}$ for which this minimum is attained, we have the decomposition of $R[G]$-modules

$$\mathcal{H}(\mathfrak{q}_1 \ldots \mathfrak{q}_r) = R((-1)^r \psi) \oplus \bigoplus_{i \geq r+1} \left(\frac{R}{\mathfrak{m}^{d_i(1,(-1)^i\psi)}} ((-1)^i \psi) \right)^2 \tag{8.5.18}$$

and we have

$$\partial^r(\kappa) = \min(L, j + \sum_{i \geq r+1} d_i(1, (-1)^i \psi)). \tag{8.5.19}$$

Proof Let n be a cycle in $\mathcal{M}_{0,F}$. By Lemma 8.5.12, we have that there is a character $\chi = \pm$ of $G = \text{Gal}(F/k)$ uniquely determined by n and there is the decomposition of

$R[G]$-modules for some $R[G]$-module $M(n)$

$$\mathcal{H}(n) \cong R(\chi) \oplus M(n) \oplus M(n).$$

We have by assumption that $\kappa \in \mathrm{KS}(T, \mathcal{F}, \mathcal{L}^{(2L-1)})$, $m_0 \in \mathcal{M}_{0,F}^{(2L-1)}$, and we have $\kappa_{m_0} \neq 0$. By Proposition 8.6.7(iii), we have that $\kappa_n \in \mathrm{St}(\mathcal{H}(T, \mathcal{F}, \mathcal{L}^{(2L-1)}))(n)$ for all $n \in \mathcal{M}_{0,F}$. Let $j \geq 0$ be the unique integer such that κ_{m_0} generates $\mathfrak{m}^j \mathrm{St}(\mathcal{H}(T, \mathcal{F}, \mathcal{L}^{(2L-1)}))(m_0)$. Then again by Proposition 8.5.7(iii), the element κ_r generates $\mathfrak{m}^j \mathrm{St}(\mathcal{H}(T, \mathcal{F}, \mathcal{L}^{(2L-1)}))(r)$ for every $r \in \mathcal{M}_{0,F}$.

Therefore as $\mathcal{H}(m) \cong R(\pm \psi) \oplus M(m) \oplus M(m)$ (Lemma 8.5.12), $\lambda(m) = \mathrm{length}(M(m))$, and as the stub Selmer module $\mathrm{St}(m)$ is a cyclic R-module, we have $\mathrm{length}_R(R\kappa_m) = L - j - \lambda(m)$, where $R\kappa_m$ is the submodule of $\mathcal{H}(m)$ generated by κ_m. Combining this with the definition of $\partial^r(\kappa)$ (Definition 8.5.11), we obtain

$$\partial^r(\kappa) = \min_n(L, j + \lambda(n) \mid n \in \mathcal{M}_{0,F}, \nu(n) = r) \tag{8.5.20}$$

where the minimum runs over all cycles $n \in \mathcal{M}_{0,F}$ such that $\nu(n) = r$. Then for $r = 0$, we obtain the required equality (8.5.19) for we have $\sum_{i \geq 1} d_i(1, (-1)^i \psi) = \mathrm{length}(M(1)) = \lambda(1)$ and we have

$$\partial^0(\kappa) = \min(L, j + \sum_{i > 0} d_i(1, (-1)^i \psi)).$$

For each pair of elements $n, \mathfrak{q} \in \mathcal{M}_{0,F}$ where $\mathfrak{q}$, is a prime divisor coprime to n, there is a diagram (8.3.24), in Proposition 8.3.23, and corresponding integers a^δ, d for $\delta = +$ and $-$ and where d is independent of δ by Proposition 8.3.23. To denote the dependence on $n, \mathfrak{q}$, we write

$$a^\delta = a^\delta(n, \mathfrak{q}), \quad d = d(n, \mathfrak{q}). \tag{8.5.21}$$

We also have the integers $\lambda^\delta(n)$ for $\delta = \pm$ defined in lemma-Definition 8.5.13 so that $\lambda^+(n) + \lambda^-(n) = \lambda(n)$.

Step 1. *Suppose that $n \in \mathcal{M}_{0,F}$. Suppose that ψ is the character of $\mathcal{H}(n)$. We have for all primes $\mathfrak{q} \in \mathcal{M}_{0,F}$ coprime to n, where $L = \mathrm{length}(R)$,*

$$\lambda^\psi(n) + L \geq \lambda^\psi(n\mathfrak{q}) \geq \lambda^\psi(n). \tag{8.5.22}$$

Let $c_1 \in R(\psi)$ be a generator of the component $R(\psi)$ of $\mathcal{H}(n)$ in the $R[G]$-module decomposition (Lemma 8.5.12) $\mathcal{H}(n) \cong R(\psi) \oplus M(n) \oplus M(n)$. By Proposition 8.3.31, we may then choose a prime $\mathfrak{q} \in \mathcal{M}_{0,F}$ coprime to n such that $\mathrm{res}_{F_\mathfrak{q}/F}(\mathfrak{m}^{L-1} c_1) \neq 0$. Then for this $\mathfrak{q}$ we have

8.5 Selmer Groups over Artinian Principal Ideal Rings

$$\lambda^\psi(n\mathfrak{q}) = \lambda^\psi(n).$$

Furthermore, for any $\mathfrak{q} \in \mathcal{M}_{0,F}$ coprime to n such that $\lambda^\psi(n\mathfrak{q}) = \lambda^\psi(n)$ then we have (notation as in (8.5.21))

$$d(n, \mathfrak{q}) = 0, a^\psi(n, \mathfrak{q}) = L$$

and also

$$\mathcal{H}^\psi(n\mathfrak{q}) \cong \bigoplus_{i \geq 1, i \text{ even}} \left(\frac{R}{\mathfrak{m}^{d_i(n,(-1)^i\psi)}} ((-1)^i \psi) \right)^2. \tag{8.5.23}$$

We have

$$\mathcal{H}(n) = H^1_{\mathcal{F}(n)}(F, T) \cong R(\psi) \oplus M(n) \oplus M(n).$$

There is a map obtained by localization at a prime $\mathfrak{q} \in \mathcal{M}_{0,F}$ which is coprime to n

$$f^\psi : \mathcal{H}^\psi(n) \longrightarrow H^1_f(F_\mathfrak{q}, T)^\psi$$

where $H^1_f(F_\mathfrak{q}, T)^\psi$ is a free R-module of rank 1 by Lemma 8.3.4. Therefore, $\ker(f^\psi)$ has colength $\leq L$ in $\mathcal{H}(n)^\psi$. Therefore,

$$\text{length}_R(\ker(f^\psi)) \geq \text{length}_R(R(\psi)) + 2\lambda^\psi(n) - L = 2\lambda^\psi(n).$$

The kernel $\ker(f^\psi)$ is equal to $\mathcal{H}^\psi_\mathfrak{q}(n) \subseteq \mathcal{H}^\psi(n)$. So we have

$$\text{length}_R(\mathcal{H}^\psi_\mathfrak{q}(n))^\psi \geq 2\lambda^\psi(n).$$

We have from Lemma 8.5.13 and in the notation of (8.5.21)

$$\lambda^\psi(n\mathfrak{q}) = \lambda^\psi(n) + L - a^\psi(n, \mathfrak{q}) - d(n, \mathfrak{q})/2. \tag{8.5.24}$$

As $d(n, \mathfrak{q}) \geq 0$, $a^\psi(n, \mathfrak{q}) \geq 0$ and $a^\psi(n, \mathfrak{q}) + d(n, \mathfrak{q}) \leq L$, this gives

$$\lambda^\psi(n) + L \geq \lambda^\psi(n\mathfrak{q}) \geq \lambda^\psi(n).$$

This proves the inequalities (8.5.22) for all primes $\mathfrak{q} \in \mathcal{M}_{0,F}$ coprime to n.

Let $c_1 \in R(\psi)$ be a generator of the component $R(\psi)$ of $\mathcal{H}(n)$ where we fix the $R[G]$-module decomposition $\mathcal{H}(n) \cong R(\psi) \oplus M(n) \oplus M(n)$. By Proposition 8.3.31, we may then choose a prime $\mathfrak{q} \in \mathcal{M}_{0,F}$ coprime to n such that $\text{res}_{F_\mathfrak{q}/F}(\mathfrak{m}^{L-1} c_1) \neq 0$.

Then the map of restriction at q

$$f^\psi : R(\psi)^\psi \oplus (M(n) \oplus M(n))^\psi \longrightarrow H^1_f(F_q, T)^\psi$$

where $H^1_f(F_q, T)^\psi$ is a free R-module of rank 1 (Lemma 8.3.4) is surjective and in particular $\ker(f^\psi)$ has colength L in $\mathcal{H}^\psi(n)$. It follows from the structure of modules over discrete valuation rings that we have an isomorphism (see, e.g. [5, lemma 5.3.2])

$$\mathcal{H}^\psi_q(n) \cong \ker(f^\psi) \cong (M(n) \oplus M(n))^\psi.$$

Therefore diagram (8.3.24) gives in the notation of (8.5.21) of that diagram

$$a^\psi(n, q) = L, d(n, q) = 0.$$

It follows from the same diagram that

$$\mathcal{H}^\psi(nq) \cong \mathcal{H}^\psi_q(n) = \ker(f^\psi) \cong (M(n) \oplus M(n))^\psi.$$

Furthermore, we have the decomposition from Lemma 8.5.12

$$\mathcal{H}(nq) \cong R(-\psi) \oplus M(nq) \oplus M(nq).$$

Therefore, we have

$$(M(nq) \oplus M(nq))^\psi \cong (M(n) \oplus M(n))^\psi$$

and the decomposition for $\mathcal{H}(nq)^\psi$ stated as (8.5.23) in Step 1 now follows from (8.5.15), and we then have

$$\lambda^\psi(nq) = \lambda^\psi(n).$$

For the last part, suppose that $q \in \mathcal{M}_{0,F}$ is such that $\lambda^\psi(nq) = \lambda^\psi(n)$. Then we have from the formula (8.5.24) that

$$L - a^\psi(n, q) - d(n, q)/2 = 0.$$

We have from the diagram (8.3.24) and Proposition 8.3.23 that $a^\psi(n, q) \geq 0, d(n, q) \geq 0$ and

$$0 \leq L - a^\psi(n, q) - d(n, q) \leq L - a^\psi(n, q) - d(n, q)/2 = 0.$$

From this, we have $a^\psi(n, q) = L, d(n, q) = 0$.

8.5 Selmer Groups over Artinian Principal Ideal Rings

It then follows from diagram (8.3.23) that $\mathcal{H}^{\mathfrak{q},\psi}(n) = \mathcal{H}^{\psi}(n)$, $\mathcal{H}^{\psi}(\mathfrak{q}n) = \mathcal{H}_{\mathfrak{q}}^{\psi}(n)$. We also have from this diagram that the cokernel of the injection

$$\mathcal{H}^{\psi}(\mathfrak{q}n) \to \mathcal{H}^{\mathfrak{q},\psi}(n)$$

is cyclic of length L. We have from this and from (8.5.15) the decomposition

$$\mathcal{H}^{\mathfrak{q},\psi}(n) = R(\chi) \oplus \bigoplus_{i\geq 1, i \text{ even}} \left(\frac{R}{\mathfrak{m}^{d_i(n,(-1)^i\chi)}}((-1)^i\chi)\right)^2.$$

It follows that there is an isomorphism of $R[G]$-modules for any $\mathfrak{q} \in \mathcal{M}_{0,F}$ for which $\lambda^{\psi}(n\mathfrak{q}) = \lambda^{\psi}(n)$.

$$\mathcal{H}^{\psi}(\mathfrak{q}n) = \bigoplus_{i\geq 1, i \text{ even}} \left(\frac{R}{\mathfrak{m}^{d_i(n,(-1)^i\chi)}}((-1)^i\chi)\right)^2.$$

which proves the decomposition (8.5.23) for any $\mathfrak{q} \in \mathcal{M}_{0,F}$ coprime to n such that $\lambda^{\psi}(n\mathfrak{q}) = \lambda^{\psi}(n)$ and completes Step 1.

Step 2. Let ψ be the character of $\mathcal{H}(n)$. Suppose that $\mathcal{H}(n)^{-\psi} = 0$. Then we have for all prime places $\mathfrak{q} \in \mathcal{M}_{0,F}$ coprime to n

$$\mathcal{H}(n\mathfrak{q})^{-\psi} \cong R(-\psi), \quad \mathcal{H}_{\mathfrak{q}}^{-\psi}(n) \cong 0, \quad \mathcal{H}(n)^{\mathfrak{q},-\psi} \cong R(-\psi)$$

$$d(n,\mathfrak{q}) = 0, \ a^{-\psi}(n,\mathfrak{q}) = 0, \ \lambda^{-\psi}(\mathfrak{q}n) = 0.$$

Let $\mathfrak{q} \in \mathcal{M}_{0,F}$ be any prime place coprime to n. As $\mathcal{H}_{\mathfrak{q}}^{-\psi}(n) \subseteq \mathcal{H}^{-\psi}(n) = 0$, we must have $\mathcal{H}_{\mathfrak{q}}^{-\psi}(n) = 0$. From diagram (8.3.24) and with the notation of the diagram (see (8.5.21)), we then have $a^{-\psi}(n,\mathfrak{q}) = 0$. Again from this diagram (8.3.24), we have that $\mathcal{H}^{\mathfrak{q},-\psi}(n)$ is an R-module of length $L = \text{length}(R)$. Furthermore, from this diagram, we have that $\mathcal{H}(n\mathfrak{q})^{-\psi}$ is a submodule of $\mathcal{H}^{\mathfrak{q},-\psi}(n)$ of colength $d(n,\mathfrak{q})$ of the module $\mathcal{H}^{\mathfrak{q},-\psi}(n)$.

But we have $\mathcal{H}(n\mathfrak{q})^{-\psi} \cong (R(-\psi) \oplus M(n\mathfrak{q}) \oplus M(n\mathfrak{q}))^{-\psi}$ which is an R-module of length $\geq L$, and it is of length equal to L if and only if $M(n\mathfrak{q})^{-\psi} = 0$. Therefore, we must have $d(n,\mathfrak{q}) = 0$ and $\mathcal{H}(n\mathfrak{q})^{-\psi} \cong R(-\psi)$ and is a free R-module of rank 1. Hence, from diagram (8.3.24), we have $\mathcal{H}(n)^{\mathfrak{q},-\psi} \cong R(-\psi)$.

It follows that

$$\lambda^{-\psi}(n\mathfrak{q}) = \lambda^{-\psi}(n) = 0$$

which completes this Step 2.

Step 3. *Suppose that ψ is the character of $\mathcal{H}(n)$ and that $n \in \mathcal{M}_{0,F}$. We have for all primes $\mathfrak{q} \in \mathcal{M}_{0,F}$ coprime to n, with the notation of (8.5.21)*

$$\lambda(n\mathfrak{q}) \geq \lambda(n) - a^{-\psi}(n, \mathfrak{q}) \geq \lambda(n) - d_1(n, -\psi).$$

Define the $R[G]$-module, given the decomposition $\mathcal{H}(n)$ given in (8.5.15),

$$N = \bigoplus_{i \geq 3, i \text{ odd}} \left(\frac{R}{\mathfrak{m}^{d_i(n,-\psi)}}(-\psi) \right)^2.$$

Then we have:

(i) *There is a prime place $\mathfrak{q} \in \mathcal{M}_{0,F}$ coprime to n such that*

$$d(n, \mathfrak{q}) = 0, a^{\psi}(n, \mathfrak{q}) = L, a^{-\psi}(n, \mathfrak{q}) = d_1(n, -\psi).$$

(ii) *Suppose that $\mathfrak{q} \in \mathcal{M}_{0,F}$ is a prime place coprime to n such that*

$$d(n, \mathfrak{q}) = 0, a^{\psi}(n, \mathfrak{q}) = L, a^{-\psi}(n, \mathfrak{q}) = d_1(n, -\psi).$$

Then we have

$$\mathcal{H}^{-\psi}(\mathfrak{q}n) \cong R(-\psi) \oplus N. \tag{8.5.25}$$

$$\mathcal{H}_{\mathfrak{q}}^{-\psi}(n) \cong R/\mathfrak{m}^{d_1(n,-\psi)}(-\psi) \oplus N \tag{8.5.26}$$

$$\mathcal{H}^{\mathfrak{q},-\psi}(n) \cong R(-\psi) \oplus \mathcal{H}_{\mathfrak{q}}^{-\psi}(n) \tag{8.5.27}$$

$$\lambda^{-\psi}(n\mathfrak{q}) = \lambda^{-\psi}(n) - d_1(n, -\psi). \tag{8.5.28}$$

Put $L = \text{length}(R)$. From Lemma 8.5.13, we have

$$\lambda(n\mathfrak{q}) = \lambda(n) + L - a^{\psi}(n, \mathfrak{q}) - a^{-\psi}(n, \mathfrak{q}) - d(n, \mathfrak{q}).$$

As $L - a^{\psi}(n, \mathfrak{q}) - d(n, \mathfrak{q}) \geq 0$ in the diagram (8.3.24) and as $a^{-\psi}(n, \mathfrak{q}) \leq d_1(n, -\psi)$ from this same diagram because $a^{-\psi}(n, \mathfrak{q})$ is the length of the longest cyclic submodule of $\mathcal{H}(n)^{-\psi}$ by the decomposition (8.5.15), we obtain the first inequality stated in the Step 3

$$\lambda(n\mathfrak{q}) \geq \lambda(n) - a^{-\psi}(n, \mathfrak{q}) \geq \lambda(n) - d_1(n, -\psi).$$

8.5 Selmer Groups over Artinian Principal Ideal Rings

(i) Suppose first that $H^1_{\mathcal{F}(n)}(F, T)^{-\psi} = 0$, i.e. $d_1(n, -\psi) = 0$. Then from Step 2, we obtain the isomorphisms (8.5.25), (8.5.26), and (8.5.27). Note also that the final equality $\lambda^{-\psi}(nq) = \lambda^{-\psi}(n) - d_1(n, -\psi)$ also holds when $H^1_{\mathcal{F}(n)}(F, T)^{-\psi} = 0$ from Step 2.

From Step 1, there is a prime $q \in \mathcal{M}_{0,F}$ coprime to n such that

$$d(n, q) = 0, a^{\psi}(n, q) = L.$$

From Step 2, we obtain

$$a^{-\psi}(n, q) = 0.$$

This proves part (i) in this case where $H^1_{\mathcal{F}(n)}(F, T)^{-\psi} = 0$.

Suppose now that $H^1_{\mathcal{F}(n)}(F, T)^{-\psi} \neq 0$ that is to say $d_1(n, -\psi) \neq 0$. We have the decomposition of $R[G]$-modules $\mathcal{H}(n) \cong R(\psi) \oplus M(n) \oplus M(n)$ by Lemma 8.5.12. Then there is a non-zero element $0 \neq c_2 \in \mathcal{H}(n)^{-\psi}$ whose annihilator in R equals $\mathfrak{m}^{d_1(n,-\psi)}$ (from the decomposition (8.5.15)) so that

$$Rc_2 \cong R/\mathfrak{m}^{d_1(n,-\psi)}.$$

As in Step 1, let $c_1 \in R(\psi)$ be a generator of a non-zero component isomorphic to $R(\psi)$ of $\mathcal{H}(n)^{\psi}$, where such a component certainly exists. By Proposition 8.3.31, we may then choose a prime $q \in \mathcal{M}_{0,F}$ such that both the following conditions hold

$$\mathrm{res}_{F_q/F}(\mathfrak{m}^{L-1} c_1) \neq 0$$

$$\mathrm{res}_{F_q/F}(\mathfrak{m}^{d_1(n,-\psi)-1} c_2) \neq 0.$$

We then have that the conclusion of Step 1 holds, and in particular, we have

$$d(n, q) = 0, a^{\psi}(n, q) = L. \tag{8.5.29}$$

Furthermore, as $\mathrm{res}_{F_q/F}(\mathfrak{m}^{d_1(n,-\psi)-1} c_2) \neq 0$, it follows from diagram (8.3.24) that

$$a^{-\psi}(n, q) = d_1(n, -\psi).$$

This completes the proof of part (i).

(ii) Suppose that $q \in \mathcal{M}_{0,F}$ is a prime place coprime to n such that

$$d(n, q) = 0, a^{\psi}(n, q) = L, a^{-\psi}(n, q) = d_1(n, -\psi). \tag{8.5.30}$$

We have by definition
$$\mathcal{H}_{\mathfrak{q}}^{-\psi}(n) = \ker(\mathcal{H}^{-\psi}(n) \to H^1(F_{\mathfrak{q}}, T)^{-\psi}).$$

We have (by (8.5.15)), where $\mathcal{H}(n) \cong R(\psi) \oplus M(n) \oplus M(n)$ by Lemma 8.5.12,
$$M(n)^{-\psi} \cong \bigoplus_{i \geq 1,\ \text{odd}} R/\mathfrak{m}^{d_i(n,-\psi)}(-\psi).$$

As $a^{-\psi}(n, \mathfrak{q}) = d_1(n, -\psi)$ is the length of the longest cyclic submodule of $M(n)^{-\psi}$, we obtain from the diagram (8.3.23)

$$\mathcal{H}_{\mathfrak{q}}^{-\psi}(n) \cong$$
$$\left(\bigoplus_{i \geq 1,\ \text{odd}} R/\mathfrak{m}^{d_i(n,(-1)^i\psi)}\right) \oplus \left(\bigoplus_{i \geq 3,\ \text{odd}} R/\mathfrak{m}^{d_i(n,(-1)^i\psi)}\right). \tag{8.5.31}$$

From the choice of $\mathfrak{q}$, we have that this kernel is a submodule of colength $d_1(n, -\psi)$ in $\mathcal{H}^{-\psi}(n) \cong (M(n) \oplus M(n))^{-\psi}$.

Put
$$N = \left(\bigoplus_{i \geq 3,\ \text{odd}} R/\mathfrak{m}^{d_i(n,(-1)^i\psi)}\right) \oplus \left(\bigoplus_{i \geq 3,\ \text{odd}} R/\mathfrak{m}^{d_i(n,(-1)^i\psi)}\right)$$

where this $R[G]$-module, $G = \mathrm{Gal}(F/k)$, is considered as a submodule of $\mathcal{H}_{\mathfrak{q}}^{-\psi}(n)$ and therefore a submodule of $\mathcal{H}^{-\psi}(\mathfrak{q}n)$ and $\mathcal{H}^{\mathfrak{q},-\psi}(n)$.

In the diagram (8.3.24), we then have $a^{-\psi}(n, \mathfrak{q}) = d_1(n, -\psi)$ by (8.5.30). Furthermore, in this diagram, the upper left arrow, for $\delta = -\psi$, has length $L - d_1(n, -\psi)$.

The submodules $\mathcal{H}_{\mathfrak{q}}^{-\psi}(n)$ and $\mathcal{H}^{-\psi}(n)$ are both $R/\mathfrak{m}^{d_1(n,-\psi)}$-modules as both are annihilated by $\mathfrak{m}^{d_1(n,-\psi)}$. Therefore, the exact sequence

$$0 \to \mathcal{H}_{\mathfrak{q}}^{-\psi}(n) \to \mathcal{H}^{-\psi}(n) \to R/\mathfrak{m}^{d_1(n,-\psi)} \to 0$$

is split, and we obtain an isomorphism of $R/\mathfrak{m}^{d_1(n,-\psi)}[G]$-modules

$$\mathcal{H}^{-\psi}(n) \cong R/\mathfrak{m}^{d_1(n,-\psi)}(-\psi) \oplus \mathcal{H}_{\mathfrak{q}}^{-\psi}(n)$$

where the inclusion $\mathcal{H}_{\mathfrak{q}}^{-\psi}(n) \subseteq \mathcal{H}^{-\psi}(n)$ becomes the inclusion of the second factor.

We similarly have the exact sequence of $R[G]$-modules

$$0 \to \mathcal{H}_{\mathfrak{q}}^{-\psi}(n) \to \mathcal{H}^{\mathfrak{q},-\psi}(n) \to R(-\psi) \to 0$$

8.5 Selmer Groups over Artinian Principal Ideal Rings

Again, this is a split sequence of $R[G]$-modules, so we have the isomorphism, where the inclusion $\mathcal{H}_{\mathfrak{q}}^{-\psi}(n) \subseteq \mathcal{H}^{-\psi,\mathfrak{q}}(n)$ is just the inclusion of the second factor

$$\mathcal{H}^{\mathfrak{q},-\psi}(n) \cong R(-\psi) \oplus \mathcal{H}_{\mathfrak{q}}^{-\psi}(n). \tag{8.5.32}$$

It then follows from (8.5.31) that

$$\mathcal{H}^{\mathfrak{q},-\psi}(n) \cong R(-\psi) \oplus R/\mathfrak{m}^{d_1(n,-\psi)}(-\psi) \oplus N. \tag{8.5.33}$$

Here $\mathcal{H}^{-\psi}(n) \cong \bigoplus_{i \geq 1, \text{ odd}} \left(R/\mathfrak{m}^{d_i(n,(-1)^i\psi)} \right)^2$ is the submodule obtained by taking the submodule $R/\mathfrak{m}^{d_i(n,-\psi)}(-\psi)$ of $R(-\psi)$ in the decomposition (8.5.33) above of $\mathcal{H}^{\mathfrak{q},-\psi}(n)$.

The diagram of (8.3.24) for the $-\psi$ components becomes the following.

$$
\begin{array}{ccc}
 & \mathcal{H}^{\mathfrak{q},-\psi}(n) & \\
 L - a^{-\psi}(n,\mathfrak{q}) \nearrow & & \nwarrow a^{-\psi}(n,\mathfrak{q}) + d(n,\mathfrak{q}) \\
\mathcal{H}^{-\psi}(n) & & \mathcal{H}^{-\psi}(n\mathfrak{q}) \\
a^{-\psi}(n,\mathfrak{q}) \nwarrow & & \nearrow L - a^{-\psi}(n,\mathfrak{q}) - d(n,\mathfrak{q}) \\
 & \mathcal{H}_{\mathfrak{q}}^{-\psi}(n) &
\end{array}
$$
(8.5.34)

The integer $d(n, \mathfrak{q})$ depends only on n, $\mathfrak{q}$, and we have $d(n, \mathfrak{q}) = 0$ by (8.5.30). Factoring out by the submodule N, we then have the injections of R-modules,

$$\frac{\mathcal{H}_{\mathfrak{q}}^{-\psi}(n)}{N} \xhookrightarrow{L-d_1} \frac{\mathcal{H}^{-\psi}(\mathfrak{q}n)}{N} \xhookrightarrow{d_1} \frac{\mathcal{H}^{\mathfrak{q},-\psi}(n)}{N}. \tag{8.5.35}$$

Furthermore, the cokernel of the first injection $\frac{\mathcal{H}_{\mathfrak{q}}^{-\psi}(n)}{N} \hookrightarrow \frac{\mathcal{H}^{-\psi}(\mathfrak{q}n)}{N}$ map has length $L - d_1(n, -\psi)$ by the above diagram (8.5.34). As $H_s^1(F_{\mathfrak{q}}, T)^{-\psi}$ is a free R-module of rank 1, it follows that the cokernel of this injection may be identified with a submodule of $H_s^1(F_{\mathfrak{q}}, T)^{-\psi}$, and hence the cokernel is a cyclic R-module of this length. Again the cokernel of the second injection $\frac{\mathcal{H}^{-\psi}(\mathfrak{q}n)}{N} \hookrightarrow \frac{\mathcal{H}^{\mathfrak{q},-\psi}(n)}{N}$ of (8.5.35) is cyclic of length $d_1(n, -\psi)$ by the diagram (8.5.34).

We claim that $\frac{\mathcal{H}^{-\psi}(\mathfrak{q}n)}{N}$ is a cyclic R-module where this can be seen as follows. We have that

$$\frac{\mathcal{H}_{\mathfrak{q}}^{-\psi}(n)}{N} \cong R/\mathfrak{m}^{d_1(n,-\psi)}(-\psi), \quad \frac{\mathcal{H}^{\mathfrak{q},-\psi}(n)}{N} \cong R(-\psi) \oplus \frac{\mathcal{H}_{\mathfrak{q}}^{-\psi}(n)}{N}$$

from (8.5.31) and (8.5.32). It follows from this and the diagram (8.5.34) that $\frac{\mathcal{H}^{-\psi}(\mathfrak{q}n)}{N}$ is an R-module generated by at most two elements.

Now $\mathcal{H}$ is odd, so that $\mathcal{H}(qn)$ is odd (Definition 8.3.20). Therefore, $\mathcal{H}(qn) \otimes_R R/\mathfrak{m}$ is an odd dimensional $R/\mathfrak{m}$-vector space. Now $\mathcal{H}(qn)^{-\psi} \otimes_R R/\mathfrak{m}$ has odd dimension over $R/\mathfrak{m}$ as $-\psi$ is the character of $\mathcal{H}(qn)$. Evidently, $N \otimes_R R/\mathfrak{m}$ has even dimension over $R/\mathfrak{m}$.

We claim that the following sequence is exact

$$0 \to N \otimes_R R/\mathfrak{m} \to \mathcal{H}^{-\psi}(qn) \otimes_R R/\mathfrak{m} \to \frac{\mathcal{H}^{-\psi}(qn)}{N} \otimes_R R/\mathfrak{m} \to 0$$

where this is obtained from the inclusion of N in $\mathcal{H}^{-\psi}(qn)$ by applying $- \otimes_R R/\mathfrak{m}$. It is only necessary to show that the left-hand map $N \otimes_R R/\mathfrak{m} \to \mathcal{H}^{-\psi}(qn) \otimes_R R/\mathfrak{m}$ is injective. Suppose then that $a \in N$ and $a \otimes 1 \in N \otimes_R R/\mathfrak{m}$ are in the kernel of the first map. Then we have that there is $b \in \mathcal{H}^{-\psi}(qn)$ such that $a = rb$ in $\mathcal{H}^{-\psi}(qn)$ where $r \in \mathfrak{m} \subset R$. But $\mathcal{H}^{-\psi}(qn) \subseteq \mathcal{H}^{q,-\psi}(n) \cong R(-\psi) \oplus \mathcal{H}_q^{-\psi}(n) \cong R(-\psi) \oplus R/\mathfrak{m}^{d_1(n,-\psi)}(-\psi) \oplus N$ from (8.5.32) and (8.5.33). It follows that $a \in \mathfrak{m}N$ so that this sequence is exact as claimed.

It follows from this exact sequence that $\frac{\mathcal{H}^{-\psi}(qn)}{N} \otimes_R R/\mathfrak{m}$ has odd dimension over $R/\mathfrak{m}$. As we have shown that $\frac{\mathcal{H}^{-\psi}(qn)}{N}$ is an R-module generated by at most two elements, it follows from the structure of modules over a discrete valuation ring that $\frac{\mathcal{H}^{-\psi}(qn)}{N}$ is generated by at most one element as an R-module and hence is a cyclic R-module.

Counting the lengths of the R-modules in the diagram of injections (8.5.35) using the isomorphisms $\frac{\mathcal{H}_q^{-\psi}(n)}{N} \cong R/\mathfrak{m}^{d_1(n,-\psi)}(-\psi)$, $\frac{\mathcal{H}^{q,-\psi}(n)}{N} \cong R(-\psi) \oplus \frac{\mathcal{H}_q^{-\psi}(n)}{N}$, it follows that this cyclic module $\frac{\mathcal{H}^{-\psi}(qn)}{N}$ has length L and is therefore isomorphic to $R(-\psi)$ and is a free R-module of rank 1.

It then follows that there is an isomorphism of $R[G]$-modules, which proves (8.5.25),

$$\mathcal{H}^{-\psi}(qn) \cong R(-\psi) \oplus N.$$

We also have (see (8.5.31), (8.5.32)), which prove (8.5.26) and (8.5.27)

$$\mathcal{H}_q^{-\psi}(n) \cong R/\mathfrak{m}^{d_1(n,-\psi)}(-\psi) \oplus N$$

$$\mathcal{H}^{q,-\psi}(n) \cong R(-\psi) \oplus \mathcal{H}_q^{-\psi}(n).$$

We then obtain the final equality (8.5.28)

$$\lambda^{-\psi}(nq) = \lambda^{-\psi}(n) - d_1(n, -\psi)$$

from (8.5.25) and (8.5.15) where this alternatively follows Lemma 8.5.13.

This completes Step 3.

Step 4. Induction Step. *Suppose that $n \in \mathcal{M}_{0,F}$ and that ψ is the character of $\mathcal{H}(n)$.*

8.5 Selmer Groups over Artinian Principal Ideal Rings

(i) *We have*

$$\min_{\mathfrak{q}} \lambda^{\psi}(n\mathfrak{q}) = \lambda^{\psi}(n)$$

$$\min_{\mathfrak{q}} \lambda^{-\psi}(n\mathfrak{q}) = \lambda^{-\psi}(n) - d_1(n, -\psi)$$

where $\mathfrak{q}$ *runs over all prime places in* $\mathcal{M}_{0,F}$ *coprime to n. These two minima are attainable simultaneously for at least one prime* $\mathfrak{q} \in \mathcal{M}_{0,F}$ *coprime to n.*

(ii) *For any* $\mathfrak{q} \in \mathcal{M}_{0,F}$ *coprime to n for which these two minima in part (i) simultaneously hold we have*

$$d(n, \mathfrak{q}) = 0, a^{\psi}(n, \mathfrak{q}) = L, a^{-\psi}(n, \mathfrak{q}) = d_1(n, -\psi).$$

(iii) *We have*

$$\min_{\mathfrak{q}} \lambda(n\mathfrak{q}) = \lambda(n) - d_1(n, -\psi)$$

where $\mathfrak{q}$ *runs over all prime places in* $\mathcal{M}_{0,F}$ *coprime to n. For any prime* $\mathfrak{q} \in \mathcal{M}_{0,F}$ *coprime to n for which*

$$\lambda(n\mathfrak{q}) = \lambda(n) - d_1(n, -\psi)$$

we then have

$$d(n, \mathfrak{q}) = 0, a^{\psi}(n, \mathfrak{q}) = L, a^{-\psi}(n, \mathfrak{q}) = d_1(n, -\psi).$$

(iv) *Fix a prime* $\mathfrak{q} \in \mathcal{M}_{0,F}$ *coprime to n for which the two minima in part (i) are attained. Then we have the $R[G]$-isomorphism*

$$\mathcal{H}(\mathfrak{q}n) \cong R(-\psi) \oplus \bigoplus_{i \geq 2} \left(\frac{R}{\mathfrak{m}^{d_i(n,(-1)^i \psi)}} ((-1)^i \psi) \right)^2. \tag{8.5.36}$$

(i) Put $L = \text{length}(R)$. Let $n \in \mathcal{M}_{0,F}$. Let ψ be the character of $\mathcal{H}(n)$. We then have the $R[G]$-module decomposition $\mathcal{H}(n) \cong R(\psi) \oplus M(n) \oplus M(n)$ by Lemma 8.5.12. By Step 1 and Step 3, we have for all primes $\mathfrak{q} \in \mathcal{M}_{0,F}$ coprime to n

$$\lambda^{\psi}(n\mathfrak{q}) \geq \lambda^{\psi}(n)$$

$$\lambda^{-\psi}(n\mathfrak{q}) \geq \lambda^{-\psi}(n) - d_1(n, -\psi).$$

It remains to show that there is a prime q coprime to n where equality holds simultaneously in these formulae.

Let $c_1 \in R(\psi)$ be a generator of the non-zero component $R(\psi)$ of $\mathcal{H}(n)$ in this decomposition of $\mathcal{H}(n)$. By Proposition 8.3.31, we may then choose a prime $q \in \mathcal{M}_{0,F}$ coprime to n such that

(a) $\operatorname{res}_{F_q/F}(\mathfrak{m}^{L-1} c_1) \neq 0.$

Then by Step 1 for this choice of q, we have

(b) $\lambda^{\psi}(nq) = \lambda^{\psi}(n)$

and if $d(n, q)$ is the integer independent of ψ defined in (8.5.21) via the diagram (8.3.24) then

(c) $d(n, q) = 0.$

Suppose first that $H^1_{\mathcal{F}(n)}(F, T)^{-\psi} = 0$. Then $d_1(n, -\psi) = 0$ and by Step 2, we have $\lambda^{-\psi}(qn) = \lambda^{-\psi}(n) = 0$ for all primes $q \in \mathcal{M}_{0,F}$ coprime to n. This with (b) proves part (i) in this case.

Suppose now that $H^1_{\mathcal{F}(n)}(F, T)^{-\psi} \neq 0$, that is to say $d_1(n, -\psi) \neq 0$. Then there is a non-zero element $0 \neq c_2 \in \mathcal{H}(n)^{-\psi}$ whose annihilator in R equals $\mathfrak{m}^{d_1(n,-\psi)}$ so that $Rc_2 \cong R/\mathfrak{m}^{d_1(n,-\psi)}$. By Proposition 8.3.31, we may then choose a prime place $q \in \mathcal{M}_{0,F}$ coprime to n such that

(d) $\operatorname{res}_{F_q/F}(\mathfrak{m}^{d_1(n,-\psi)-1} c_2) \neq 0$

and also that (a) holds simultaneously.

By Lemma 8.5.13, we have

$$\lambda^{-\chi}(nq) = \lambda^{-\chi}(n) - a^{-\psi}(n, q) - d(n, q)/2.$$

By (c) and (d), we then obtain the equality

$$\lambda^{-\psi}(nq) = \lambda^{-\psi}(n) - d_1(n, -\psi).$$

This with (b) proves part (i).

(ii) Suppose that $q \in \mathcal{M}_{0,F}$ is any prime place coprime to n for which these two minima simultaneously of (i) hold that is to say we have

$$\lambda^{\psi}(nq) = \lambda^{\psi}(n)$$

8.5 Selmer Groups over Artinian Principal Ideal Rings

$$\lambda^{-\psi}(n\mathfrak{q}) = \lambda(n) - d_1(n, -\psi).$$

From Lemma 8.5.13, we have the following relations, with the notation of (8.5.21),

$$\lambda^{\psi}(n\mathfrak{q}) = \lambda^{\psi}(n) + L - a^{\psi}(n, \mathfrak{q}) - d(n, \mathfrak{q})/2$$

$$\lambda^{-\psi}(n\mathfrak{q}) = \lambda^{-\psi}(n) - a^{-\psi}(n, \mathfrak{q}) - d(n, \mathfrak{q})/2$$

As $\lambda^{\psi}(n\mathfrak{q}) = \lambda^{\psi}(n)$, the first relation gives

$$L - a^{\psi}(n, \mathfrak{q}) - d(n, \mathfrak{q})/2 = 0.$$

But by the diagram (8.5.34) (or (8.3.24)), we have $L - a^{\psi}(n, \mathfrak{q}) - d(n, \mathfrak{q}) \geq 0$ from which it follows that

$$a^{\psi}(n, \mathfrak{q}) = L, \quad d(n, \mathfrak{q}) = 0.$$

The second relation then shows that

$$a^{-\psi}(n, \mathfrak{q}) = d_1(n, -\psi).$$

This completes part (ii)

(iii) As

$$\lambda(\mathfrak{q}n) = \lambda^{\psi}(\mathfrak{q}n) + \lambda^{-\psi}(\mathfrak{q}n)$$

this follows immediately from parts (i) and (ii).

(iv) Let $\mathfrak{q} \in \mathcal{M}_{0,F}$ be a prime coprime to n and for which the two minima in part (i) are attained simultaneously. Then by part (ii), we have

$$d(n, \mathfrak{q}) = 0, a^{\psi}(n, \mathfrak{q}) = L, a^{-\psi}(n, \mathfrak{q}) = d_1(n, -\psi).$$

The decomposition (8.5.15) gives that

$$\mathcal{H}(n) \cong R(\chi) \oplus \bigoplus_{i \geq 1} \left(\frac{R}{\mathfrak{m}^{d_i(n,(-1)^i\chi)}}((-1)^i\chi) \right)^2$$

From the diagram (8.3.24)) and that $a^{\psi}(n, \mathfrak{q}) = L$, we obtain

$$\mathcal{H}^{\psi}_{\mathfrak{q}}(n) = (\bigoplus_{i \geq 1} \left(\frac{R}{\mathfrak{m}^{d_i(n,(-1)^i\chi)}}((-1)^i\chi) \right)^2)^{\psi}.$$

As $d(n, \mathfrak{q}) = 0, a^\psi(n, \mathfrak{q}) = L$, we obtain again from the diagram (8.3.24)

$$\mathcal{H}^\psi(\mathfrak{q}n) = \mathcal{H}^\psi_\mathfrak{q}(n) \cong (\bigoplus_{i \geq 1} \left(\frac{R}{\mathfrak{m}^{d_i(n, (-1)^i \chi)}}((-1)^i \chi) \right)^2)^\psi.$$

On the other hand, from the equalities $d(n, \mathfrak{q}) = 0, a^\psi(n, \mathfrak{q}) = L, a^{-\psi}(n, \mathfrak{q}) = d_1(n, -\psi)$ and from Step 3(ii), we obtain

$$\mathcal{H}^{-\psi}(\mathfrak{q}n) \cong R(-\psi) \oplus N.$$

The decomposition of $\mathcal{H}(\mathfrak{q}n)$ stated in (8.5.36) of part (iv) of the Step now follows from those of these two components $\mathcal{H}^{-\psi}(\mathfrak{q}n)$ and $\mathcal{H}^\psi(\mathfrak{q}n)$, which completes this Step 4.

Step 5. *Let ψ be the character of $\mathcal{H}$. We have then*

$$\min_{\mathfrak{q}_1 \ldots \mathfrak{q}_r} \lambda(\mathfrak{q}_1 \ldots \mathfrak{q}_r) = \lambda(1) - \sum_{i=1}^{r} d_i(1, (-1)^i \psi). \tag{8.5.37}$$

Here the minimum runs over all cycles $\mathfrak{q}_1 \ldots \mathfrak{q}_r$ which are products of r distinct primes in $\mathcal{M}_{0,F}$. Furthermore, for any cycle $\mathfrak{q}_1 \ldots \mathfrak{q}_r \in \mathcal{M}_{0,F}$ for which this minimum is attained we have the decomposition of $R[G]$-modules

$$\mathcal{H}(\mathfrak{q}_1 \ldots \mathfrak{q}_r) = R((-1)^r \psi) \oplus \bigoplus_{i \geq r+1} \left(\frac{R}{\mathfrak{m}^{d_i(1, (-1)^i \psi)}}((-1)^i \psi) \right)^2. \tag{8.5.38}$$

We prove these formulae (8.5.37) and (8.5.38) by induction on r. For $m \in \mathcal{M}_{0,F}$, write $\nu(m) = r$ when m is the product of r distinct prime factors in $\mathcal{M}_{0,F}$.

From the decomposition (8.5.15), the formula (8.5.37) as well as the decomposition (8.5.38) hold trivially when $r = 0$.

Fix $r \in \mathbb{N}$, and assume that these formulae (8.5.37) and (8.5.38) hold for r.

Select a cycle $n \in \mathcal{M}_{0,F}$ with $\nu(n) = r$ and for which

$$\lambda(n) = \min_{\mathfrak{q}_1 \ldots \mathfrak{q}_r} \lambda(\mathfrak{q}_1 \ldots \mathfrak{q}_r).$$

By the induction hypothesis, we have $\lambda(n) = \lambda(1) - \sum_{i=1}^{r} d_i(1, (-1)^i \psi)$ and that the decomposition (8.5.38) holds for $n = \mathfrak{q}_1 \ldots \mathfrak{q}_r$.

We have by Step 4(iii)

$$\min_\mathfrak{q} \lambda(n\mathfrak{q}) = \lambda(n) - d_1(n, -\psi).$$

8.5 Selmer Groups over Artinian Principal Ideal Rings

where q runs over all prime places in $\mathcal{M}_{0,F}$ coprime to n. Fix any prime $\mathfrak{q} \in \mathcal{M}_{0,F}$ coprime to n for which this minimum is attained. Then by Step 4(i), (iii), and (iv), we have

$$\lambda^{\psi}(n\mathfrak{q}) = \lambda^{\psi}(n)$$

$$\lambda^{-\psi}(n\mathfrak{q}) = \lambda^{-\psi}(n) - d_1(n, -\psi).$$

and also an isomorphism of $R[G]$-modules, where $G = \mathrm{Gal}(F/k)$ and χ is the character of $\mathcal{H}(n)$, from (8.5.36) of Step 4

$$\mathcal{H}(n\mathfrak{q}) \cong R(-\chi) \oplus \bigoplus_{i \geq 2} \left(\frac{R}{\mathfrak{m}^{d_i(n,(-1)^i \chi)}} ((-1)^i \chi) \right)^2. \tag{8.5.39}$$

Comparing the decomposition $\mathcal{H}(n) = R(\chi) \oplus \bigoplus_{i \geq 1} \left(\frac{R}{\mathfrak{m}^{d_i(n,(-1)^i \chi)}} ((-1)^i \chi) \right)^2$ of (8.5.15) with the induction hypothesis (8.5.38) where we put $n = \mathfrak{q}_1 \ldots \mathfrak{q}_r$, we obtain

$$d_i(n, (-1)^i \chi) = d_{i+r}(1, (-1)^{i+r} \psi) \quad (i \geq 1)$$

$$\chi = (-1)^r \psi.$$

The formula (8.5.39) then becomes

$$\mathcal{H}(n\mathfrak{q}) \cong R((-1)^{r+1} \psi) \oplus \bigoplus_{i \geq r+2} \left(\frac{R}{\mathfrak{m}^{d_i(1,(-1)^i \chi)}} ((-1)^i \psi) \right)^2.$$

Hence the formula (8.5.38) holds for $r + 1$ and $n\mathfrak{q}$, and hence the numerical formula (8.5.37) holds for $n\mathfrak{q}$ and $r + 1$. It follows by induction that the numerical formula (8.5.37) holds for all $r \in \mathbb{N}$.

We have also shown that if n is a cycle with $v(n) = s$ for which $\lambda(n) = \min_{m, v(m)=s} \lambda(m)$ and the decomposition (8.5.38) holds for n then for any prime $\mathfrak{q}$ coprime to n for which $\lambda(n\mathfrak{q}) = \min_{m, v(m)=s+1} \lambda(m)$ then the decomposition (8.5.38) holds for $\mathcal{H}(n\mathfrak{q})$. It follows that the decomposition (8.5.38) holds for any cycle n with $v(n) = r + 1$ and any $r \geq 0$ for which $\lambda(n)$ is minimal among all cycles with $v(n) = r + 1$. This completes Step 5.

Step 6. End of proof.

The equality (8.5.17) follows from (8.5.37) of Step 5. The decomposition (8.5.18) follows from (8.5.38) also of Step 5. The formula (8.5.19) follows (8.5.20), and the formula already proved (8.5.17) which completes the proof of the theorem. □

Corollary 8.5.40 *Under the hypotheses of Theorem 8.5.16, suppose that* $\kappa \in$ **KS**$(T, \mathcal{F}, \mathcal{L})$ *and* $\kappa_1 \neq 0$. *Then we have with the notation of Definition 8.5.11*

$$\partial^0(\kappa) \geq \partial^1(\kappa) \geq \partial^2(\kappa) \geq \ldots \geq 0$$

$$e_0(\kappa) \geq e_2(\kappa) \geq e_4(\kappa) \ldots \geq 0$$

$$e_1(\kappa) \geq e_3(\kappa) \geq e_5(\kappa) \ldots \geq 0.$$

Furthermore we have, putting $e_i = e_i(\kappa)$,

$$H^1_{\mathcal{F}}(F, T) \cong R(\psi) \oplus \bigoplus_{i \geq 0} \left(\frac{R}{\mathfrak{m}^{e_i}} ((-1)^{i+1}\psi) \right)^2.$$

Proof As $\kappa_1 \neq 0$, then we have $\partial^0(\kappa) < L$, where $L = \text{length}(R)$. So by Theorem 8.5.16, we have $\partial^r(\kappa) = j + \sum_{i>r} d_i(1, (-1)^i \psi))$ for all r, whence the corollary from the decomposition (8.5.15). □

8.6 Selmer Groups over Discrete Valuation Rings

(8.6.1) In this section, we apply the results of the previous section Sect. 8.5 to determine the structure of Selmer groups of Kolyvagin systems over discrete valuation rings of Heegner point and CM point type.

In the next section Sect. 8.7, this is applied to elliptic curves, but in this section, we consider an abstract Kolyvagin system of the required type.

(8.6.2) The notation is the same as in the section Sect. 8.4. We then have that

- k is a global field equipped with the special set of places ∞;
- F is an imaginary quadratic field extension of k with respect to ∞;
- R is a coefficient ring (see (4.1.2)) which is a mixed characteristic complete discrete valuation ring with local parameter π and maximal ideal $\mathfrak{m}$ and where the residue characteristic l of R is different from 2 and the characteristic of k;
- T is an R-representation of G_F;
- $(T, \mathcal{F}, \mathcal{L})$ is a Selmer triple on F of the Heegner point and CM point type given by Definition 7.3.3, (7.3.4), (7.3.2)), and for the above imaginary quadratic extension F/k and the integers $d_\mathfrak{q}$ for all $\mathfrak{q} \in \mathcal{L}$, defined in (7.3.4).

We assume for the rest of this section Sect. 8.6 that

- $(T, \mathcal{F}, \mathcal{L})$ is such that both $(T^*, \mathcal{F}^*, \mathcal{L})$ and $(T, \mathcal{F}, \mathcal{L})$ satisfy the hypotheses H.0-H.5 of Sect. 8.1 where $T \cong T^*$ by Remark 8.1.11;
- $\tau \in G_k$ is the element given by hypothesis H.4(b);

8.6 Selmer Groups over Discrete Valuation Rings

- $\mathcal{L}$ contains all but finitely many primes of F lying over primes of k unramified and inert in F/k;
- $\kappa \in \mathbf{KS}(T, \mathcal{F}, \mathcal{L})$ is a Kolyvagin system for $(T, \mathcal{F}, \mathcal{L})$.

For any $i \in \mathbb{N}$, define

$$R^{(i)} = R/\mathfrak{m}^i, \ T^{(i)} = T/\mathfrak{m}^i T, \ \mathcal{L}^{(i)} = \mathcal{L}_i(T), \ \mathcal{M}^{(i)}_{0,F} = \mathcal{M}_{0,F}(\mathcal{L}^{(i)})$$

where the last means that $\mathcal{M}^{(i)}_{0,F}$ is the set of squarefree cycles formed of primes $\mathfrak{q}$ of $\mathcal{L}^{(i)}$ lying over primes $\mathfrak{p}$ of k which are inert and and unramified in F/k (see (7.3.4)).

As in (7.3.4), $\mathcal{M}_{i,F}, i \in \mathbb{N}$, is the set of squarefree cycles formed of distinct primes $\mathfrak{q}$ of $\mathcal{L}_i$ lying over primes $\mathfrak{p}$ of k which are inert and and unramified in F/k and by convention the unit cycle 1 belongs to $\mathcal{M}_{i,F}, i \in \mathbb{N}$.

(8.6.3) If M is an R-module with an action by τ where τ restricted to M has order 2, denote by M^+, M^- the isotypical components of M on which τ acts like $+1, -1$, respectively. Write

$$\mathcal{H}^a_b(c) = H^1_{\mathcal{F}^a_b(c)}(F, T), \ \overline{\mathcal{H}}^a_b(c) = H^1_{\mathcal{F}^a_b(c)}(F, T/\mathfrak{m}T)$$

$$\mathcal{H}^{(i),a}_b(c) = H^1_{\mathcal{F}^a_b(c)}(F, T/\mathfrak{m}^i T).$$

If anyone of a, b, c is equal to 1, then it will be omitted from the notation.

Recall also from Definition 8.4.9 that in view of the decomposition $\mathcal{H}^{(i)}(n) \cong R^{(i),\epsilon} \oplus M^{(i)}(n) \oplus M^{(i)}(n)$ of $\mathcal{H}^{(i)}(n) = H^1_{\mathcal{F}(n)}(F, T^{(i)})$ (Proposition 8.4.7), put

$$\lambda^{(i)}(n) = \text{length}_R(M^{(i)}(n)), \ \text{St}^{(i)}(n) = \mathfrak{m}^{\lambda^{(i)}(n)} \mathcal{H}^{(i)}(n)$$

where $\text{St}^{(i)}(n)$ is the stub Selmer module for $T^{(i)}$ (see Definition 8.3.19).

(8.6.4) For the next proposition, let $\text{fract}(R)$ be the field of fractions of R, $\mathcal{D} = \text{fract}(R)/R$ and $A = T \otimes_R \mathcal{D}$. For a given Selmer structure $\mathcal{F}$ on T, by propagating $\mathcal{F} \otimes \text{fract}(R)$ from $T \otimes \text{fract}(R)$ to A, we obtain a Selmer structure again denoted by $\mathcal{F}$ on A.

Proposition 8.6.5 *Suppose that $\kappa_1 \neq 0$. Then there is a finite torsion $R[G]$-module M, where $G = \text{Gal}(F/k)$, together with a decomposition of $R[G]$-modules*

$$H^1_{\mathcal{F}}(F, A) \cong \mathcal{D}(\psi) \oplus M \oplus M. \tag{8.6.6}$$

Here $\mathcal{D}(\psi)$ is the R-divisible submodule isomorphic to $\mathcal{D}$ on which G acts via a character $\psi = \pm$. There is a canonical isomorphism, for all i, $H^1_{\mathcal{F}}(F, T^{(i)}) \cong H^1_{\mathcal{F}}(F, A)[\mathfrak{m}^i]$ and the module $H^1_{\mathcal{F}}(F, T)$ is a free R-module of rank 1, $R[G]$-isomorphic to $R(\psi)$, and is the π-adic Tate module of the torsion module $H^1_{\mathcal{F}}(F, A)$, where π is a local parameter of R.

Proof Since $H^1_{\mathcal{F}}(F,T) \cong \varprojlim H^1_{\mathcal{F}}(F, T^{(i)})$ and $\kappa^{(i)}$ is the reduction mod $\mathfrak{m}^i$ of κ and $\kappa_1 \neq 0$, we must have that $\kappa_1^{(i)}$ is non-zero for i sufficiently large. Fix such an i. Taking $n=1$ in Proposition 8.4.14, we have

$$\kappa_1^{(i)} \in \mathrm{St}^{(i)}(1)$$

and in particular $\mathrm{St}^{(i)}(1) \neq 0$.

We have an equality of torsion Galois modules

$$A = \bigcup_{j \geq 1} T^{(j)},$$

and Lemma 8.1.16 then implies that there are $R[G]$-isomorphisms

$$H^1_{\mathcal{F}}(F, T^{(j)}) \cong H^1_{\mathcal{F}}(F, A[\mathfrak{m}^j]) \cong H^1_{\mathcal{F}}(F, A)[\mathfrak{m}^j] \qquad (j \geq 1),$$

and we obtain from the decomposition of Lemma 8.5.12 that there is a decomposition of $R[G]$-modules for some character $\psi = \pm$

$$H^1_{\mathcal{F}}(F, A)[\mathfrak{m}^i] \cong (R/\mathfrak{m}^i)(\psi) \oplus M^{(i)} \oplus M^{(i)}$$

with $\mathrm{length}_R(M^{(i)}) < i$ because $\mathrm{St}^{(i)}(1) \neq 0$.

For all $j \geq i$, we have

$$H^1_{\mathcal{F}}(F, A)[\mathfrak{m}^j] \cong (R/\mathfrak{m}^j)(\psi) \oplus M^{(j)} \oplus M^{(j)}$$

with $\mathrm{length}_R(M^{(j)}) < j$, by lemma 8.4.12(iii), and we therefore have for all $j \geq i$

$$H^1_{\mathcal{F}}(F, T^{(j)})[\mathfrak{m}^i] \cong H^1_{\mathcal{F}}(F, A)[\mathfrak{m}^i] \cong (R/\mathfrak{m}^i)(\psi) \oplus M^{(j)}[\mathfrak{m}^i] \oplus M^{(j)}[\mathfrak{m}^i]$$

where $\mathrm{length}\, M^{(j)}[\mathfrak{m}^i] < i$ for all $j \geq i$. It follows from the structure of finitely generated R-modules that we have $\mathrm{length}\, M^{(j)} < i$, and hence $M^{(j)}$ is a finite torsion $R[G]$-module whose $R[G]$-isomorphism class is independent of j for all sufficiently large j. Therefore for some finite torsion $R[G]$-module M, there is an $R[G]$-isomorphism

$$H^1_{\mathcal{F}}(F, A) \cong \mathcal{D}(\psi) \oplus M \oplus M. \tag{8.6.7}$$

Since $H^1_{\mathcal{F}}(F, T^{(i)}) \cong H^1_{\mathcal{F}}(F, A)[\mathfrak{m}^i]$, we obtain

$$H^1_{\mathcal{F}}(F, T) = \varprojlim H^1_{\mathcal{F}}(F, A)[\mathfrak{m}^i]$$

8.6 Selmer Groups over Discrete Valuation Rings

where the transition morphisms $H^1_{\mathcal{F}}(F, A)[\mathfrak{m}^{i+1}] \to H^1_{\mathcal{F}}(F, A)[\mathfrak{m}^i]$ are multiplication by π, where π is a local parameter of R; that is to say the Selmer group $H^1_{\mathcal{F}}(F, T)$ is the π-adic Tate module of the R-torsion module $H^1_{\mathcal{F}}(F, A)$. From the decomposition (8.6.7), $H^1_{\mathcal{F}}(F, T)$ is therefore a free R-module of rank 1 and so is $R[G]$-isomorphic to $R(\psi)$. □

Definition 8.6.8 Suppose that $(T, \mathcal{F}, \mathcal{L})$ is odd (see Definition 8.4.9). For $\kappa \in \mathrm{KS}(T, \mathcal{F}, \mathcal{L})$, we define

$$\partial^\infty(\kappa) = \min_{i \in \mathbb{N},\, n \in \mathcal{M}^{(i)}_{0,F}} \{\max_a \{a \mid \kappa_n^{(i)} \in \mathfrak{m}^a \mathcal{H}^{(i)}(n)\}\}.$$

Here the interior maximum runs over all $a \in \mathbb{N} \cup \{\infty\}$ and $\partial^\infty(\kappa) \in \mathbb{N} \cup \{\infty\}$. Note that $\partial^\infty(\kappa) = \infty$ if and only if $\kappa = 0$.

We say that κ is *primitive* if $\partial^\infty(\kappa) = 0$. Note that κ being primitive is the same as $\kappa^{(1)} \neq 0$.

We similarly define the element $d(\kappa) \in \mathbb{N} \cup \{\infty\}$, where the outer minimum runs over all $i \in \mathbb{N}$ and over all $n \in \mathcal{M}^{(i)}_{0,F}$ such that $\rho(n) = 1$, where $\rho(n)$ is given in Definition 8.3.7,

$$d(\kappa) = \min_{i \in \mathbb{N},\, n \in \mathcal{M}^{(i)}_{0,F},\, \rho(n)=1} \left\{ \max_a \{a \mid \kappa_n^{(i)} \in \mathfrak{m}^a \mathcal{H}^{(i)}(n)\} \right\}.$$

The only difference between $\partial^\infty(\kappa)$ and $d(\kappa)$ is that in the latter, the $n \in \mathcal{M}^{(i)}_{0,F}$ are restricted to run over cycles with $\rho(n) = 1$.

Lemma 8.6.9 *Let $\kappa \in \mathrm{KS}(T, \mathcal{F}, \mathcal{L})$ and suppose that $l > 4$. Suppose that $(T, \mathcal{F}, \mathcal{L})$ is odd. Then for all integers $i \geq 1$ and all $n \in \mathcal{M}^{(2i-1)}_{0,F}$, the element $d(\kappa) \in \mathbb{N} \cup \{\infty\}$ satisfies $\kappa_n^{(i)} R^{(i)} = \mathfrak{m}^{d(\kappa)} \mathrm{St}^{(i)}(n)$ where $\mathrm{St}^{(i)}(n)$ is the stub Selmer module (see (8.6.3).*

Proof By Theorem 8.3.35, for every $i \geq 1$, there is an integer $0 \leq d_i \leq i$ such that if $n \in \mathcal{M}^{(i)}_{0,F}$ and $\rho(n) = 1$, then we have the isomorphisms of submodules of $\mathcal{H}^{(i)}(n)$

$$\kappa_n^{(i)} R^{(i)} \cong \mathfrak{m}^{d_i} R^{(i)} \cong \mathfrak{m}^{d_i} \mathcal{H}^{(i)}(n).$$

Since $\mathcal{M}^{(i+1)}_{0,F} \subseteq \mathcal{M}^{(i)}_{0,F}$, we evidently have $d_i \leq d_{i+1}$ and also that if $d_i < i$, then $d_i = d_{i+1} = d_{i+2} = \ldots$. We let $d(\kappa) = \lim_{i \to \infty} d_i$ where it is possible that $d(\kappa) = +\infty$; it can be checked that this $d(\kappa)$ coincides with that given in Definition 8.6.8. It follows that if $n \in \mathcal{M}^{(i)}_{0,F}$ and $\rho(n) = 1$, then we have

$$\kappa_n^{(i)} R^{(i)} \cong \mathfrak{m}^{d(\kappa)} R^{(i)}$$

for all $i \geq 1$.

Put, for $n \in \mathcal{M}_{0,F}^{(2i-1)}$ (as in Definition 8.3.7),

$$\rho(n) = \dim_{R/\mathfrak{m}} \overline{\mathcal{H}}(n), \quad \rho^\delta(n) = \dim_{R/\mathfrak{m}} \overline{\mathcal{H}}(n)^\delta \quad (\delta = \pm).$$

By Proposition 8.4.14, if $i \geq 1$ and $n \in \mathcal{M}_{0,F}^{(2i-1)}$, then we have $\kappa_n^{(i)} \in \mathrm{St}^{(i)}(n) \otimes \Gamma_n$. We will prove by induction on $r(n) = \max(\rho^+(n), \rho^-(n))$ that if $n \in \mathcal{M}_{0,F}^{(2i-1)}$, then $\kappa_n^{(i)} R^{(i)} = \mathfrak{m}^{d(\kappa)} \mathrm{St}^{(i)}(n)$. This evidently holds from the first paragraph of this proof when $\rho(n) = 1$. We may then assume that $\rho(n) > 1$. As we have that $\kappa_n^{(i)} \in \mathrm{St}^{(i)}(n)$, we need only to consider the case where $\mathrm{St}^{(i)}(n) \neq 0$.

From Corollary 8.3.33 and as $n \in \mathcal{M}_{0,F}^{(2i-1)}$, we may choose a prime cycle $\mathfrak{q} \in \mathcal{M}_{0,F}^{(2i-1)}$ coprime to n so that $n\mathfrak{q} \in \mathcal{M}_{0,F}^{(2i-1)}$ and where $r(n\mathfrak{q}) < r(n)$ and

$$\mathrm{res}_{F_\mathfrak{q}/F}(\mathrm{St}^{(i)}(n)[\mathfrak{m}]) \neq 0. \tag{8.6.10}$$

Write $\mathrm{length}_R(N)$ for the length of an artinian R-module N. For any $i \geq 1$, we have by (8.6.10), where $\kappa_n^{(i)} R^{(i)}$ is the submodule of $\mathrm{St}^{(i)}(n) \otimes \Gamma_n$ generated by $\kappa_n^{(i)}$,

$$\mathrm{length}_R(\kappa_n^{(i)} R^{(i)}) = \mathrm{length}_R(\mathrm{res}_{F_\mathfrak{q}/F}(\kappa_n^{(i)} R^{(i)})).$$

By the Kolyvagin system relations (7.1.8), this equals

$$\mathrm{length}_R(\mathrm{res}_{F_\mathfrak{q}/F}(\kappa_{n\mathfrak{q}}^{(i)} R^{(i)})).$$

By the induction hypothesis, this equals

$$\mathrm{length}_R(\mathrm{res}_{F_\mathfrak{q}/F}(\mathfrak{m}^{d(\kappa)} \mathrm{St}^{(i)}(n\mathfrak{q}))).$$

By Proposition 8.3.25, this equals

$$\mathrm{length}_R(\mathrm{res}_{F_\mathfrak{q}/F}(\mathfrak{m}^{d(\kappa)} \mathrm{St}^{(i)}(n))).$$

Again by (8.6.10), this equals

$$\mathrm{length}_R(\mathfrak{m}^{d(\kappa)} \mathrm{St}^{(i)}(n)).$$

That is to say, we have then shown that $\mathrm{length}_R(\kappa_n^{(i)} R^{(i)}) = \mathrm{length}_R(\mathfrak{m}^{d(\kappa)} \mathrm{St}^{(i)}(n))$. This means that $\kappa_n^{(i)} R^{(i)} = \mathfrak{m}^{d(\kappa)} \mathrm{St}^{(i)}(n)$ which completes the induction. □

Proposition 8.6.11 *Suppose that $\kappa \in \mathbf{KS}(T, \mathcal{F}, \mathcal{L})$ is a non-zero Kolyvagin system and that $l > 4$. Suppose that $(T, \mathcal{F}, \mathcal{L})$ is odd. Then we have $d(\kappa) \geq \partial^\infty(\kappa)$. Furthermore, we have $d(\kappa) = 0$ if and only if κ is primitive.*

8.6 Selmer Groups over Discrete Valuation Rings

Proof By the definition of $d(\kappa)$ and $\partial^\infty(\kappa)$ (Definition 8.6.8), it is immediate that $d(\kappa) \geq \partial^\infty(\kappa)$. For the second claim, it suffices to prove $\partial^\infty(\kappa) = 0$ implies that $d(\kappa) = 0$. That $\partial^\infty(\kappa)$ is zero is equivalent to $\kappa^{(1)} \neq 0$. So let $n \in \mathcal{M}_{0,F}^{(1)}$ be such that $\kappa_n^{(1)} \neq 0$. By Lemma 8.6.9, we have $0 \neq \kappa_n^{(1)} R^{(1)} = \mathfrak{m}^{d(\kappa)} \mathrm{St}^{(1)}(n)$. Hence we have $\mathfrak{m}^{d(\kappa)} \mathrm{St}^{(1)}(n) \neq 0$ and this implies that $d(\kappa) = 0$. □

Remark 8.6.12 The Kolyvagin system $\kappa \in \mathbf{KS}(T, \mathcal{F}, \mathcal{L})$ induces a Kolyvagin system $\kappa^{(i)}$ in $\in \mathbf{KS}(T^{(i)}, \mathcal{F}, \mathcal{L}^{(i)})$ for all $i \geq 0$ by taking κ mod $\mathfrak{m}^i$. Associated with κ and $\kappa^{(i)}$ are the numerical invariants $d(\kappa)$, $\partial^\infty(\kappa)$ (Definition 8.6.8) and $\partial^r(\kappa^{(i)})$, $\partial^r(\kappa)$ (below in Definition 8.6.15). The relation between these is not entirely clear.

We have $d(\kappa) = \partial^\infty(\kappa)$ for the case of Kolyvagin systems obtained from Heegner points and CM points on elliptic curves over global fields where this follows from the formula (8.6.14) in Theorem 8.6.13 combined with Theorem 8.7.25 in the next section Sect. 8.7 and under the hypotheses of the latter theorem. Furthermore, under the hypotheses of Theorem 8.6.19, we obtain $\partial^r(\kappa) = \lim_{i \to +\infty} \partial^r(\kappa^{(i)})$.

See also Exercise 8.6.26.

Theorem 8.6.13 *Suppose there is a Kolyvagin system $\kappa \in \mathbf{KS}(T, \mathcal{F}, \mathcal{L})$ with $\kappa_1 \neq 0$. Assume that $l > 4$. In the decomposition of $R[G]$-modules given by Proposition 8.6.5*

$$H_\mathcal{F}^1(F, A) \cong \mathcal{D}(\psi) \oplus M \oplus M$$

where M is a finite $R[G]$-module, we have

$$\mathrm{length}_R(M) = \mathrm{length}_R(H_\mathcal{F}^1(F, T)/R.\kappa_1) - d(\kappa). \qquad (8.6.14)$$

In particular, we have $\mathrm{length}_R(M) \leq \mathrm{length}_R(H_\mathcal{F}^1(F, T)/R.\kappa_1))$ with equality if and only if κ is primitive.

Proof Proposition 8.6.5 gives the decomposition of $R[G]$-modules $H_\mathcal{F}^1(F, A) \cong \mathcal{D}(\psi) \oplus M \oplus M$ where M is a finite $R[G]$-module and $\psi = \pm$ is a character of $\mathrm{Gal}(F/k)$, which gives the first part. It follows from this decomposition that the Selmer triple $(T, \mathcal{F}, \mathcal{L})$ has odd rank (Definitions 8.3.20 and 8.4.9). Again from Proposition 8.6.5 and this decomposition, we have that there is a canonical $R[G]$-module isomorphism, for all i, $H_\mathcal{F}^1(F, T^{(i)}) \cong H_\mathcal{F}^1(F, A)[\mathfrak{m}^i]$ and the module $H_\mathcal{F}^1(F, T)$ is a free R-module of rank 1 and is the π-adic Tate module of the R-torsion module $H_\mathcal{F}^1(F, A)$, where π is a local parameter of R, and so is $R[G]$-isomorphic to $R(\psi)$.

For all sufficiently large i, we then obtain the $R[G]$-module decomposition

$$H_\mathcal{F}^1(F, T/\mathfrak{m}^i T) = \mathcal{H}^{(i)}(1) \cong (R/\mathfrak{m}^i)(\psi) \oplus M \oplus M.$$

By Lemma 8.6.9, we have $\kappa_1^{(i)} R^{(i)} = \mathfrak{m}^{d(\kappa)} \mathrm{St}^{(i)}(1)$. Put

$$\lambda = \mathrm{length}_R(M)$$

where this equals $\lambda^{(i)}(1)$, for i sufficiently large (see (8.4.10)).

From Lemma 8.1.16, there are canonical $R[G]$-isomorphisms

$$H^1_{\mathcal{F}}(F, T^{(i)}) \cong H^1_{\mathcal{F}}(F, A[\mathfrak{m}^i]) \cong H^1_{\mathcal{F}}(F, A)[\mathfrak{m}^i]$$

, and by Proposition 1.3.10, there is an $R[G]$-isomorphism

$$H^1_{\mathcal{F}}(F, T) \cong \varprojlim_i H^1_{\mathcal{F}}(F, T^{(i)}).$$

As $H^1_{\mathcal{F}}(F, T)$ is a free R-module of rank 1 and is the π-adic Tate module of the R-torsion module $H^1_{\mathcal{F}}(F, A)$, by Proposition 8.6.5, it follows that the map

$$H^1_{\mathcal{F}}(F, T)//\mathfrak{m}^i H^1_{\mathcal{F}}(F, T) \longrightarrow H^1_{\mathcal{F}}(F, T/\mathfrak{m}^i T)$$

is injective and taking $i > d(\kappa) + \lambda$ this injectivity combined with $\kappa_1^{(i)} R^{(i)} = \mathfrak{m}^{d(\kappa)} \mathrm{St}^{(i)}(1)$ shows that we have $\kappa_1 R = \mathfrak{m}^{\lambda + d(\kappa)} H^1_{\mathcal{F}}(F, T)$. Hence, we have

$$\mathrm{length}_R(M) = \mathrm{length}_R(H^1_{\mathcal{F}}(F, T)/R.\kappa_1) - d(\kappa).$$

The inequality stated in the final sentence of the theorem follows immediately, and this is an equality if and only if $d(\kappa) = 0$ and where this latter is equivalent to κ being primitive from Proposition 8.6.11. □

Definition 8.6.15 The Kolyvagin system $\kappa \in \mathbf{KS}(T, \mathcal{F}, \mathcal{L})$ induces a Kolyvagin system $\kappa^{(i)}$ in $\in \mathbf{KS}(T^{(i)}, \mathcal{F}, \mathcal{L}^{(i)})$ for all $i \geq 0$ by taking κ mod $\mathfrak{m}^i$.

Recall from Definition 8.5.11 that we have for $\kappa^{(i)}$ and $r \in \mathbb{N}$, where $i = \mathrm{length}_R R/\mathfrak{m}^i$,

$$\partial^r(\kappa^{(i)}) = \min\{i - \mathrm{length}(R^{(i)} \kappa_m^{(i)}) | \, m \in \mathcal{M}_{0,F}^{(i)}, \nu(m) = r\}.$$

Here $\nu(m)$ denotes the number of distinct prime divisors of the cycle $m \in \mathcal{M}_{0,F}$.

The elementary divisors of $\kappa^{(i)}$ are defined by

$$e_j(\kappa^{(i)}) = \partial^j(\kappa^{(i)}) - \partial^{j+1}(\kappa^{(i)}) \qquad (i \geq 0).$$

Define for κ

$$\partial^r(\kappa) = \max(j : \kappa_m \in \mathfrak{m}^j H^1_{\mathcal{F}(m)}(F, T/I_m T) \otimes \Gamma_m | \, m \in \mathcal{M}_{0,F}, \nu(m) = r\}.$$

8.6 Selmer Groups over Discrete Valuation Rings

The elementary divisors of κ are similarly defined by

$$e_j(\kappa) = \partial^j(\kappa) - \partial^{j+1}(\kappa).$$

(8.6.16) To prepare for the next Theorem 8.6.19, suppose that $\kappa \in \mathbf{KS}(T, \mathcal{F}, \mathcal{L})$ and $\kappa_1 \neq 0$. From Proposition 8.6.5, there is a finite $R[G]$-module M such that

$$H^1_{\mathcal{F}(n)}(F, A) \cong \mathcal{D}(\psi) \oplus M \oplus M.$$

where $\psi = \pm$ is a character of $\mathrm{Gal}(F/k)$. The $R[G]$-module M can be decomposed into cyclic $R[G]$-submodules, so that we then have the decomposition into cyclic $R[G]$-modules

$$H^1_{\mathcal{F}(n)}(F, A) = \mathcal{D}(\psi) \oplus \bigoplus_{h \geq 1} \left(\frac{R}{\mathfrak{m}^{d_h(n, (-1)^h \psi)}}((-1)^h \psi) \right)^2 \tag{8.6.17}$$

where each component $R/\mathfrak{m}^{d_h(n,(-1)^h\psi)}((-1)^h\psi)$ is in the $(-1)^h\psi$-eigenspace for all h and where the integers $d_h = d_h(n, (-1)^h\psi)$ may be assumed to satisfy

$$d_1 \geq d_3 \geq d_5 \geq \ldots \geq 0 \text{ and } d_2 \geq d_4 \geq d_6 \geq \ldots \geq 0. \tag{8.6.18}$$

The character ψ and these integer sequences are uniquely determined by $H^1_{\mathcal{F}}(F, A)$.

For all $i \in \mathbb{N}$, the R-module $\mathrm{St}^{(i)}(n) = \mathfrak{m}^{\lambda^{(i)}(n)} \mathcal{H}^{(i)}(n)$ is the stub Selmer module for $(T^{(i)}, \mathcal{F}, \mathcal{L}^{(i)})$ over $R^{(i)}$ for any n (see (8.6.3)). Then for all $i \in \mathbb{N}$, we have a Kolyvagin system $\kappa^{(i)}$ obtained by taking the reduction of κ mod $\mathfrak{m}^i$ (see Remark 8.6.12).

As the unit cycle $1 \in \mathcal{M}^{(i)}_{0,F}$ for all i, it follows from Proposition 8.4.14 that $\kappa^{(i)}_1 \in \mathrm{St}^{(i)}(1) \otimes \Gamma_1$ for all i. As $\mathrm{St}^{(i)}(1)$ is a cyclic R-module, there is a unique integer $9 \leq j(i) \leq i$ such the $\kappa^{(i)}_1$ generates $\mathfrak{m}^{j(i)} \mathrm{St}^{(i)}(1)$ for all i.

Theorem 8.6.19 *Suppose that $\kappa \in \mathbf{KS}(T, \mathcal{F}, \mathcal{L})$, where $\kappa_1 \neq 0$, and that $l > 4$. Then there is an unique integer $j \in \mathbb{N}$ such that $\kappa^{(i)}_1$ generates $\mathfrak{m}^j \mathrm{St}^{(i)}(\mathcal{H})(1) \otimes \Gamma_1$ for all i (see (8.6.16)). Also we have*

$$\partial^r(\kappa) = \lim_{i \to +\infty} \partial^r(\kappa^{(i)}).$$

Furthermore, we have

$$\partial^0(\kappa) \geq \partial^1(\kappa) \geq \partial^2(\kappa) \geq \ldots \geq 0$$

$$e_0(\kappa) \geq e_2(\kappa) \geq e_4(\kappa) \ldots \geq 0, \quad e_1(\kappa) \geq e_3(\kappa) \geq e_5(\kappa) \ldots \geq 0$$

and we have (see (8.6.17) and (8.6.18)) where $d_i = d_i(1, (-1)^i \psi)$

$$\partial^r(\kappa) = j + \sum_{i \geq r+1} d_i.$$

Finally, there is the decomposition, putting $e_i = e_i(\kappa)$,

$$H^1_{\mathcal{F}}(F, A) \cong \mathcal{D}(\psi) \oplus \bigoplus_{i \geq 0} \left(\frac{R}{\mathfrak{m}^{e_i}}((-1)^{i+1}\psi) \right)^2.$$

Here the component $\mathcal{D}(\psi)$ is the divisible subgroup of $H^1_{\mathcal{F}}(F, A)$ and has an action by $\mathrm{Gal}(F/k)$ where the character of this subgroup is $\psi = \pm$.

Proof From Proposition 8.6.5 replacing $\mathcal{F}$ by $\mathcal{F}(n)$, we have the decomposition of $R[G]$-modules

$$H^1_{\mathcal{F}(n)}(F, A) \cong \mathcal{D}(\psi) \oplus M(n) \oplus M(n) \tag{8.6.20}$$

where $M(n)$ is a finite $R[G]$-module and $\mathcal{D}(\psi)$ is the R-divisible submodule of $H^1_{\mathcal{F}(n)}(F, A)$ and which forms a single eigencomponent under $\mathrm{Gal}(F/k)$ and where its character is ψ.

This decomposition shows that the Selmer triple $(T, \mathcal{F}, \mathcal{L})$ has odd rank (Definitions 8.3.20 and 8.4.9). Again from Proposition 8.6.5 and Lemma 8.1.16, we have that there is a canonical isomorphism of $R[G]$-modules, for all i,

$$H^1_{\mathcal{F}(n)}(F, T^{(i)}) \cong H^1_{\mathcal{F}(n)}(F, A)[\mathfrak{m}^i] \tag{8.6.21}$$

and the module $H^1_{\mathcal{F}(n)}(F, T)$ is a free R-module of rank 1 and is the π-adic Tate module of the R-torsion module $H^1_{\mathcal{F}(n)}(F, A)$, where π is a local parameter of R.

We obtain from (8.6.20) and (8.6.21) that there are isomorphisms of $R[G]$-modules, for all i,

$$H^1_{\mathcal{F}(n)}(F, T^{(i)}) \cong (R/\mathfrak{m}^i)(\psi) \oplus (M(n) \oplus M(n))[\mathfrak{m}^i]. \tag{8.6.22}$$

From the decomposition (8.6.22), for $n \in \mathcal{M}_{0,F}$ and $i \in \mathbb{N}$, the $R[G]$-module $H^1_{\mathcal{F}(n)}(F, T^{(i)})$ decomposes as a sum of cyclic $R[G]$-modules, where ψ is the character of $H^1_{\mathcal{F}(n)}(F, T^{(i)})$,

$$H^1_{\mathcal{F}(n)}(F, T^{(i)}) = R^{(i)}(\psi) \oplus \bigoplus_{h \geq 1} \left(\frac{R}{\mathfrak{m}^{d_h(i,n,(-1)^h\psi)}}((-1)^h \psi) \right)^2 \tag{8.6.23}$$

8.6 Selmer Groups over Discrete Valuation Rings

where each component $R/\mathfrak{m}^{d_h(i,n,(-1)^h\psi)}((-1)^h\psi)$ is in the $(-1)^h\psi$-eigenspace for all h and where the integer $d_h(i,n,(-1)^h\psi)$ depends on i for all h and where we may assume that, by permuting the components and writing $d_h = d_h(i,n,(-1)^h\psi)$,

$$i \geq d_1 \geq d_3 \geq d_5 \geq \ldots$$

$$i \geq d_2 \geq d_4 \geq d_6 \geq \ldots.$$

These conditions uniquely determine the integers $d_h(i,n,(-1)^h\psi) \in \mathbb{N}$ in the decomposition of (8.6.23).

Comparing the decompositions (8.6.17) and (8.6.23) using the isomorphism (8.6.21), it follows that we have for all i

$$d_h(i,n,(-1)^h\psi) = \min(i, d_h(n,(-1)^h\psi)). \tag{8.6.24}$$

We have already remarked that $\kappa_1^{(i)}$ lies in the stub Selmer module $\mathfrak{m}^{j(i)}\mathrm{St}^{(i)}(\mathcal{H})(1)$ for all i (see (8.6.16)). Let $j(i) \in \mathbb{N}$, where $0 \leq j(i) \leq i$, be the unique integer such that $\kappa_1^{(i)}$ generates $\mathfrak{m}^{j(i)}\mathrm{St}^{(i)}(\mathcal{H})(1)$ for all i (see (8.6.16)).

We have that $\kappa_1^{(i)} \equiv \kappa_1 \bmod \mathfrak{m}^i$ hence $\kappa_1^{(i)} \neq 0$ for all $i \gg 0$ as $\kappa_1 \neq 0$. It follows that $j(i) < i$ for all $i \gg 0$. We have by definition (see (8.6.3)) in view of the decomposition $\mathcal{H}^{(i)}(n) \cong R^{(i),\epsilon} \oplus M^{(i)}(n) \oplus M^{(i)}(n)$ of $\mathcal{H}^{(i)}(n) = H^1_{\mathcal{F}(n)}(F, T^{(i)})$ (Proposition 8.4.7), that

$$\lambda^{(i)}(n) = \mathrm{length}_R(M^{(i)}(n)), \quad \mathrm{St}^{(i)}(n) = \mathfrak{m}^{\lambda^{(i)}(n)}\mathcal{H}^{(i)}(n)$$

where $\mathrm{St}^{(i)}(n)$ is the stub Selmer module for $T^{(i)}$.

By (8.6.20) and (8.6.21), we have

$$M^{(i)}(1) \cong M(1)[\mathfrak{m}^i]$$

so that $\lambda^{(i)}(1) = \lambda(1)$ for all $i \gg 0$.

We have that $H^1_{\mathcal{F}}(F, T)$ is an $R[G]$-module which is free R-module of rank 1 and is the π-adic Tate module of $H^1_{\mathcal{F}}(F, A)$ (Proposition 8.6.5). We have $\kappa_1 \in H^1_{\mathcal{F}}(F, T)$ and a commutative diagram of $R[G]$-modules

$$\begin{array}{ccc} H^1_{\mathcal{F}}(F, A)[\mathfrak{m}^{i+1}] & \xrightarrow{f_{i+1}} & \mathcal{H}^{(i+1)}(1) \\ \pi \downarrow & & \downarrow \\ H^1_{\mathcal{F}}(F, A)[\mathfrak{m}^i] & \xrightarrow{f_i} & \mathcal{H}^{(i)}(1) \end{array}$$

Here the maps f_i, f_{i+1} are the isomorphisms (8.6.21) and the map labelled π is that obtained by multiplication by π.

By definition of $j(i)$ and as $\lambda^{(i)}(1) = \lambda(1)$ for all $i \gg 0$, we obtain that $\kappa_1^{(i)} R^{(i)} \cong \mathfrak{m}^{j(i)+\lambda(1)} H_{\mathcal{F}}(F, T^{(i)})$ for $i \gg 0$. It follows that length $\kappa_1^{(i)} R^{(i)} = i - j(i) - \lambda(1)$ for $i \gg 0$. The above commutative diagram gives $\pi f_{i+1}^{-1} R\kappa^{(i+1)} = f_i^{-1} R\kappa^{(i)}$. Hence we have length $\kappa_1^{(i+1)} R^{(i+1)} = i + 1 - j(i+1) - \lambda(1)$ for $i \gg 0$ from which we obtain $j(i) = j(i+1)$ for $i \gg 0$. Hence, we have that $j(i)$ is independent of i for all $i \gg 0$. Write j for this common value of $j(i)$ for $i \gg 0$. This gives the first statement of the theorem that there is a unique integer $j \in \mathbb{N}$ such that $\kappa_1^{(i)}$ generates $\mathfrak{m}^j \mathrm{St}^{(i)}(\mathcal{H})(1) \otimes \Gamma_1$ for all i.

Then the formula (8.5.19) of Theorem 8.5.16 applied to $T^{(i)}$ and $\kappa_1^{(i)}$ combined with $j(i) = j$ for $i \gg 0$ and the equalities (8.6.24) shows that for $i \gg 0$, we have

$$\partial^r(\kappa^{(i)}) = \min(i, j + \sum_{h \geq r+1} \min(i, d_h(1, (-1)^h \psi))). \tag{8.6.25}$$

Fix $s \geq 0$, and let $\delta = \partial^s(\kappa)$. Then there is $n \in \mathcal{M}_{0,F}$ with $\nu(n) = s$ such that $\kappa_n \notin \mathfrak{m}^{\delta+1} H^1_{\mathcal{F}(n)}(F, T/I_n T)$ and $\kappa_n \in \mathfrak{m}^{\delta} H^1_{\mathcal{F}(n)}(F, T/I_n T)$.

We have by Proposition 8.4.14 that $\kappa_n^{(\delta+1)} \in \mathrm{St}^{(\delta+1)}(n)$. Furthermore, the stub Selmer module $\mathrm{St}^{(\delta+1)}(\mathcal{H})(n)$ for $(T^{(\delta+1)}, \mathcal{F}, \mathcal{L}^{(\delta+1)})$ is a free $R^{(\delta+1)}/\mathfrak{m}^{\delta+1-\lambda(n)}$-module of rank 1, where we write $\lambda(n)$ in place of $\lambda^{(\delta+1)}(n)$. It follows that as $R^{(\delta+1)} \kappa_n^{(\delta+1)}$ is a submodule of this free $R^{(\delta+1)}/\mathfrak{m}^{\delta+1-\lambda(n)}$-module, we must have

$$\kappa_n^{(\delta+1)} \in \mathfrak{m}^{\delta+1-\mathrm{length}(R^{(h+1)}\kappa_n^{(\delta+1)})} \mathcal{H}^{(\delta+1)}(n).$$

This combined with $\kappa_n \notin \mathfrak{m}^{\delta+1} H^1_{\mathcal{F}(n)}(F, T/I_n T)$ shows that we have $\partial^s(\kappa^{(\delta+1)}) \leq \delta$. From (8.6.25), it follows that $\partial^s(\kappa^{(i)}) \leq \delta = \partial^s(\kappa)$ for all $i \geq \delta$.

On the other hand, since $\kappa_n \in \mathfrak{m}^{\delta} H^1_{\mathcal{F}(n)}(F, T/I_n T)$ for every $n \in \mathcal{M}_{0,F}$ with $\nu(n) = s$, by definition, we have $\partial^s(\kappa^{(i)}) \geq \delta$ for every $i \geq \delta$. This combined with the previous paragraph shows that $\partial^s(\kappa) = \sup_{i \geq 0}(\partial^s(\kappa^{(i)}))$. Since $\partial^s(\kappa^{(i)})$ is a non-decreasing function of i by (8.6.25), it follows that $\partial^s(\kappa) = \lim_{i \to +\infty} \partial^s(\kappa^{(i)})$ as stated in the theorem.

The rest of Theorem 8.6.19 follows from (8.6.17), (8.6.24), (8.6.25) and this limit. □

Exercise 8.6.26 Let $\kappa \in \mathbf{KS}(T, \mathcal{F}, \mathcal{L})$ and assume the hypotheses of Lemma 8.6.9 hold. Show that $d(\kappa) = \infty$ if and only if $\kappa = 0$. [Hint: If $d(\kappa) = \infty$, we must have $d_i = i$ for all i in the notation of the proof of Lemma 8.6.9. Then use Proposition 8.3.36.]

8.7 Shafarevich-Tate Groups of Elliptic Curves

In this section, we apply the previous results to the Shafarevich-Tate groups of elliptic curves over global fields via Kolyvagin systems.

8.7 Shafarevich-Tate Groups of Elliptic Curves

For an elliptic curve over a global field, we require a Selmer triple $(T_l, \mathcal{F}, \mathcal{L})$ obtained from the elliptic curve as well as a Heegner system given by Heegner points and CM points and finally a Kolyvagin system obtained from this Heegner system.

The Selmer group structure of elliptic curves can be determined using Kolyvagin systems and Euler systems. This has already been done in the case of elliptic curves over the rational number field by Kolyvagin via Euler systems of Heegner points and similarly for elliptic curves over global function fields in [5] again by Euler systems arising from Heegner points on Drinfeld modular curves.

In this section, using Kolyvagin systems, we prove uniformly a decomposition of the Selmer groups and hence the Shafarevich-Tate groups for elliptic curves over totally real number fields parametrized by Shimura curves, or classical modular curves when the ground field is $\mathbb{Q}$, and also for elliptic curves over global function fields parametrized by Drinfeld modular curves.

(8.7.1) The notation for Kolyvagin systems of Heegner point and CM point Kolyvagin systems holds in this section where this is now restricted to the case of elliptic curves (see Definition 7.3.3, (7.3.4)). In particular, we let

- k be a global field;
- F be a quadratic Galois field extension of k;
- $R = \mathbb{Z}_l$ be the l-adic completion of $\mathbb{Z}$ where l is a prime number different from the characteristic of k;
- E/k be an elliptic curve where it is assumed that if k has characteristic zero then E/k does not have potential complex multiplication;
- T_l be the l-adic Tate module of E;
- $\mathcal{F}$ be the classical Selmer structure on T_l over F as well as on the finite group schemes $E[n]/k$ for all $n \in \mathbb{N}$ coprime to the characteristic of k (see (5.5.7)–(5.5.11), Definition 7.3.3);
- $S(\mathcal{F})$ be the finite exceptional set of places consisting of the finite set of places of F where T is ramified together with places of F above l, if F is a number field, and the Archimedean places of F, if any;
- $\mathcal{L}$ be a subset of Σ_F disjoint from $S(\mathcal{F})$.

Modularity of E/k

(8.7.2) We assume throughout this section Sect. 8.7 that the field extension F/k is one of the following.

(a) The field k is a totally real number field; F is a totally imaginary quadratic extension field of F. The field k is equipped with a special set of places, namely, the set ∞_k of (equivalence classes of) Archimedean places of k;
(b) The field k is a global field of positive characteristic equipped with a special set ∞_k of places, F/k is an imaginary quadratic extension with respect to ∞_k.

Despite these two classes of fields (a), (b) distinguished by characteristic, we often clump together the rational number field case $k = \mathbb{Q}$ with the positive characteristic case as these are similar to each other regarding Heegner points, whereas the general Shimura curve case for totally real number fields is rather different.

(8.7.3) (Function field and rational number field case). If either (i) k has positive characteristic and E/k has split (Tate) multiplicative reduction at the place in ∞_k or (ii) k has characteristic zero and $k = \mathbb{Q}$, then by Theorem 2.6.2, there is a finite surjective morphism of k-schemes

$$\pi : X_0^{\text{Ell.Sp.}}(\mathfrak{J}) \to E$$

where $\mathfrak{J}$ is the conductor of E/k with any place in ∞_k removed. This map is only determined up to translation in E. Nonetheless, there is a unique choice of map π by "rigidification by by ξ" as in paragraph (3.4.4).

(8.7.4) (Totally real number field case). Suppose now that k is a totally real algebraic number field. Let H is an open compact subgroup of $\widehat{B}^*$, where B is a quaternion algebra over k of the type in Sect. 2.3. Let N_H^*/k be the projective Shimura curve defined by H (see (2.3)).

Suppose that there is a finite surjective morphism of k-schemes

$$\pi : N_H^* \longrightarrow E.$$

The map π is defined only up to translation in the group scheme E. If the subscheme N_H of N_H^* is non-compact, then N_H and N_H^* are classical modular curves defined over $\mathbb{Q}$, and the rigidification of ξ is already given in the previous paragraph.

Suppose then that N_H is compact. Then again π may be "rigidified by ξ" as in (3.4.11) by using a "Hodge class". In general, it is not known when such a parametrization $\pi : N_H^* \longrightarrow E$ of an elliptic curve by a Shimura curve exists.

(8.7.5) We say that E/k is *modular* if there is a finite surjective morphism of k-schemes

$$\pi : X_0^{\text{Ell.Sp.}}(\mathfrak{J}) \to E \text{ or } \pi : N_H^* \longrightarrow E$$

as in (8.7.3) and (8.7.4) above where π is rigidified by ξ.

We assume for the rest of this section Sect. 8.7 that E/k is modular in this sense that a morphism of the above type exists and the fields F/k are of one of the above types in (8.7.2) and E/k does not have potential complex multiplication if k is a number field.

Heegner Points and CM Points on E/k and Axioms for Heegner Systems

(8.7.6) (Function field and rational number field case). Heegner points can be defined on E as in Definition 3.4.5 where a Heegner point on E is written $(a, \mathfrak{J}_1, c, \pi)$. It is here

8.7 Shafarevich-Tate Groups of Elliptic Curves

assumed that $\mathfrak{J}$ is the conductor of E/k without the component at ∞_k and also that $\mathfrak{J}$ and F/k satisfy the *Heegner condition* that all prime ideal components of $\mathfrak{J}$ split completely in the field extension F/k (see (3.2.2)) and that $\mathfrak{J} = \mathfrak{J}_1 \mathfrak{J}_2$ as in (3.2.3). This is a necessary condition to construct the Heegner points.

By Proposition 6.6.5, the Heegner points $x_c = (1, \mathfrak{J}_1, c, \pi) \in E(F[c])$ form a Heegner system (Sect. 6.5) where c runs over a set $\mathcal{C}$ consisting of all squarefree cycles on k which are coprime to a finite set of excluded primes. The finite excluded set of primes is given explicitly in (6.6.2), Definition 6.7.6 and Proposition 6.7.7.

(8.7.7) (Totally real number field and compact Shimura curve case). By Proposition 6.7.7, we have the CM point $x_c = \pi x(c) \in E$ on the elliptic curve E and defined over the field $F_c = F(x(c))$ (see Definitions 3.3.4, 3.4.12, 3.4.16, 3.4.21, and Proposition 6.7.7).

By Proposition 6.7.7, these CM points x_c form a Heegner system (Sect. 6.5), where c runs over all elements of a set $\mathcal{C}$ consisting of all squarefree cycles on k which are coprime to a finite set of excluded primes. The finite excluded set of primes from $\mathcal{C}$ is given explicitly in Definition 6.7.6 and Proposition 6.7.7.

Note that x_1 is a CM point given by $x_1 = \pi x(1) \in E(F(x(1)))$.

The Selmer Triple on E

(8.7.8) Let $\mathcal{E}_1$ be the set of all prime numbers coprime to the characteristic of k, and for each $l \in \mathcal{E}_1$, let $T_l = T_l(E)$ be the l-adic Tate module of E/k.

Let $(T_l, \mathcal{F}, \mathcal{L})$ be the Selmer triple given by E and the quadratic field extension F/k; that is to say (see also (5.5.7)–(5.5.11) or Definition 7.3.3 and take $J = E$ in the notation there)

- T_l is the l-adic Tate module of E/k as a G_k-module and hence by restriction as a G_F-module;
- $S(\mathcal{F})$ is the finite set of places of the field F where T_l is ramified, i.e. the places of bad reduction of $E \times_k F$, together with the places above l (if any) and the archimedean places (if any);
- $\mathcal{F}$ is the classical Selmer structure for the elliptic curve $E \times_k F$ with exceptional set $S(\mathcal{F})$ (see Sect. 5.5);
- $\mathcal{L}$ is a subset of Σ_F disjoint from $S(\mathcal{F})$.

Given this $(T_l, \mathcal{F}, \mathcal{L})$ and $l \in \mathcal{E}_1$, and the imaginary quadratic extension F/k, define the integers d_q, ideals I_n, groups Γ_n, sets of primes $\mathcal{L}_i$, and the sets of cycles $\mathcal{M}_{i,F}$ exactly as in Definition 7.3.3 and (7.3.4).

The set $\mathcal{M}_{i,k}$ of cycles, for any $i \in \mathbb{N}$, is defined to be the set of cycles on k lying below the cycles in $\mathcal{M}_{i,F}$ on F where we recall that any prime divisor dividing a cycle in $\mathcal{M}_{i,F}$ lies over a prime of k unramified and inert in F/k (see (7.4.4) and (7.4.5) for a precise definition). There is a canonical bijection $\mathcal{M}_{i,F} \cong \mathcal{M}_{i,k}$ obtained by lowering a cycle on F to a cycle on k (see (7.4.5)).

Derived Cohomology Classes

(8.7.9) There is a subset $\mathcal{E}_2 \subseteq \mathcal{E}_1$ obtained by removing a finite set of prime numbers from $\mathcal{E}_1$, with the following property for all prime numbers $l \in \mathcal{E}_2$:
For all $c \in \mathcal{M}_{1,k}$ (see (8.7.8)), there is the derived cohomology class

$$\theta_c \in H^1(F, T_l/I_{c^\sharp}T_l) \otimes \Gamma_{c^\sharp}$$

where the component in $\Gamma_{c^\sharp}$ arises because of the choices of generators ϕ_q we have made for each prime q dividing $c^\sharp$ (see Definition 7.4.17). The finite set of excluded primes of $\mathcal{E}_1$ is provided by Ax.7 of Sect. 6.5 (see Definition 7.4.17).

The Heegner Point and CM Point Kolyvagin System κ for E/k

(8.7.10) Take the Heegner system x_c constructed in (8.7.6) and (8.7.7) where c ranges over all cycles in a set $\mathcal{C}$ consisting of all squarefree cycles on k which are coprime to a finite exceptional set of prime places. The finite excluded set of prime places from $\mathcal{C}$ is given explicitly in (6.6.2), Definition 6.7.6, and Proposition 6.7.7.

Let $\mathcal{E}_3 \subset \mathcal{E}_2$ be the subset of prime numbers obtained by removing from $\mathcal{E}_2$ the finitely many primes dividing 2 and the primes dividing the order of Pic(O_k), where O_k is the ring of integers of k with respect to ∞_k (see Theorem 7.6.14).

Take $i = 1$ in the definition of a Kolyvagin system (Definition 7.1.9). Let $\theta_c \in H^1(F, T_l/I_{c^\sharp}T_l) \otimes \Gamma_{c^\sharp}$ for all $c \in \mathcal{M}_{1,k}$ be the derived cohomology classes (8.7.9) obtained from the Heegner system x_c.

By Theorem 7.6.14 taking $l \in \mathcal{E}_3$, there is a Kolyvagin system κ for $(T_l, \mathcal{F}, \mathcal{L})$ such that $\kappa_1 = \theta_1$ and which is obtained from the collection of the θ_c for all $c \in \mathcal{M}_{1,k}$ by applying automorphisms to the cohomology groups $H^1(F, T/IT)$ where I runs over ideals of R.

This Kolyvagin system $\kappa \in \mathbf{KS}(T_l, \mathcal{F}, \mathcal{L})$ is of Heegner point and CM point type for E/k (Definition 7.3.3 and (7.3.4)). Note that $\mathcal{E}_3$ contains all but finitely many prime numbers.

Hypotheses H.0-H.5 Sect. 8.1

(8.7.11) By Proposition 8.1.18, there is a set of prime numbers $\mathcal{E}_4$ such that the hypotheses H.0-H.5 hold for the Selmer triple $(T_l, \mathcal{F}, \mathcal{L})$ of (8.7.8) for all $l \in \mathcal{E}_4$.

The set $\mathcal{E}_4$ contains all but finitely many prime numbers, if k is a number field, and it has positive Dirichlet density, if k is a function field. We then put

$$\mathcal{E} = \mathcal{E}_3 \cap \mathcal{E}_4, \tag{8.7.12}$$

so that again, the set $\mathcal{E}$ contains all but finitely many prime numbers, if k is a number field, and it has positive Dirichlet density, if k is a function field.

8.7 Shafarevich-Tate Groups of Elliptic Curves

In conclusion, for all $l \in \mathcal{E}$, we have a Kolyvagin system $\kappa \in \mathbf{KS}(T_l, \mathcal{F}, \mathcal{L})$ for $i = 1$ (Definition 7.1.9) and such that the hypotheses H.0-H.5 hold for the Selmer triple $(T_l, \mathcal{F}, \mathcal{L})$ of (8.7.8).

The Shafarevich-Tate Group of the Elliptic Curve E/k

(8.7.13) Let E/k be the given elliptic curve as in (8.7.1). The module $H^1_{\mathcal{F}}(F, E[n])$, for any positive integer n coprime to the characteristic of k, is the classical n-Selmer group which fits into the exact sequence where $\text{III}(E/F)$ is the Shafarevich-Tate group of $E \times_k F/F$ (see (5.5.9))

$$0 \longrightarrow E(F)/nE(F) \longrightarrow H^1_{\mathcal{F}}(F, E[n]) \longrightarrow \text{III}(E/F)_n \longrightarrow 0.$$

Letting n run over all powers of a prime number l coprime to the characteristic of k and taking the direct limit, we obtain the exact sequence

$$0 \longrightarrow E(F) \otimes_{\mathbb{Z}} \frac{\mathbb{Q}_l}{\mathbb{Z}_l} \longrightarrow H^1_{\mathcal{F}}(F, T_l \otimes_{\mathbb{Z}_l} \frac{\mathbb{Q}_l}{\mathbb{Z}_l}) \longrightarrow \text{III}(E/F)_{l^\infty} \longrightarrow 0. \qquad (8.7.14)$$

For the Tate module T_l of E and its corresponding Selmer module $H^1_{\mathcal{F}}(F, T_l)$, we have the exact sequence (from (5.5.11))

$$0 \longrightarrow E(F) \otimes_{\mathbb{Z}} \mathbb{Z}_l \longrightarrow H^1_{\mathcal{F}}(F, T_l) \longrightarrow \varprojlim_n \text{III}(E/F)_{l^n} \longrightarrow 0. \qquad (8.7.15)$$

If $\text{III}(E/F)$ is a finite group, this last exact sequence becomes an isomorphism $H^1_{\mathcal{F}}(F, T_l) \cong E(F) \otimes \mathbb{Z}_l$.

The Main Theorems

(8.7.16) We asume for the rest of this section Sect. 8.7 that:

- F/k is a field extension of one of the types (a) or (b) in (8.7.2);
- E/k is a modular elliptic curve in the sense of (8.7.5);
- $\mathcal{E}$ is the set of prime numbers of (8.7.12) which contains all but finitely many prime numbers if k is a number field and is a set of positive Dirichlet density if k has positive characteristic;
- $\mathcal{L}$ of the Selmer triple $(T_l, \mathcal{F}, \mathcal{L})$ contains all but finitely prime places of F which lie over primes of k which are inert and unramified in F/k;
- κ_c for $c \in \mathcal{M}_{1,F}$ is the Kolyvagin system constructed in (8.7.11) for the Selmer triple $(T_l, \mathcal{F}, \mathcal{L})$ for any $l \in \mathcal{E}$ such that $\kappa_1 = \theta_1$. Furthermore, the hypotheses H.0-H.5 hold for the Selmer triple $(T_l, \mathcal{F}, \mathcal{L})$ for all $l \in \mathcal{E}$ by (8.7.11).

(8.7.17) For the next Theorem 8.7.18, we put

$\mathcal{D} = \mathbb{Q}_l/\mathbb{Z}_l$;
$A = T_l \otimes_R \mathcal{D}$.

By propagating the Selmer structure $\mathcal{F} \otimes \mathbb{Q}_l$ from $T_l \otimes \mathbb{Q}_l$ to $A = T_l \otimes_{\mathbb{Z}_l} \mathcal{D}$, we obtain a Selmer structure again denoted by $\mathcal{F}$ on A.

Theorem 8.7.18 *Fix any $l \in \mathcal{E}$, and let $(T_l, \mathcal{F}, \mathcal{L})$ be the Selmer triple as above (8.7.8). Let κ be the Heegner point and CM point Kolyvagin system for $(T_l, \mathcal{F}, \mathcal{L})$ of (8.7.11). Assume the following.*

- $l > 4$;
- *the Heegner point and CM point $N_{F_1/F}x_1$ has infinite order in $E(F)$ (see (8.7.6), (8.7.7))*;
- $d(\kappa) \in \mathbb{N} \cup \{\infty\}$ *is the element attached to the Kolyvagin system κ in Definition 8.6.8.*

Then $H^1_{\mathcal{F}}(F, T_l)$ is a free R-module of rank 1 and there is a finite $R[G]$-module M such that

$$H^1_{\mathcal{F}}(F, A) \cong \mathbb{Q}_l/\mathbb{Z}_l \oplus M \oplus M$$

with

$$\text{length}_R(M) = \text{length}_R(H^1_{\mathcal{F}}(F, T_l)/R\kappa_1) - d(\kappa). \tag{8.7.19}$$

In particular, we have $\text{length}_R(M) \leq \text{length}_R(H^1_{\mathcal{F}}(F, T_l)/R.\kappa_1))$ with equality if and only if κ is primitive. Furthermore, $\text{III}(E/F)_{l^\infty}$ is finite, and there is an isomorphism of $R[G]$-modules

$$\text{III}(E/F)_{l^\infty} \cong M \oplus M.$$

Proof By (8.7.11), for all $l \in \mathcal{E}$, the Selmer triple $(T_l, \mathcal{F}, \mathcal{L})$ satisfies hypotheses H.0-H.5. As E is an elliptic curve, there is a canonical isomorphism of G_k-modules $T_l \cong T_l^*$ where T_l^* is the dual of T_l given by the Weil pairing. Furthermore, as $\mathcal{F}$ is the classical Selmer structure on T_l, the dual Selmer structure $\mathcal{F}^*$ on $T_l^* \cong T_l$ coincides with $\mathcal{F}$ under this isomorphism, and hence $(T_l^*, \mathcal{F}^*, \mathcal{L})$ also satisfies hypotheses H.0-H.5.

We have $\kappa_1 = \theta_1$ by construction of κ where θ_1 is a derived cohomology class (see (8.7.10)). We have that θ_1 is the following, distinguishing two cases for the field k.

- (Function field/rational number field) In the case where k is a function field or is the rational number field (see (8.7.6))

8.7 Shafarevich-Tate Groups of Elliptic Curves

$$\theta_1 = N_{F_1/F} x_1 \in E(F), \quad \text{where} \quad x_1 = (1, \mathfrak{J}_1, 1, \pi) \in E(F[1])$$

where x_1 is a Heegner point, $F_1 = F[1]$ is the Hilbert class field of F when $k = \mathbb{Q}$ and is the corresponding ring class field $F_1 = F[1]$ when k is a function field (see (6.6.4)), and $N_{F_1/F}$ is the norm from F_1 to F.

- (Number field $\neq \mathbb{Q}$) When k is a number field distinct from $\mathbb{Q}$, then we have

$$\theta_1 = N_{F_1/F} x_1 \in E(F),$$

Here $x_1 \in E(F_1)$ is a CM point given by $x_1 = \pi x(1) \in E(F_1)$ (see (8.7.7)).

In both cases, the derived cohomology classes θ_c satisfy $\theta_c \in H^1(F, T_l/I_{c\sharp} T_l) \otimes \Gamma_{c\sharp}$. Furthermore, we have $I_1 = 0$ (see (7.1.4)), so we have $\theta_1 \in H^1(F, T_l) \otimes \Gamma_1$. Therefore as θ_1 is the image of $N_{F_1/F} x_1$ under the injective Kummer map $\delta : E(F) \to H^1(F, T_l)$ (see (6.5.2)) and $N_{F_1/F} x_1 \in E(F)$ has infinite order, we have that $\theta_1 \neq 0$.

The hypotheses of Theorem 8.6.13 then hold and the equality (8.7.19) as well as the inequality follow from this theorem. The final isomorphism for $\text{III}(E/F)_{l^\infty}$ follows from the exact sequence (8.7.14) and the decomposition of $H^1_{\mathcal{F}}(F, A) \cong \mathcal{D}(\psi) \oplus M \oplus M$ of Proposition 8.6.5. □

Corollary 8.7.20 *Assume the hypotheses of Theorem 8.7.18 hold and $\mathcal{E}$ is the set of prime numbers of the theorem. Let $\text{III}(E/F)_{\mathcal{E}}$ be the subgroup of the torsion group $\text{III}(E/F)$ of elements of order only divisible by primes in $\mathcal{E}$.*

Then the subgroup $\text{III}(E/F)_{\mathcal{E}}$ is a finite group where $\mathcal{E}$ contains all except finitely many prime numbers, if k is a number field, and $\text{III}(E/F)$ is a finite group, if k is a function field. Furthermore, $E(F) \otimes_{\mathbb{Z}} \mathbb{Q}$ is a 1-dimensional $\mathbb{Q}$-vector space.

Proof For each prime number $l \in \mathcal{E}$, let $^{(l)}\kappa$ be the Heegner point and CM point Kolyvagin system for l-adic Tate module $T_l = T_l(E/k)$ as in (8.7.10) and (8.7.11) (see also Sect. 7.6). We write $\mathfrak{m}_l$ for the maximal ideal of $\mathbb{Z}_l$ and $A^{(l)} = T_l \otimes_{\mathbb{Z}_l} \mathbb{Q}_l/\mathbb{Z}_l$.

By definition, the element $^{(l)}\kappa_1$ is the image in $H^1(F, T_l)$ of the point $N_{F_1/F} x_1 \in E(F)$ via the injective Kummer map (see (6.5.2), see also Proposition 7.6.9)

$$E(F) \otimes_{\mathbb{Z}} \mathbb{Z}_l \longrightarrow H^1(F, T_l).$$

As the point $N_{F_1/F} x_1$ has infinite order in $E(F)$, it follows that $^{(l)}\kappa_1$ is non-zero for all prime numbers $l \in \mathcal{E}$.

Write $T_l^{(i)} = T_l/\mathfrak{m}_l^i T_l$ for $i \in \mathbb{N}$. We have from Lemma 8.1.16 that there are isomorphisms

$$H^1_{\mathcal{F}}(F, T_l^{(i)}) \cong H^1_{\mathcal{F}}(F, A^{(l)}[\mathfrak{m}_l^i]) \cong H^1_{\mathcal{F}}(F, A^{(l)})[\mathfrak{m}_l^i]$$

and
$$H^1_{\mathcal{F}}(F, T_l) \cong \varprojlim_i H^1_{\mathcal{F}}(F, T_l^{(i)})$$

where this last isomorphism is from Proposition 1.3.10.

It follows from Theorem 8.7.18 that for all $l \in \mathcal{E}$, we have

$$H^1_{\mathcal{F}}(F, A^{(l)}) \cong \mathbb{Q}_l/\mathbb{Z}_l \oplus M^{(l)} \oplus M^{(l)} \tag{8.7.21}$$

where $M^{(l)}$ is a finite abelian l-group depending on l, $H^1_{\mathcal{F}}(F, T_l)$ is a free $\mathbb{Z}_l$ module of rank 1, and where

$$\text{length}_{\mathbb{Z}_l}(M^{(l)}) \leq \text{length}_{\mathbb{Z}_l}(H^1_{\mathcal{F}}(F, T_l)/\mathbb{Z}_l.^{(l)}\kappa_1).$$

From the exact sequence (8.7.15) and that $H^1_{\mathcal{F}}(F, T_l)$ is a free R module of rank 1 and (Theorem 8.7.18) and that $N_{F_1/F}(x_1)$ has infinite order, we obtain that $E(F) \otimes_{\mathbb{Z}} \mathbb{Z}_l$ is a free $\mathbb{Z}_l$-module of rank 1. Again from this and from the exact sequence (8.7.15), it follows that we have $\varprojlim_n \mathrm{III}(E/F)_{l^n}$ is a finite group. Hence, we have that $\varprojlim_n \mathrm{III}(E/F)_{l^n}$ must be zero. Again from the exact sequence (8.7.15), we obtain that $H^1_{\mathcal{F}}(F, T_l)$ is equal to its submodule $E(F) \otimes_{\mathbb{Z}} \mathbb{Z}_l$.

We then have the isomorphisms of groups for all $l \in \mathcal{E}$

$$H^1_{\mathcal{F}}(F, T_l)/\mathbb{Z}_l.^{(l)}\kappa_1 \cong \frac{E(F)}{\mathbb{Z}N_{F_1/F}x_1} \otimes \mathbb{Z}_l. \tag{8.7.22}$$

As $E(F) \otimes_{\mathbb{Z}} \mathbb{Z}_l$ is a free $\mathbb{Z}_l$-module of rank 1 and $E(F)$ is a finitely generated abelian group, it follows that $\frac{E(F)}{\mathbb{Z}N_{F_1/F}x_1}$ is a finite abelian group. Therefore,

$$H^1_{\mathcal{F}}(F, T_l)/\mathbb{Z}_l.^{(l)}\kappa_1$$

is a finite l-group for all $l \in \mathcal{E}$, and this group is zero for all but finitely many $l \in \mathcal{E}$.

As $M^{(l)}$ is a finite abelian group depending on l and whose length is less than or equal to $\text{length}_{\mathbb{Z}_l}(H^1_{\mathcal{F}}(F, T_l)/\mathbb{Z}_l^{(l)}\kappa_1)$, we obtain that $M^{(l)}$ is zero for all but finitely many $l \in \mathcal{E}$.

We have

$$\mathrm{III}(E/F)_{\mathcal{E}} \cong \prod_{l \in \mathcal{E}} \mathrm{III}(E/F)_{l^\infty}$$

and then from the exact sequence (8.7.14) and the decomposition (8.7.21), it follows that $\mathrm{III}(E/F)_{\mathcal{E}}$ is a finite group. If k is a function field and because $\mathcal{E}$ contains at least one element, then it follows [33] that $\mathrm{III}(E/F)$ is a finite group. □

8.7 Shafarevich-Tate Groups of Elliptic Curves

Remarks 8.7.23

(i) Kato and Trihan [33] demonstrate that if k is a global function field of characteristic p and A/k is an abelian variety then $\mathrm{III}(E/F)$ is finite if and only if $\mathrm{III}(E/F)_{l^\infty}$ is finite for one prime number l including the possibility $l = p$. Furthermore, if $\mathrm{III}(E/F)_{l^\infty}$ is finite for one prime number l, they also demonstrate that the conjectures of Birch and Swinnerton Dyer are true for A/k. See also [3].

(ii) That Heegner points on elliptic curves over $\mathbb{Q}$ have infinite order can be expressed in terms of the non-vanishing of first derivatives of L-series of these curves by the Gross-Zagier formula [23].

(8.7.24) For the next theorem, let $\mathcal{D} = \mathbb{Q}_l/\mathbb{Z}_l$, where $R = \mathbb{Z}_l$, and $A = T_l \otimes_{\mathbb{Z}_l} \mathcal{D}$. For the given classical Selmer structure $\mathcal{F}$ on T_l, by propagating $\mathcal{F} \otimes \mathbb{Q}_l$ from $T \otimes \mathbb{Q}_l$ to A, we obtain a Selmer structure again denoted by $\mathcal{F}$ on A.

We write $G = \mathrm{Gal}(F/k)$. For a character $\psi = \pm$ of G over R, write $\mathcal{D}(\psi) = \mathcal{D} \otimes_R R(\psi)$.

For all $l \in \mathcal{E}$, there is the corresponding Heegner point and CM point Kolyvagin system $\kappa \in \mathbf{KS}(T_l, \mathcal{F}, \mathcal{L})$ of (8.7.11). For all $i \in \mathbb{N}$, we have a Kolyvagin system $\kappa^{(i)}$ obtained by taking the reduction of κ mod $\mathfrak{m}^i$ (see Remark 8.6.12). The R-module $\mathrm{St}^{(i)}(n) = \mathfrak{m}^{\lambda^{(i)}(n)} H^1_{\mathcal{F}(n)}(F, T_l^{(i)}/I_n T_l^{(i)})$ is the stub Selmer module for $(T_l^{(i)}, \mathcal{F}, \mathcal{L}^{(i)})$ over $R^{(i)}$ for any n and any i (see (8.6.3)).

As the unit cycle $1 \in \mathcal{M}^{(i)}_{0,F}$ for all i, it follows from Proposition 8.4.14 that $\kappa^{(i)}_1 \in \mathrm{St}^{(i)}(1) \otimes \Gamma_1$ for all i. As $\mathrm{St}^{(i)}(1)$ is a cyclic R-module, there is a unique integer $0 \leq j(i) \leq i$ such the $\kappa^{(i)}_1$ generates $\mathfrak{m}^{j(i)}\mathrm{St}^{(i)}(1)$ for all i. By Theorem 8.6.19, there is an integer $j \in \mathbb{N}$ independent of i such that $\kappa^{(i)}_1$ generates $\mathfrak{m}^j \mathrm{St}^{(i)}(1)$ for all i.

Theorem 8.7.25 *Suppose that $l \in \mathcal{E}$, $l > 4$. Let $\kappa \in \mathbf{KS}(T_l, \mathcal{F}, \mathcal{L})$ be the corresponding Heegner point and CM point Kolyvagin system of (8.7.11). Suppose that $N_{F_1/F}(x_1)$ has infinite order in $E(F)$.*

Then we have $\kappa_1 \neq 0$, and there is a finite $R[G]$-module $M(n)$ for any $n \in \mathcal{M}_{0,F}$ such that

$$H^1_{\mathcal{F}(n)}(F, A) \cong \mathcal{D}(\psi) \oplus M(n) \oplus M(n). \tag{8.7.26}$$

where $\psi = \pm$ is a character of $\mathrm{Gal}(F/k)$. The $R[G]$-module $M(n)$ can be decomposed into cyclic $R[G]$-submodules, so that we have

$$H^1_{\mathcal{F}(n)}(F, A) = \mathcal{D}(\psi) \oplus \bigoplus_{h \geq 1}\left(\frac{R}{\mathfrak{m}^{d_h(n,(-1)^h\psi)}}((-1)^h\psi)\right)^2 \tag{8.7.27}$$

where each component $R/\mathfrak{m}^{d_h(n,(-1)^h\psi)}((-1)^h\psi)$ is in the $(-1)^h\psi$-eigenspace for all h and where the integers $d_h = d_h(n, (-1)^h\psi)$ may be assumed to satisfy

$$d_1 \geq d_3 \geq \ldots \geq 0 \text{ and } d_2 \geq d_4 \geq d_6 \geq \ldots \geq 0. \tag{8.7.28}$$

There is an integer $j \in \mathbb{N}$ be such that κ_m generates $\mathfrak{m}^j \operatorname{St}(\mathcal{H})(m)$ for all $m \in \mathcal{M}_{0,F}$. Furthermore, we have (see Definition 8.6.15 for $e_i(\kappa)$ and $\partial^r(\kappa)$)

$$\partial^0(\kappa) \geq \partial^1(\kappa) \geq \ldots \geq 0$$

$$e_0(\kappa) \geq e_2(\kappa) \geq e_4(\kappa) \ldots \geq 0, \quad e_1(\kappa) \geq e_3(\kappa) \geq e_5(\kappa) \ldots \geq 0$$

and we have, where $d_i = d_i(1, (-1)^i \psi)$ in the notation of (8.7.27)

$$\partial^r(\kappa) = j + \sum_{i \geq r+1} d_i.$$

We finally have the decompositions, putting $e_i = e_i(\kappa)$ and where $\text{III}(E/F)_{l^\infty}$ is a finite group,

$$H^1_{\mathcal{F}}(F, A) \cong \mathcal{D}(\psi) \oplus \bigoplus_{i \geq 0} \left(\frac{R}{\mathfrak{m}^{e_i}}((-1)^{i+1}\psi) \right)^2$$

$$\text{III}(E/F)_{l^\infty} \cong \bigoplus_{i \geq 0} \left(\frac{R}{\mathfrak{m}^{e_i}}((-1)^{i+1}\psi) \right)^2.$$

Proof The decomposition of (8.7.26) follows from Proposition 8.6.5. The formula (8.7.27) then evidently follows and where the exponents can be rearranged to satisfy (8.7.28). The character ψ and these integer sequences d_i satisfying (8.7.28) are uniquely determined by $H^1_{\mathcal{F}(n)}(F, A)$ and ψ is independent of n.

The element $\kappa_1 = \theta_1$ is the image of $N_{F_1/F}(x_1)$ under the injective Kummer map $\delta : E(F) \to H^1(F, T_l)$ (see (6.5.2)). As $N_{F_1/F}(x_1) \in E(F)$ has infinite order in $E(F)$, we have that $\kappa_1 \neq 0$.

The existence of $j \in \mathbb{N}$ such that κ_m generates $\mathfrak{m}^j \operatorname{St}(\mathcal{H})(m)$ is explained in (8.7.24) and follows from Theorem 8.6.19 and Proposition 8.5.7(ii).

The rest of the theorem can now be read off from Theorem 8.6.19 and Theorem 8.7.18 where the decomposition of $\text{III}(E/F)_{l^\infty}$ follows from that of $H^1_{\mathcal{F}}(F, A)$ and the exact sequence (8.7.14). □

Remark 8.7.29 Theorem 8.7.25 covers the decompositions of Shafarevich-Tate groups already known for the cases of elliptic curves over $\mathbb{Q}$ parametrized by classical modular

8.7 Shafarevich-Tate Groups of Elliptic Curves

curves due to Kolyvagin [36, 47] and those elliptic curves over global fields parametrized by Drinfeld modular curves [5].

The character ψ in the decomposition (8.7.26) of the theorem equals $+$ if the point $N_{F_1/F}(x_1) \in E(F)$ belongs to $E(k)$ and equals $-$ otherwise.

In the case where $k = \mathbb{Q}$ and F is an imaginary quadratic field, the point $N_{F_1/F}(x_1) \in E(F)$ has infinite order if and only if $L'(E/F, 1) \neq 0$ where $L(E/F, s)$ is the L-function of E/F by the theorem of Gross-Zagier [23].

Reference

1. Agashe, A., Stein, W.: Visibility of Shafarevich-Tate groups of abelian varieties. J. Number Theory **97**, 171–185 (2002)
2. Artin,E., Tate, J.: Class Field Theory. AMS Chelsea Publishing, AMS, Providence (2008)
3. Bauer, W.: On the conjecture of Birch and Swinnerton-Dyer for abelian varieties over function fields in characteristic $p > 0$. Invent. Math. **108**, 263–287 (1992)
4. Brown M.L.: Heegner Modules and Elliptic Curves. Springer Lecture Notes in Mathematics No. 1849. Springer Verlag, Berlin-Heidelberg-New York (2004)
5. Brown, M.L,: Structure of Tate-Shafarevich groups of elliptic curves over global function fields. Kyoto J. Math. **55**(4), 687–772 (2015)
6. Breuil, C., Conrad, B., Diamond, F., Taylor, R.: The Modularity of elliptic curves over $\mathbb{Q}$: wild 3-adic exercises. J. A.M.S. **14**, 843–939 (2001)
7. Bloch, S., Kato, K.: L-functions and Tamagawa numbers of motives. In: The Grothendieck Festschrift, vol. 1, pp. 333–400. Birkhäuser, Boston-Basel-Berlin (1990)
8. Česnavičius, K.: p-Selmer growth in extensions of degree p. To appear in the Journal of the L.M.S.
9. Česnavičius, K.: Topology on cohomology of local fields. To appear in Forum of Mathematics, Sigma
10. Cassels, J.W.S, Fröhlich, (eds.): Algebraic Number Theory. Thompson Book Company, Washington DC (1967)
11. Cremona, J.E., Mazur, B.: Visualizing elements in the Shafarevich-Tate group. Exp. Math. **9**, 13–28 (2002)
12. Curtis, C.W., Reiner, I.: Methods of Representation Theory, vol. 1. John Wiley and Sons, New York (1981)
13. Colliot-Thélène, J-L., Skorobogatov, A.N.: The Brauer-Grothendieck Group. Preprint (2019)
14. Cornut, C. Vatsal, V.: Non-triviality of Rankin-Selberg L-functions and CM points. In: L-Functions and Galois Representations (Durham, July 2004), LMS Lecture Note Series 320, pp.121–186. Cambridge University Press, Cambridge (2007)
15. de Jong, A.J.: A result of Gabber. Preprint. http://www.math.columbia.edu/~dejong/papers/2-gabber.pdf
16. Deligne, P., Husemöller, D.: Survey of Drinfel'd Modules. Contemporary Mathematics, vol. 67, pp. 25–91. AMS (1987)
17. Deligne, P., Rapoport, M.: Les schémas des modules de courbes elliptiques. In: Modular Functions in One Variable II, Springer Verlag Lecture Notes in Mathematics, No. 349, pp. 143–316. Berlin-Heidelberg-New York (1973)

18. Edixhoven, B.: Rational elliptic curves are modular (after Breuil, Conrad, Diamond and Taylor). Séminaire Bourbaki no 871, 1999–2000
19. Flach, M.: A generalisation of the Cassels-Tate pairing. J. Reine Angew. Math. **412**, 113–127 (1990)
20. Flach, M.: A finiteness theorem for the symmetric square of an elliptic curve. Invent. Math. **109**, 307–328 (1992)
21. Freitas, N., Le Hung, B.V., Siksek, S.: Elliptic curves over real quadratic fields are modular. Invent. Math. **201**, 159–206 (2015)
22. Gross B.H., Kolyvagin's work on modular elliptic curves. In: Coates, J.H., Taylor, M.J. (eds.) *L*-Functions and Arithmetic, pp. 235–256. Cambridge University Press, Cambridge (1990)
23. Gross B.H., Zagier D.: Heegner points and derivatives of *L*-series. Invent. Math. **84**, 225–320 (1986)
24. González-Avilés, C.D.: Arithmetic duality theorems for 1-motives over function fields. Preprint
25. Gekeler, E-U.: Drinfeld Modular Curves. Lecture Notes in Mathematics 1231. Springer, Berlin-Heidelberg-New York-Tokyo (1986)
26. Grothendieck, A.: Le groupe de Brauer I. In: Dix exposś sur la cohomologie des schémas, pp. 46–65. North Holland (1968)
27. Grothendieck, A. Le groupe de Brauer III. In: Dix exposś sur la cohomologie des schémas, pp. 88–188. North Holland (1968).
28. Gekeler, E-U., Reversat, M.: Jacobians of Drinfeld modular curves. J. Reine Angew. Math. **476**, 27–93 (1996)
29. Hayes, D.R.: Explicit class field theory in global function fields. In Studies in Algebra and Number Theory, volume 6 of Advanced Mathematical Studies, pp. 173–217. Academic Press, New York (1979)
30. Howard, B.: The Heegner point Kolyvagin system. Comp. Math. **141**(6), 1439–1472 (2004)
31. Jannsen, U.: Continuous étale cohomology. Math. Ann. **280**, 207–245 (1988)
32. Kato, K.: p-adic Hodge theory and values of zeta functions of modular forms. In: Cohomologie p-adiques et applications arithmétiques III. Astérisque **295**, 117–290 (2004)
33. Kato, K., Trihan, F.: On the conjectures of Birch and Swinnerton-Dyer in characteristic $p > 0$. Invent. Math. **153**, 537–592 (2003)
34. Kolyvagin V.A., Euler systems. In: Cartier, P., et al. (eds.) The Grothendieck Festschrift, vol. II, pp. 435–483. Progress in Mathematics 8. Birkhäuser, Boston (1990)
35. Kolyvagin V.A.: Finiteness of $E(\mathbb{Q})$ and Ш $(E/\mathbb{Q})$ for a class of Weil curves. Math. USSR **32**(3) (1989)
36. Kolyvagin V.A.: On the structure of Shafarevich-Tate groups. In: Proceedings of USA-USSR Symposium on Algebraic Geometry, Chicago 1989. Lecture Notes in Mathematics No. 1479. Springer, Berlin-Heidelberg-New York (1991)
37. Kolyvagin V.A.: On the structure of Selmer groups. Math. Ann. **291**, 253–259 (1991)
38. Katz, N.M., Mazur, B.: Arithmetic Moduli of Elliptic Curves. Princeton University Press, Princeton (1985)
39. Kresch, A., Tschinkel, Y.: Effectivity of Brauer-Manin obstructions. https://arxiv.org/abs/math/0612665
40. Kühne, L.: Intersections of class fields. Preprint
41. Kurihara, M.: Refined Iwasawa theory and Kolyvagin systems of Gauss sum type. Proc. L.M.S. **104**, 728–769 (2012)
42. Lang, S.: Cyclotomic Fields. Springer, New York, Berlin, Heidelberg, Tokyo (1983)
43. Lang, S.: Complex Multiplication I and II. Graduate Texts in Mathematics, No. 121. Springer, New York, Berlin, Heidelberg, Tokyo (1990)

44. Lang, S., Tate, J.: Principal homogeneous spaces over abelian varieties. Am. J. Math. **80**, 659–684 (1958)
45. Masser, D.W.: Division values of elliptic functions. Bull. Lond. Math. Soc **9**, 49–53 (1977)
46. Mazur, B.: An introductory lecture on Euler systems. https://pdfs.semanticscholar.org/3074/eb4b95256b4b6688be9ca19506b247adabf4.pdf
47. Mccallum, W.G.: Kolyvagin's work on Shafarevich-Tate groups. In: Coates, J.H. (eds), Taylor, M.J. (eds.) L-Functions and Arithmetic, pp. 295–316. Cambridge University Press, Cambridge (1990)
48. Milne, J.S.: Arithmetic Duality Theorems, Perspectives in Mathematics. Academic Press, Boston (1986)
49. Milne, J.S.: Canonical models of (mixed) Shimura varieties and automorphic vector bundles. In: Clozel, L., Milne, J.S. (eds.) Automorphic Forms, Shimura Varieties, and L-Functions, Vol. I. Proceedings of the Conference Held at the Iniversity of Michigan, 1988. Academic Press, New York (1990)
50. Milne, J.S.: Introduction to Shimura Varieties. https://www.jmilne.org/math/xnotes/svi.pdf
51. Milne, J.S.: Etale Cohomology. Princeton University Press, Princeton (1980)
52. Milne, J.S.: Elements of order p in the Tate-Shafarevich group. Bull. LMS **2**, 293–296 (1970)
53. Milne, J.S.: Algebraic Groups. (Preliminary version, finally published by Cambridge University Press (2017))
54. Mazur, B., Rubin, K.: Kolyvagin Systems. Memoirs of the A.M.S. No. 799 (2004)
55. Morgan, A., Smith, A.: The Cassels-Tate pairing for finite galois modules. https://arxiv.org/pdf/2103.08530.pdf
56. Nekovar J.: The Euler system method for CM points on Shimura curves. In: L-functions and Galois Representations (Durham, July 2004), LMS Lecture Note Series 320, pp. 471–547. Cambridge University Press, Cambridge (2007)
57. Neukirch, J.: Class Field Theory. Springer Verlag, Berlin-Heidelberg-New York (1986).
58. Neukirch, J., Schmidt, A., Wingberg, K.: Cohomology of Number Fields, 2nd edn. Springer, Berlin-Heidelberg-New York (2015)
59. Oort, F., Tate, J.: Group schemes of finite order. Ann. sci. Ecole. Norm. Sup. **3**, 1–21 (1970)
60. Raynaud, M.: Caractéristique d'Euler-Poincaré d'un faisceau et cohomologie des variétés abéliennes. In: Grothendieck, A. (ed.) Dix Exposés sur la Cohomologie des Schémas, pp. 12–30. North Holland, Amsterdam (1967)
61. Raynaud, M.: Schémas en groupes de type $(p, \ldots, p)$. Bull. Soc. Math. France **102**, 241–280 (1974)
62. Rubin, K: Euler Systems. Annals of Mathematics Studies, vol. 147. Princeton University Press, Princeton (2000)
63. Rubin, K.: Kolyvagin's system of Gauss sums. In: van der Geer, G., et al. (eds.) Arithmetic Algebraic Geometry. Progress in Mathematics, vol. 89, pp. 309–324. Birkhäuser, Boston (1991)
64. Rubin, K.: Elliptic curves with complex multiplication and the conjecture of Birch and Swinnerton Dyer. In: Viola, C. (ed.) Arithmetic Theory of Elliptic Curves. Spinger Verlag Lecture Notes in Mathematics, vol. 1716, pp. 167–234 (1999)
65. Rubin, K.: Euler systems and modular elliptic curves. In: Galois Representations in Arithmetic Algebraic Geometry. London Mathematical Society. Lecture Notes Series, vol. 254, pp. 351–368 (1998)
66. Serre, J-P.: Corps Locaux. Hermann, Paris (1968)
67. Serre, J-P.: Propriétés galoisiennes des points d'ordre fini des courbes elliptiques. Invent. Math. **15**, 47–122 (1972)
68. Serre, J-P.: Cohomologie Galoisienne. Lecture Notes in Mathematics, No. 5, 4th edn. Springer, Berlin-Heidelberg-New York (1973)

69. Serre, J-P.: Lectures on the Mordell-Weil theorem. Translated and edited by Martin Brown from notes by Michel Waldschmidt, 3rd edn. Springer, Berlin (1997)
70. Scholl, A.J.: An introduction to Kato's Euler systems. In: Galois Representations in Arithmetic Algebraic Geometry, L.M.S. Lecture Notes Series, vol. 254, pp. 379–460. Cambridge University Press, Cambridge (1998)
71. Shatz, S.S.: Cohomology of artinian group schemes over local fields. Ann. Math. **79**, 411–449 (1964)
72. Shatz, S.S.: Groups schemes, formal groups and p-divisible groups. In: G., Silverman, J.H. (eds.) Arithmetic Geometry Cornell, pp. 29–78. Springer, Berlin (1986)
73. Shimura, G.: Introduction to the Arithmetic Theory of Automorphic Functions. Publ. Math. Soc. Japan No. 11, Tokyo (1971)
74. Silverman, J.H.: The Arithmetic of Elliptic Curves. Graduate Texts in Mathematics No. 106, 2nd edn. Springer, Heidelberg, London, New York (2008)
75. Silverman, J.H.: Advanced Topics in the Arithmetic of Elliptic Curves. Graduate Texts in Mathematics No. 151, . Springer, Heidelberg, London, New York (1994)
76. Skorobogatov, A.: Torsors and Rational Points. Cambridge University Press., Cambridge (2001)
77. Serre, J-P., Tate, J.: Good reduction of abelian varieties. Ann. Math. **88**, 492–517 (1968)
78. Tate, T.: Relations between K_2 and galois cohomology. Invent. Math. **36**, 257–274 (1976)
79. The Stacks Project. https://stacks.math.columbia.edu/browse
80. Vignéras, M.F.: Arithmétique des Algèbres des Quarternions. Lecture Notes in Mathematics No. 800. Springer, Berlin-Heidelberg-New York (1980)
81. Wiles, A.: Modular elliptic curves and Fermat's last theorem. Ann. Math. **141**, 443–551 (1995)
82. Zanarella, M.: On Howard's main conjecture and the Heegner point Kolyvagin system (2019). Arxiv:1908.09197v1
83. Zhang, S.: Heights of Heegner points on Shimura curves. Ann. Math. **153**, 27–147 (2001)

Index

B
Brauer group, 114
Brauer-Manin obstruction, 114–117

C
Cartier dual, 74
Cassels-Tate pairing, 230–237
Character of $\mathcal{H}(m)$, 268
Classical descent method, 132
CM points - Shimura curves, 31
  on elliptic curves, 60
Coefficient ring, 69
Cohomological dimension, 15
Conductor of Galois representations, 24–24
  of an elliptic curve, 23–26
  Swan conductor, 24
Continuous cochain cohomology, 5–16
  corestriction, 7
  inflation, 6
  inflation-restriction sequence, 9
  restriction, 6
  of $\hat{\mathbb{Z}}$, 10–15
Corestriction homomorphism of galois cohomology, 16
Core vertex, 249–254
  definition, 249
Cycle, 2
  support of, 2

D
Director of a field, 166
Divisor class group, 119
Derived cohomology classes, 193–198
  definition, 198

Drinfeld module - definition over a field, 27
  characteristic of, 27
  definition over a scheme, 28
  Drinfeld upper half-plane, 40

E
Elliptic curve, 23
  conductor of, 23–26
  isotrivial, 43
  modularity, 41–42
  modularity theorem, 41, 56
  Ogg's conductor formula, 25
  torsion points, 42–45
Elliptic space, 36
  I-cyclic subgroup of an elliptic space, 37
Euler system
  of cyclotomic units, 169
  definition, 139–141
  of elliptic units, 170
  is a General Euler System, 141–143
  general (*see* General Euler system)
  of Heegner points, 168
  of Kato, 174
  module of Euler systems, 140
  morphisms, 144–152
  of Stickelberger elements, 171

F
Finite-singular comparison map, 97–103
  fppf topology, fppf cohomology, 72
Frobenius element, arithmetic and geometric, 2
  normal form, 83–97

G

Galois cohomology, 15
   corestriction, 16
   inflation, 16
   restriction, 16
General Euler System
   of CM points, 160–165
   definition, 136
   dimension, 137
   of dimension zero, 137–139, 147–152
   of Heegner points, 152–160
   of a Heegner system, 165–169
   morphisms, 144–152
Generalized dihedral group, 19, 52
Global duality, 107–110
   local duality, 81

H

Hasse principle, 111
   Brauer Manin obstruction, 116
   failure for elliptic curves and cubic surfaces, 112
   failure for torsors, 118
   F-obstruction, 112
   for quadric hypersurfaces, 112
   Reichardt-Lind curve, 116
Heegner condition, 49
Heegner point, 48–51
   on elliptic curves, 55–67
      definition, 57
   Euler system, 165–169
   general Euler system, 152–160
   Heegner condition, 49
   on $\mathbf{Y}_0^{\text{Ell.Sp.}}(I)$, definition, 50
Heegner point and CM point Kolyvagin system, 191
Hub of a simplicial sheaf, 262
H.0-H.5 hypotheses, 218
   for the classical Selmer structure, 228

I

I-cyclic subgroup, definition, 37
Isotrivial elliptic curve, 43
Iwasawa theory, 187

K

Kolyvagin operators, 195–197
Kolyvagin system, 177–188
   definition, 181
   as global sections of simplicial sheaves, 185
   Heegner point and CM point Kolyvagin system definition, 191
   Heegner point, 190–216
   module of Kolyvagin systems, 181
   primitive Kolyvagin system, 287
   relaxed, 184
Kummer sequence and Kummer homomorphism, 119, 122, 119, 164

L

l-adic representation – definition, 70
   dual (for a local field), 71, 75, 77
   Tate twist, 70
L-function, 305
Local condition, 79
   cartesian, 80, 81, 218, 227
   dual, 81
   finite, 81, 82
   functorial over a category, 80
   L-transverse, 80, 179
   propagated, 79, 226
   relaxed, 80
   singular quotient, 82, 99
   strict, 80
   unramified, 80
Local duality, 72–79
Locally cyclic simplicial sheaf, 262

M

Modularity of elliptic curves, 41–42
Modulator of a CM point, 52
Module scheme, 74
Moduli schemes $\mathbf{Y}_0^{\text{Ell.Sp.}}(I)$, $\mathbf{Y}_0(N)$, $\mathbf{Y}_0^{\text{Drin}}(I)$, 38
   analytic form, 40–41
   compactified $X_0^{\text{Ell.Sp.}}(I)$, $X_0(N)$ $X_0^{\text{Drin}}(I)$, 39–39
   cusps, 39
   Hecke operators, 39, 58
   rigidification by ξ, 56, 59

Index

N
Normal form, 83–97

O
Odd/even Selmer triple, 245, 258
Ogg's conductor formula, 25
Orders, 19
   in quadratic field extensions, 17

P
Partial order on Selmer structures, 106
Poitou-Tate local duality, 75, 81
   global duality, 107–110
Pontrjagin dual, 4, 108
Propagated local condition, 79, 226

Q
Quotient category of a representation Quot(T), 71

R
Reichardt-Lind curve, 116
Relaxed Kolyvagin system, 184
Representation, definition, 70
   dual (for a local field), 71, 75, 77
   dual (for a global field), 71, 108
   l-adic representation, definition, 70
   unramified, 97
   over local principal ideal ring, 21–21
   quotient category of a representation, 71
Restriction homomorphism of galois cohomology, 16
Rigidification by ξ
   classical/Drinfeld modular curves, 56
   Shimura curves, 59
Ring class field, 17
   modulator of CM point, 52
Ring of integers with respect to a special set of places, 4

S
Selmer module, 106–107
   classical, of an abelian variety, 122,123
Selmer structure, 105
   canonical, 106
   classical Selmer structure on abelian varieties, 117–133
   dual, 106, 230
   local condition, 79
   modified $\mathcal{F}_a^b(c)$, 107
   partial order on Selmer structures, 106
   Selmer module, 106
   Selmer sheaf, 185
   Selmer triple, 178
   stub Selmer module, 245, 258, 243–249
   stub Selmer sheaf, 255
Selmer triple, 178
   for Heegner point systems, 190
   liftable to a discrete valuation ring, 264, 269
   odd, even, 245, 258
Shafarevich-Tate group
   definition, 118
   generalities, 118–133
   n-torsion is finite, 131
   Reichardt-Lind curve, 116
Shatz duality, 76–79
Shimura curves, 28–36, 58–67, 160–165, 165–169
   of classical type, 55
   CM points, 31
   CM point on elliptic curves, 59
   compactification, 30–31
   decomposition of the Jacobians $J(M_H^*)$, $J(N_H^*)$, 35–36
   Hecke algebra, 33
   Hecke operator, 33
   Hecke correspondence, 31
   Hodge class, 59
   quaternion algebra, 28
   rigidification by ξ, 59
   the system M_H, M_H^*, N_H, N_H^*, 30
Shimura-Taniyama-Weil conjecture, 42
Simplicial sheaves, 185
   hub, 262
   Kolyvagin systems, 185
   locally cyclic, 262
   Selmer sheaf, 185
   stub Selmer sheaf, 255
   surjective path, 262
Singular quotient, 82, 99
Special set of places, 3
   imaginary quadratic extension with respect to, 3
   ring of integers with respect to, 4

Standard topology on cohomology groups, 72–74
Stub Selmer module, 243–249
    definition, 245
    for quotients of a representation, 258
Stub Selmer sheaf - definition, 255
Strict cohomological dimension, 15
Surjective path of a simplicial sheaf, 262

T
Torsion points on elliptic curves, 42–45
Torsor, definition, 113

Transverse condition, L-transverse local condition, 80, 179

V
Visibility, 132

W
Weak approximation, 111
    Brauer Manin obstruction, 116
    failure for elliptic curves and cubic surfaces, 112
    F-obstruction, 112

The manufacturer's authorised representative in the EU is Springer Nature Customer Service Centre GmbH, Europaplatz 3, 69115 Heidelberg, Germany. If you have any concerns regarding our products, please contact ProductSafety@springernature.com

Printed and bound by CPI Group (UK) Ltd, Croydon, CR0 4YY
26/03/2026
02078980-0001